Climate Law and Developing Countries

NEW HORIZONS IN ENVIRONMENTAL AND ENERGY LAW

Series Editors: Kurt Deketelaere, *Professor of Law, University of Leuven, Belgium and University of Dundee, Scotland and* Zen Makuch, *Reader in Law, Barrister, Imperial College, London, UK*

Environmental law – including the pressing considerations of energy law and climate change – is an increasingly important area of legal research and practice. Given the growing interdependence of global society and the significant steps being made towards environmental protection and energy efficiency, there are few people untouched by environmental and energy lawmaking processes.

At the same time, environmental and energy law is at a crossroads. The command and control methodology that evolved in the 1960s and 1970s for air, land and water protection may have reached the limit of its environmental protection achievements. New life needs to be injected into our environmental protection regimes – perhaps through the concept of sustainability in its environmental, economic and social forms. The same goes for energy policy and law, where liberalisation, environmental protection and security of supply are at the centre of attention. This important series seeks to press forward the boundaries of environmental and energy law through innovative research into environmental and energy law, doctrine and case law. Adopting a wide interpretation of environmental and energy law, it includes contributions from both leading and emerging international scholars.

Climate Law and Developing Countries

Legal and Policy Challenges for the World Economy

Edited by

Benjamin J. Richardson

Osgoode Hall Law School, York University, Canada

Yves Le Bouthillier

Faculty of Law, University of Ottawa, Canada

Heather McLeod-Kilmurray

Faculty of Law, University of Ottawa, Canada

Stepan Wood

Osgoode Hall Law School, York University, Canada

NEW HORIZONS IN ENVIRONMENTAL AND ENERGY LAW

Edward Elgar

Cheltenham, UK • Northampton, MA, USA

Published by
Edward Elgar Publishing Limited
The Lypiatts
15 Lansdown Road
Cheltenham
Glos GL50 2JA
UK

Edward Elgar Publishing, Inc.
William Pratt House
9 Dewey Court
Northampton
Massachusetts 01060
USA

A catalogue record for this book
is available from the British Library

Library of Congress Control Number: 2009933405

ISBN 978 1 84844 226 9

Printed and bound by MPG Books Group, UK

Contents

Contributors

Sumudu Atapattu

Sumudu Atapattu is the Associate Director of the Global Legal Studies Center at the University of Wisconsin Law School, United States. She teaches International Environmental Law and her book entitled *Emerging Principles of International Environmental Law* was published by Transnational Publishers in 2006. Dr Atapattu holds an LLM and a PhD from the University of Cambridge, and she is an Attorney-at-Law of the Supreme Court of Sri Lanka. Dr Atapattu is also the Lead Counsel for Poverty and Human Rights at the Center for International Sustainable Development Law in Montreal, Canada. Before she came to the United States, she was an Associate Professor at the Faculty of Law, University of Colombo, Sri Lanka.

Deepa Badrinarayana

Deepa Badrinarayana is an Assistant Professor at Chapman University School of Law, United States. She holds a Doctorate in Juridical Studies in Environmental Law and a Master's in Environmental Law from Pace Law School, as well as a BA and LLB(Hons) from the National Law School of India University. Professor Badrinarayana is also a member of the IUCN Commission on Environmental Law. Between 2005 and 2006, she was a Visiting Scholar at the Center for Global Legal Studies, Columbia Law School. Earlier, she was a Research Officer for a Government of India–World Bank Environmental Capacity-Building Project, at the National Law School of India.

Kees Bastmeijer

Kees Bastmeijer is Professor of Nature Conservation and Water Law at the Law Faculty of Tilburg University, The Netherlands. His research interests include nature protection law, the role of law in protecting the global commons, transboundary environmental impact assessment, legal aspects of sustainable tourism and corporate social responsibility. His publications on the international management of the Polar Regions illustrate his special interest in the role of law in protecting the world's last wildernesses. Before his appointment at Tilburg University, Professor

Bastmeijer worked from 1989 to 1999 as a legal policy adviser at the Dutch Ministry of Environmental Protection.

Yves Le Bouthillier

Professor Yves Le Bouthillier teaches international law, including international environmental law, in the French common law program of the Faculty of Law of the University of Ottawa. He is also the current director of the IUCN Academy of Environmental Law the Secretariat of which is located at the University of Ottawa. From 2000 to 2002 he was scholar in residence at the Canadian Department of Foreign Affairs and International Trade where he worked on international environmental law instruments.

Carolina Dutra

A qualified lawyer from Brazil, Carolina was recently a graduate student specialising in biofuels and climate change law. She completed a Master's in International Law and Environmental Law at the Catholic University of Santos. Carolina is the recipient of a scholarship from the Foundation for Scientific Research of Sao Paulo State.

Saja Erens

Saja Erens, LLM, MPhil is a PhD researcher at the Tilburg Graduate Law School, Tilburg University, The Netherlands. Her research centres on the field of European and international nature protection law. She is particularly interested in the transnational legal development and future challenges nature conservation law is faced with (e.g., climate change). Her current project deals with ecological compensation in special protection and conservation areas in the European Union, with a special focus on the cross-border dimensions of ecological compensation (i.e., from a perspective that transcends the borders between states, legislation, and scientific disciplines).

Robert Fowler

Rob Fowler is an Adjunct Professor in the Law School at the University of South Australia, having retired in late 2008 after teaching and researching in the field of environmental law in Australia for over 30 years. He continues to teach environmental law part-time at the University of South Australia and also at the University of Adelaide and the Australian National University. He is the Chair of the IUCN Academy of Environmental Law and represents the Oceania region on its Governing Board. Rob is also Chair of the South Australian Environmental Defender's Office, a public interest environmental law firm, and serves as a member of the

Governing Board of the Australian Conservation Foundation, a national environmental NGO in Australia. He is also a member of the IUCN Environmental Law Commission, and participates in two of its Specialist Groups, on soils and forests.

William Hare

William Hare is a physicist and environmental scientist with more than 20 years of experience related to climate change and stratospheric ozone depletion. He is a scientist at the Potsdam Institute for Climate Impact Research (PIK), where he works on assessing climate impacts and risks from greenhouse gas emissions. He is also a Director of Climate Analytics, a non-profit organisation which, together with PIK, runs the PREVENT project providing scientific support to Small Island Developing States and Least Developed Country climate negotiators. William was a lead author for the Intergovernmental Panel on Climate Change's *Climate Change 2007: Mitigation of Climate Change* component of its Fourth Assessment Report (AR4) and a Topic Leader for the AR4 Synthesis Report. In 2008, he was awarded *honoris causa* a Doctorate of Science from Murdoch University, Western Australia for his contribution to the climate change field.

Emmanuel B. Kasimbazi

Dr Kasimbazi is a Senior Lecturer in Law at the Faculty of Law, Makerere University, Uganda, as well as a legal consultant and advocate. He holds an LLM from the University of Calgary, Canada, and PhD from the University of Kwazulu-Natal, South Africa. Dr Kasimbazi is chairman of the East African Association of Impact Assessment and a member of the IUCN Commission on Environmental Law, and has consulted for many international and national agencies including the World Bank, African Development Bank, United Nations Environmental Programme (UNEP), Nile Basin Initiative, Food and Agriculture Organisation (FAO), Germany Technical Cooperation (GTZ), and various bilateral aid agencies including the United States Agency for International Development (USAID).

Eric Kwa

Dr Kwa is a Senior Lecturer in the Faculty of Law at the University of Papua New Guinea, and former Dean of his Faculty. He specialises in constitutional and environmental law. Holding a PhD from the University of Auckland, Dr Kwa has published widely and contributed to numerous workshops, particularly in relation to indigenous peoples and environmental law. Among several works, Dr Kwa has authored *Natural*

Resources Law of Papua New Guinea (2001) and *Twenty Years of the Papua New Guinea Constitution* (2003). Dr Kwa has undertaken a range of consultancy services, including to AusAid and UNDP, and in Papua New Guinea he has been instrumental in drafting several environmental statutes including the Biosafety and Biotechnology Bill and the Provincial Forest Conservation Bill.

Marie-Pierre Lanfranchi

Marie-Pierre Lanfranchi is a Senior Lecturer in Public Law at the University Paul Cézanne, Aix-Marseille III (CERIC, National Centre for Scientific Research), France. She teaches in and publishes on public international law, World Trade Organisation law, and international environmental law. Marie-Pierre Lanfranchi is responsible for the CERIC research programme funded by the National Agency for Research on 'Emerging Countries (BICS) in the Governance of Globalization'.

Jolene Lin

Jolene Lin is Assistant Professor at the Faculty of Law, University of Hong Kong. She teaches International Environmental Law and a course on Law, Economics, Regulation and Development. Her research interests include climate change law, administrative law and environmental law. She is the International Environmental Law editor of the *Singapore Year Book of International Law* as well as an associate member of the Asia-Pacific Centre for Environmental Law (APCEL), at the National University of Singapore.

Kirsten Macey

Kirsten Macey has followed the international climate negotiations since 2004 and is currently working as a policy officer with Climate Analytics. She has over eight years of experience consulting for non-government organisations and the public sector on international and national climate policy issues including those related to reducing emissions from deforestation; land use, land use change and forestry; carbon capture and storage; the Clean Development Mechanism; fluorinated gases; and energy efficiency. Kirsten holds a BSc in Environmental Sciences majoring in planning and science policy.

Sandrine Maljean-Dubois

Sandrine Maljean-Dubois is a Senior Researcher at the University Paul Cézanne, Aix-Marseille III (CERIC, National Centre for Scientific Research), France. She has edited several books on international and European environmental law. She has published numerous articles in this field, especially relating to the protection of biodiversity and biosecurity, climate change, the articulation between international environmental law and WTO law, non-compliance mechanisms and the role of the judiciary. In 2008, Sandrine Maljean-Dubois was the Director of the French section of the Centre for Studies and Research in International Law and International Relations at the Hague International Academy and she supervised the study of the theme 'implementation of environmental law'.

Heather McLeod-Kilmurray

Dr Heather McLeod-Kilmurray is an Assistant Professor in the Environmental Law Group at the Faculty of Law, University of Ottawa. She was law clerk at the Federal Court of Appeal, and practised at Environment Canada, and at the Sheffield City Council. Her PhD focused on the effect of environmental ethics and principles on judicial approaches to procedures such as class actions, injunctions and public interest standing. She has published on genetically modified organisms, ecofeminism and toxic torts and co-authored the chapter on Canada for *The Role of the Judiciary in Environmental Governance: Comparative Perspectives* (Kluwer) with Professor Jamie Benidickson. She teaches Environmental Law, Climate Change and Legal Change, Administrative Law and Tort Law.

Damilola S. Olawuyi

Damilola S. Olawuyi obtained an LLB (1st Class Honours) from the Igbinedion University, Nigeria, and a BL (1st Class Honours) from the Nigerian Law School, Abuja. Among his other qualifications, he holds an LLM from the University of Calgary, Canada, and an LLM from Harvard Law School. Damilola has practised and taught law in Nigeria and has published widely in the field of international environmental law. He has most recently been an instructor on the Environmental Law Programme of the United Nations Institute for Training and Research. Damilola is a Fellow of the Canadian Institute of Resources Law and the Alberta Law Foundation, and is a member of various professional societies including the Environmental Law Alliance Worldwide.

Marjan Peeters

Marjan is a Professor at the Department of Public Law at Maastricht University, The Netherlands. She publishes regularly on various topics of environmental law, with a special interest in legal aspects of climate change, and market-based regulatory instruments. She leads the elective Master's course 'International and European Environmental Law' and she has experience with teaching projects in Indonesia. In 2006, she edited with Professor Kurt Deketelaere the volume *EU Climate Change Policy: the Challenge of New Regulatory Initiatives*. In 2008, she co-edited with Professor Michael Faure the book *Climate Change and European Emissions Trading: Lessons for Theory and Practice.* In early 2008, Marjan became a member of the Governing Board of the IUCN Academy of Environmental Law, and since April 2008 she has held the position of Professor of Environmental Policy and Law at her University.

Benjamin J. Richardson
Benjamin J. Richardson is a Professor at Osgoode Hall Law School, Canada, and previously held academic appointments at the law schools of the Universities of Manchester (UK) and Auckland (New Zealand). Before he became an academic, he worked for the NSW National Parks and Wildlife Service in Australia and the IUCN in Kenya and Nepal. Professor Richardson is also a scholar of aboriginal law, and during 2006–08 he was co-director of Osgoode's award-winning Intensive Program in Aboriginal Law. Among his publications, Professor Richardson has authored or co-edited six books, including *Indigenous Peoples and the Law* and *Socially Responsible Investment Law*. Professor Richardson has co-chaired the Research Committee of the IUCN Academy of Environmental Law since June 2007.

Francesco Sindico
Francesco Sindico is a Lecturer at the School of Law of the University of Surrey, Guildford, England, where he also is Deputy Director of the Environmental Regulatory Research Group. He is also Associate Editor of the *Carbon and Climate Law Review* (CCLR). Francesco's main research interest lies in the interplay between climate and trade, which has led him to co-edit a special issue of the *CCLR* on 'Climate Change in a Global Economy'. He has also worked on other climate and environmental law related topics, including the linkage between climate and security, and the legal aspects of the environmental protection of transboundary aquifers.

Claire Stockwell
Claire Stockwell is a consultant in the field of climate change law and policy. Claire has attended the international climate law negotiations since 2003 and has worked with many organisations, civil society and youth groups, including the Potsdam Institute for Climate Impact Research, the United Nations Institute for Training and Research, and Greenpeace International. In 2008, she was awarded a Gordon Global Fellowship to research technology transfer in the context of the climate change regime.

Claire holds an International Master's in Environmental Policy from Roskilde University in Denmark and a BSc from McGill University in Canada. She hopes to complete her LLB degree before the end of the negotiations on the post-2012 climate regime!

Solange Teles da Silva

Solange is a Brazilian environmental lawyer, who obtained her PhD at Paris I University – Panthéon-Sorbonne. Presently, she is a Professor of Law at the State University of Amazonas (Manaus, Brazil) and at the Mackenzie Presbyterian University, and is coordinator of a major research project on 'Law, Natural Resources and Environmental Conflicts: the Amazon Cooperation Treaty'. Among her many affiliations, Solange is a researcher for the National Council for Scientific and Technological Development, and a researcher at Centre d'Etudes et de Recherches en droit de l'environnement, de l'aménagement et de l'urbanisme, of Paris I University – Pantheon-Sorbonne. Further, she is the Director of Publications at the Institute of Law and Citizenship Studies (Brazil), and the Associate Editor of *Fundação Getulio Vargas Law Review*. Solange has been a Visiting Professor at Valencia University (Spain), Paris South University (France), and Georgia State University (United States).

Jonathan Verschuuren

Professor Dr Jonathan Verschuuren is a Full Professor of International and European Environmental Law at Tilburg University, The Netherlands. Presently he is also the Vice Dean for research as well as the Director of the Tilburg Graduate Law School. His research mainly focuses on the role of international and European Union environmental law in legal practice. In 1993, Verschuuren received his PhD degree *cum laude* on the basis of a dissertation on the *Constitutional Right to Environmental Protection*. Since then, he has authored more than 100 publications in the field of environmental law, including several books and articles in leading journals. He has been a Visiting Professor at the Universities of Connecticut (United States), North West/Potchefstroom (South Africa), and Leuven (Belgium).

Christina Voigt

Dr Christina Voigt, LLM, is post-doctoral Research Fellow and Lecturer at the Department of Public and International Law, University of Oslo, Norway. She obtained the First (Universität Passau, 1996) and Second (1999) German Legal State Examination, holds a Master of Laws in Environmental Law degree, LLM (Envir), from the University of Auckland, New Zealand (2002), and received a Doctorate in Law from the University of Oslo in 2007. She teaches international and national environmental law, climate change law and policy as well as public international law. Her current post-doctoral research focuses on 'Safeguarding the Environmental Integrity of the Global Carbon Market'.

Angela Williams
Dr Angela Williams is a Lecturer in Law at the University of Sussex, United Kingdom. Her primary research interests are in the area of climate change law, in particular examining issues of equity, justice and development. Her current research project contemplates ideas of climate justice, from exploring how justice informs current climate change law to critiquing the emerging body of climate change based jurisprudence, and recommending how concepts of justice might be better promoted in a new international agreement. Her earlier publications examined climate change displacement, the role of sustainable development within international environmental law, and legal conceptualisations of sustainable tourism.

Stepan Wood
Dr Stepan Wood is an Associate Professor at Osgoode Hall Law School, York University, where he teaches environmental law, climate change law and property law, and coordinates the J.D./Master in Environmental Studies joint degree program. His current research focuses on voluntary international standards for environmental management and social responsibility. His publications cover a range of topics including sustainability, globalization, environmental regulation, the ISO 14001 environmental management system standard, and international environmental law. He is co-editor (with Benjamin J. Richardson) of *Environmental Law for Sustainability* (Hart, 2006) and co-author (with Stephen Clarkson) of *A Perilous Imbalance: The Globalization of Canadian Law and Governance* (U.B.C. Press, forthcoming). He is a long-time Canadian delegate to ISO Technical Committee 207 (Environmental Management), a member of the Canadian national committee on ISO 26000 (Social Responsibility), and a member of the Bar of New York.

Acknowledgements

This book is indebted to the assistance and co-operation of many people and organisations. It is based on a selection of papers presented at the IUCN Academy of Environmental Law's international conference on 'Climate Law and Developing Countries Post-2012: North and South Perspectives', held at the University of Ottawa over September 26–28, 2008 and organised jointly by the IUCN Academy, the Faculty of Law of the University of Ottawa and Osgoode Hall Law School at York University. The organising committee for the conference was composed of the four co-editors of this book as well as Carolyn Farquhar, manager of the IUCN Academy of Environmental Law, and Robert Kibugi, doctoral student at the Faculty of Law of the University of Ottawa. The work of the committee was supported by others, including professors of the Environmental Law Group, the Institute for the Environment, as well as numerous law students and members of the administrative staff of the Law Faculty. Stephanie Cummins, from RD Creative, also played a key role coordinating the logistics for the conference.

The success of the conference is also due to the generosity of many donors, who are gratefully acknowledged: Canadian International Development Agency, International Development Research Centre, Law Foundation of Ontario, Social Sciences and Humanities Research Council of Canada, Ontario Power Generation, the University of Ottawa, as well as the law firms Heenan Blaikie, Bennett Jones, Macleod Dixon and Osler, Hoskin & Harcourt.

Finally, we also appreciate the support and patience of the staff of Edward Elgar Publishing, and their efficiency in preparing the book for timely publication.

1. Introduction: climate law and developing countries

Benjamin J. Richardson, Yves Le Bouthillier, Heather McLeod-Kilmurray and Stepan Wood*

1. INTRODUCTION

> While all regions will eventually feel the effects of climate change, it will have a disproportionately harmful effect on developing countries – and in particular poor communities who are already living at or close to the margins of survival. Changes in the climate will amplify the existing challenges posed by tropical geography, a heavy dependence on agriculture, rapid population growth, and a limited capacity to cope with an uncertain climate. The world is already likely to fall short of the Millennium Development Goals for 2015 in many regions of the world. Climate change threatens the long-term sustainability of development progress. . . . The challenge now is to limit the damage, both by mitigation and adaptation.
>
> (Stern, 2007, pp. 92–93)

Climate change is one of the most pressing challenges of our time. It is a global problem, but experienced very differently in the so-called developed and developing worlds. While the academic literature on climate change and law is vast, it contains a significant gap. Little attention has been devoted to current and future issues concerning climate law in developing countries. Furthermore, there is very little published work on this topic by developing country legal scholars. This book begins to fill that gap. The chapters were originally presented at an international conference on *Climate Law in Developing Countries Post-2012*, co-sponsored by the IUCN Academy of Environmental Law, the University of Ottawa Faculty of Law and Osgoode Hall Law School, and held in Ottawa on September 26–28, 2008.[1] The Academy was established in 2003 with the support of the IUCN (International Union for the Conservation of Nature) to further the development of environmental law and policy on a global scale.[2] It is a world-wide network of more than 110 universities and institutions, dedicated to advancing knowledge of environmental law as an essential means of achieving a sustainable world. Its vision is to build sustained capacity in legal education and advance conceptual understanding and implementation of environmental law, particularly in developing countries. It pursues its vision by delivering programmes aimed at enhancing university teaching capacity in environmental law, generating global research programmes that support national and international environmental policy making, and convening major international conferences and exchanges. Its Secretariat is located at the University of Ottawa Faculty of Law.

The conference that inspired this book attracted leading experts on climate change from the North and the South, with the aim of enhancing understanding of the legal challenges that developing countries face in mitigating and adapting to climate change while meeting their social and economic needs. It also addressed how developing countries can position themselves in the negotiations to develop a new international legal regime at the expiry of the Kyoto Protocol. The book is part of the ongoing mandate of the IUCN Academy of Environmental Law to generate collaborative research on the most pressing issues in environmental law, with a view to enabling changes in national and international environmental policy agendas and also environmental law teaching. This book is the first major publication to result from the Academy's research stream in Climate Law, established in 2007. It is intended to serve as a catalyst for further collaborative research in this area. It will be of particular use to scholars, policy makers and negotiators in developing countries involved in climate law-related reforms in their countries, by providing substantive legal and policy recommendations for governments and other participants in the climate change debate, domestically and globally.

The contributions to this book were selected to cover a variety of topics and perspectives, and to ensure broad geographical diversity. The book is structured around several themes: Climate Justice; Expanding the Scope of the Climate Change Legal Regimes; The Clean Development Mechanism and Mitigation Strategies; and Climate Policy Bridging the North and South. This opening chapter sets the scene by introducing the phenomenon of climate change, considering the problem of climate change law in the context of developing countries, situating this book in the broader legal academic literature, and summarizing the main points of the chapters. A book on this subject is vulnerable to dating quickly, given the dynamic nature of legal developments concerning climate change, including current negotiations for a new international legal instrument to replace or revise the Kyoto Protocol. The contributors to this volume focus on generic themes and ongoing concerns, such as debates regarding climate justice and North–South collaboration, in the hope that this book will provide a more enduring analysis of the subject of climate law and developing countries.

2. GLOBAL CLIMATE CHANGE AND ITS IMPACTS

Any lingering scientific uncertainty in relation to the anthropogenic causes of recent global climate change was resolved by the most recent report of the Intergovernmental Panel on Climate Change (IPCC), created in 1988 by the World Meteorological Organization (WMO) and the United Nations Environment Program (UNEP) (Ott, 2007, p. 9). The IPCC concludes that '[w]arming of the climate system is unequivocal, as is now evident from observations of increases in global average air and ocean temperatures, widespread melting of snow and ice and rising global average sea level' (IPCC, 2007d, p. 30). The Panel also concludes that anthropogenic sources of greenhouse gases (GHGs), which 'alter the energy balance of the climate system', are 'drivers of climate change' (ibid., p. 37). The IPCC identified five main 'reasons for concern' about climate change: 'risks to unique and threatened systems' ('such as polar and high mountain communities and ecosystems'); 'risks of extreme weather events'; 'distribution of impacts and vulnerabilities' ('those in the weakest economic position are often the most vulnerable to climate change' as are those in 'specific groups such as the poor and elderly'); 'aggregate impacts' and 'risks of large-scale singularities' (such as 'sea level rise contribution from thermal expansion alone' and from melting of ice sheets) (ibid., p. 19). These precise effects will vary across geographical locations, and generally include increased temperatures, threats to species, reduced crop productivity, changes in wind and precipitation patterns, sea level rise, water temperature rise, coastal flooding and erosion, extreme weather events, health impacts such as malnutrition and the spread of infectious disease, and the concurrent impacts on economic and social well-being that these effects entail. Global warming threatens permanent disruption of ecosystems and biodiversity (Millennium Ecosystem Assessment, 2005) and, ultimately, the survival of life on earth.

The IPCC warns that rises in global temperature in excess of 2 degrees Celsius over the next century – the forecast upper limit that would avoid 'dangerous climate change' – should be strenuously avoided. The keynote speaker at the conference on *Climate Law in Developing Countries*, Professor William Rees, originator of 'ecological footprint' analysis, catalyzed conference participants by echoing these dire scientific warnings and pointing out that, since the last IPCC report, global warming trends – such as the retreat of Arctic sea ice – have accelerated even faster than the Panel's worst case scenarios. Climate science is extremely important to negotiators of the post-Kyoto agreements since they must set concrete targets – in terms of the ultimate levels of atmospheric GHG concentrations to achieve, the emission reductions required to achieve these levels, and the timeframe in which they are to be achieved.

The IPCC has indicated that atmospheric GHG concentrations should be stabilized at 450 parts per million (ppm) CO_2 equivalent.[3] In terms of timescale, it is important to recall that CO_2 emissions already in the atmosphere will continue to accumulate even if we were able to stop any further emissions immediately. And since signing the Kyoto Protocol, in which Annex I countries agreed to reduce their emissions to fixed percentages below 1990 levels, the emissions of almost all these countries have increased, often dramatically. The IPCC indicates that '[g]lobal GHG emissions due to human activities have grown . . ., with an increase of 70% between 1970 and 2004'. (IPCC, 2007d, p. 5). Canada, for example, undertook to reduce its emissions to 6 per cent below 1990 levels by 2012, 'yet by 2007, Canada's emissions r[o]se to 27% over 1990 levels. Under a business-as-usual scenario, Canada's emissions would reach more than double 1990 levels by the year 2050' (Donner, 2007, p. 1).

Any post-Kyoto agreement must therefore tackle these numbers: the stabilization level, the required emission reductions, and the timeframe in which to achieve them. The Working Group considering future commitments to include in the post-Kyoto agreements is guided by the IPCC's Fourth Assessment Report of 2007. That report advocates a reduction in developed country GHG emissions ranging from 25–40 per cent below 1990 levels by 2020, on the assumption that emissions reductions of this magnitude should limit the rise in global temperatures to 2 degrees Celsius; anything beyond that change would likely be very dangerous for the climate. Some environmental groups and commentators argue it is necessary for Annex I countries to achieve 80–90 per cent reductions below 1990 levels by 2050.[4] Many experts, such as Canadell and others (2007), believe the IPCC's worst-case scenario predictions are optimistic, given recent evidence such as the pace of Arctic and Antarctic ice sheet melting, declining ocean and terrestrial carbon sinks, and accelerating CO_2 emissions, which all indicate that climate 'tipping points' may be approaching faster than we think, making action even more urgent. A growing number of leading climate scientists warns that the IPCC's stabilization target of 450 ppm is too high, and argues that it should be set at the more precautionary level of 350 ppm at most (e.g., Hansen, et al., 2008).[5] As Rob Fowler points out in chapter 8 of this book, '[t]his extends the mitigation strategy from a "zero carbon" (Epstein, et al., 2008; Makhijani, 2007) to a "negative carbon" scenario', since as George Monbiot has observed, to achieve a 350 ppm target, 'we are talking at a minimum of a 100% cut in [CO_2 emissions] and it looks like it might have to go to 110% or 115%'.[6] Of course, whether the stabilization is set at 450 ppm or 350 ppm, this goal will require strenuous efforts on the part not only of Annex I nations, but of all nations, particularly such rapidly industrializing economies as China and India.

The IPCC's 2007 Report also contains positive messages, in that it concludes that with appropriate adaptation and mitigation strategies, the dire direction in which humanity is currently headed could be changed. The term 'mitigation' in this context refers to efforts to reduce GHG emissions, deforestation and other drivers of climate change. Mitigation may be achieved either by reducing emissions of GHGs from sources such as fossil fuel combustion, or by enhancing their removal by carbon sinks. Forests, soils and oceans act as carbon sinks, removing carbon from the atmosphere. 'Adaptation' refers to efforts aimed at 'adapting and reducing [our] vulnerability to the impacts of weather- and climate-related events' (IPCC, 2007d, p. 56). Adaptation is a crucial but relatively neglected priority, given that the impacts of climate change are already being felt and the planet is already 'pre-committed' to continued global warming for the short term at least, while GHGs already in the atmosphere exert their effects (Flannery, 2005).

Economist Nicholas Stern (2007, p. xvi) has estimated mitigation costs at around 1 per cent of GDP. He concludes that these are 'small relative to the costs and risks of climate change that will be avoided' and that tackling climate change can lead to economic opportunities, environmental protection, and energy security (ibid.). He concludes that 'there is still time to avoid the worst impacts of climate change if strong collective action starts now' (2007, p. xxvii). Appropriate legal frameworks and agreements governing action on climate change are essential elements of this collective global action.

3. CLIMATE LAW IN THE CONTEXT OF DEVELOPING COUNTRIES

Understanding 'Developing Countries'

This book is primarily concerned with climate law relating to *developing countries*. It considers this topic from numerous angles, including domestic climate law within developing countries, regional responses to climate change involving developing countries, developing countries' strategies in global climate change negotiations, the impacts of developed country laws and policies upon developing countries, and the future of the Clean Development Mechanism (CDM). While the chapters present a great diversity of perspectives on these issues, they take the dichotomy between 'developing' and 'developed' countries as their starting point. There is always a risk of oversimplification in the use of these terms. Yet even while adopting this conventional terminology, the book demonstrates amply the irreducible heterogeneity of what is often referred to as the Global South and the highly complicated roles played by 'developing' and 'developed' nations in contributing and responding to the problem of climate change.

The vocabulary of 'developing countries' and variations such as the 'Third World' emerged in the 1950s in the wake of the decolonization process that swept much of Africa, Asia, Latin America and Oceania (Escobar, 1995). Such terms are not merely convenient geo-political descriptions; they are imbued with more loaded, nuanced meanings. They have been used to imply the supposed undeveloped and primitive state of these 'peripheral' parts of the world, and to denote a state of becoming more like the supposedly more advanced nations at the 'centre' of the global economy. Conversely, such labels have also become unifying symbols for postcolonial states in their struggle to achieve real economic and political independence (Rajan and Mohanram, 1995). Since the 1980s, academics and activists alike have also increasingly spoken of 'the South' or the 'Global South', to denote the acute political and economic divisions between the wealthy industrialized countries, known collectively as 'the North' (due to their predominant location in the Northern latitudes), and the poorer countries of the South (mostly occupying the southern part of the globe). The persistent vast disparities in living standards between their peoples, and the resulting lack of equilibrium in the political and economic power between them, have led some critics such as McClintock (1995) to view the rhetoric of postcolonialism as 'prematurely celebratory'.

As the various contributors to this book demonstrate, these categorical labels elide substantial economic, social and environmental differences among developing countries, which are relevant to international debates about which countries are most vulnerable to global warming and how to distribute the responsibilities to mitigate it. The assumptions of international climate law and other environmental law regimes built on crude distinctions between the North and South may therefore be problematic.

Indeed, the cultural, economic and geographical disparities among developing countries, and their capacity to address climate change, tend to be as great as those between the North and South. The rapidly industrializing economies of South and East Asia, such as South Korea, China and, to a lesser extent, Thailand and Malaysia have attained levels of affluence that are the envy of many of their Southern peers (Hughes, 2009; Kim, 1998). Rapid industrialization has also wrought a plethora of environmental problems (McElroy, Nielsen and Lydon, 1998; Richardson, 2005). By contrast, many African nations, particularly in the sub-Saharan region, lag substantially on social and economic indicators relating to poverty and public health (United Nations Economic Commission for Africa, 2006). Some of the largest developing countries – such as China, India, Brazil and

Indonesia – are rapidly emerging as leading GHG emitters, whether in the form of escalating deforestation or fossil fuel combustion. Others – such as small island states, states with large low-lying areas and states with rapidly advancing deserts – are bearing the brunt of the impacts of a changing climate without having made significant contributions to the problem. With yet others relying heavily on petroleum exports for their national livelihoods, a wide divergence of interests and positions among developing countries on the topic of climate change should come as no surprise.

Apart from differences among developing nations, there is also great variation in climate change-related roles, vulnerabilities, capacities, interests and positions *within* such countries. Minority groups in some countries continue to wage struggles to achieve a measure of self-determination and autonomy, including Tibetans, Kurds, Tamils and many others who were unable to participate in the decolonization process (Otto, 1996). Some of these struggles involve resistance to development pressures that exacerbate global warming, such as deforestation of the traditional lands of indigenous peoples in the Amazon (Watson, 2008). Others contribute to these same climate change-inducing pressures as marginalized groups seek to join the mainstream development bandwagon.

Many of the variations among and within Southern states are directly relevant to the problem of climate change. Small island states in the South Pacific and other oceans are home to communities that are among the most vulnerable to the impacts of climate change including rising sea levels and devastation of marine ecosystems. The low-lying countries of Tuvalu and Kiribati may be entirely inundated, requiring the wholesale evacuation of their populations, as Angela Williams and Eric Kwa explain in their chapters in this book. The proliferation of regional blocs, such as the Alliance of Small Island States (AOSIS), illustrates the growing cleavages in the developing world on some environmental and economic issues. AOSIS is a coalition of small-island and low-lying coastal countries that have similar development challenges and environmental concerns, especially their vulnerability to the adverse impacts of climate change.[7] On the other hand, Arab oil-producing states have tended to collaborate in opposing stringent international standards to mitigate global warming (Carrell, 2003).

Developing Country Issues in International Climate Law

The evolution of international climate change law, like other fields of international environmental law, has been shaped greatly by the political struggles between the North and South and tensions within these geo-political groupings (Anand, 2004; Thomas, 1992). Although developing countries overwhelmingly ratified the United Nations Framework Convention on

Climate Change (UNFCCC)[8] of 1992 and the Kyoto Protocol[9] of 1997, their commitment was secured principally on the basis that the industrialized countries of the North would take primary responsibility for reducing global GHG emissions, as well as furnishing the financial and technological resources to enable the South to develop sustainably without heavy reliance on fossil fuels (Weizsäcker, et al., 1999). Informed by the principles of 'intragenerational equity' and 'common but differentiated responsibility' (Weiss, 1989), this approach in the UNFCCC and Kyoto Protocol has been justified on the basis that the industrialized countries have been the main contributors to global warming and remain the predominant GHG polluters (Rajamani, 2006). The Protocol, which came into effect in 2005, set binding limitations on the amount of CO_2 and related GHGs that the countries of the OECD (Organisation for Economic Co-operation and Development) and the former Soviet Union and Eastern Europe (collectively known as Annex 1 Parties under the UNFCCC) may emit, with a target that they reduce their emissions on average by 5.2 per cent below 1990 levels in the first commitment period of 2008–2012. The only substantial obligations agreed to by non-Annex 1 Parties, under Article 10, are to take certain measures to improve the quality of the reporting of their anthropogenic emissions and to 'formulate, implement, publish and regularly update national . . . programmes containing measures to mitigate climate change and measures to facilitate adequate adaptation to climate change'.

While this might have appeared to be a good deal for developing nations, by focusing on the responsibilities of Annex 1 states the international climate law negotiations tended to neglect issues of greater interest to some Southern governments, such as supporting climate adaptation in their vulnerable communities (Najam, Huq and Sokona, 2003). Furthermore, the Kyoto Protocol's method for assigning GHG emission targets has created a potentially detrimental precedent for developing countries, which increasingly are expected to have their own targets (Agarwal, et al., 1999). By setting the first commitment period targets as a percentage of 1990 emissions, rather than on the basis of emissions per capita, the Kyoto Protocol's allocation formula favours countries with large recent GHG emissions over those whose recent emissions are small. As countries with relatively low per capita GHG emissions, developing countries would be relatively disadvantaged by such a precedent. For example, the average Canadian emits just under 15 tons of CO_2 annually while the average person in China emits slightly less than three tons (International Energy Agency, 2008). Concomitantly, the Kyoto Protocol's focus on minimizing the burden of implementation on the Annex 1 Parties through 'flexibility mechanisms' such as emissions trading encourages the movement of resources to the more lucrative investment opportunities in the global carbon market, rather than towards meeting the challenges of promoting sustainable development and adaptation to climate impacts for the most at risk peoples unable to participate in such markets.

The principal mechanism created by the Kyoto Protocol for involving developing countries in climate change mitigation is the CDM. As explained in this book's chapters by Christina Voigt, Emmanuel Kasimbazi and Damilola S. Olawuyi, the CDM has yet to achieve its potential for facilitating collaboration between developing and developed countries in financing climate-friendly projects in the South. The majority of projects have been skewed towards a small number of countries, such as China, and many such projects would probably have been undertaken regardless of the CDM (thus failing to satisfy the 'additionality' requirement of the CDM).

Adaptation to global warming is another acute concern of developing countries, due to constraints in their adaptive capacity, as Jolene Lin and Eric Kwa explain in their chapters. The IPCC advised in its 2007 Fourth Assessment Report that the adverse impacts of climate change will probably fall hardest on the developing world (IPCC, 2007a). The CDM imposes a levy on projects to raise funds to assist developing countries to meet these adaptation challenges, but the Adaptation Fund only became operational in 2008. Some other funding mechanisms tied to the Kyoto regime have been established. The Protocol obliges developed nations to 'provide new and additional financial resources to meet the agreed full costs incurred by developing country Parties'[10] and to 'provide such financial resources, including for the transfer of technology, needed by the developing country Parties to meet the agreed full incremental costs of advancing the implementation'.[11] At the sixth UNFCCC Conference of the Parties (COP) at Bonn in 2001, the Parties agreed to establish several intergovernmental funds including the Special Climate Change Fund (SCCF) for capacity building, technology transfer, certain sector-specific projects, and economic diversification in developing countries, and the Least Developed Countries Fund to assist such countries adapt to climate change.[12] Yet, the funds remain woefully inadequate to enable developing countries to shift to a low carbon economy, while policy instruments to encourage private capital to invest responsibly for the same cause are lacking under current international law (Richardson, 2009).

From the perspective of the North, a significant lacuna in current international climate law is the lack of controls on deforestation, as Stockwell and others explain later in this book. Massive forest clearing in Brazil, Indonesia and other equatorial countries, sometimes supposedly in the name of climate-friendly biofuel production, accounts for some

20 per cent of CO_2 emissions and destroys globally important carbon sinks (IPCC, 2007b). Presently, the CDM does not encompass projects to protect such forests and manage them sustainably, and there are few economic incentives for Southern states to protect such a global resource in the interests of posterity.

Consequently, as will be discussed later in this chapter, many policy issues central to the position of developing countries remain to be negotiated in the agreement which will eventually replace the commitments in the Kyoto Protocol after 2012.

Climate Governance Constraints and Opportunities within Developing Countries

Regardless of how international climate law formally recognizes the position of the South, many developing countries are severely constrained in their abilities to take action domestically to mitigate or adapt to climate change. Throughout much of the South, environmental law and policy have ostensibly begun to reach maturity, manifested by extensive legislative enactments, judicial activism and the establishment of specialist environmental management agencies. Yet these advances in environmental governance often remain seriously handicapped by acute shortages of financial and technical resources, reliance on poorly designed regulations, and more general problems associated with the fragility of the rule of law in those countries plagued by corruption and civil strife (Andreen, 2000; Peerenboom, 2002). Consequently, implementation of CDM projects, control of deforestation, and regulation of industrial GHG emissions are at best sporadic.

For many years, environmental problems and their ineffectual regulation in the South were attributed largely to exogenous influences associated with the international economy. During the 1980s and early 1990s, the misguided policies of the multilateral development banks including the World Bank were heavily criticized. Their penchant for large-scale, capital intensive projects that wrought significant social and environmental harm to local communities, as well as their structural adjustment programs designed to impose a neoliberal model of privatization, deregulation and market liberalization, were viewed by many as simply means to create new patterns of dependency in an economic system dominated by the West (Gathii, 1998–99, 2000; Martens, 1989; Rich, 1994). Other obstacles identified by critics include transnational corporations, Western aid programs, the international trade regime's unequal terms of trade for Southern economies, and the foreign debt burden on the poorest countries (Amin, 1976; Hayter and Watson, 1985; Klein, 2000; Sharma, 2006). Such an economic context could thus constrain environmental policy choices domestically, particularly regulations that inhibit exploitation of natural resources in order to meet international market demands (Chen, 2009).

Other research stresses endogenous factors in explaining weak and ineffective environmental governance in the South (Boer, Ramsay and Rothwell, 1998; Harashima, 2000; Kameri-Mbote and Cullet, 1995). Environmental law regimes in many developing countries suffer from a fragmented and incoherent regulatory structure, comprising community-based management, colonial-era statutes still on the books, and more recent legislation and policies modelled on inappropriate Western precedents (Richardson, 2000). Much environmental legislation is more akin to a form of resource management law, to provide for the 'orderly' exploitation of forests, fisheries, minerals and other natural resources. It also habitually takes the form of command-and-control regulation, seeking to control behaviour through sticks rather than carrots. Because of their institutional and legal weaknesses, many states have seen the solution to poor compliance as simply to centralize environmental administration (Wunsch and Olowu, 1990). Since the mid-1990s, some states including Kenya, India and Mexico have enacted more strategic environmental law regimes including the creation of specialized national environmental agencies and comprehensive environmental protection statutes (Boer, 1999). In addition, some nations including Brazil, the Philippines, South Africa and Indonesia have added environmental protection clauses to their national constitutions as an expression of their commitment (Bruch, Coker and VanArsdale, 2001). Such provisions generally have yet to raise substantially the status of environmental protection compared to competing economic and social priorities.

Consequently, the implementation of robust climate change-related laws within developing countries faces some daunting challenges. To abate GHG emissions from surging industrialization or rampant land use changes in the South requires overcoming the same institutional and governance challenges that have impeded the implementation of environmental law generally. Chapters in this book by Solange Teles da Silva and Carolina Dutra, on Brazil, and Emmanuel Kasimbazi, concerning Uganda, illustrate these difficulties in relation to building a green biofuels industry and a sustainable forestry sector respectively. Elegantly drafted laws alone will not suffice. As Wang (2002, p. 28) explains, 'law becomes effective by social forces and pressures interested in and working for its implementation. Without a proper institutional setting, the law will remain a fig-leaf, pretending action without changing social reality'. The failure of the 'law and development' movement of the 1960s and 1970s shattered the belief that if laws are reformed and legal institutions strengthened, nothing can hold back the triumph of the rule of law (Trubek and Santos, 2006). While the 'Good Governance' movement promoted by the international development community takes a broader view of how legal reform may succeed, including the importance of capacity-building, it still suffers from some of the mistakes of earlier approaches to take account of local customs and traditions and to pay attention to social and economic justice (Gathii, 1999).

To suggest that the story of climate law and policy in developing countries is solely one of constraints and frustration would be misleading, however. One of the goals of this book is to show that climate law in developing countries is also a story of opportunity and innovation. Increasingly, developing country governments are devoting conscious efforts to the development of climate law and policy at national and local levels. They are experimenting with a range of policy approaches and instruments, from CDM projects to GHG emission limits, clean technology subsidies, anti-desertification campaigns and initiatives to protect biodiversity in the face of climate change. Some of these measures are inspired by the pursuit of cost-savings, efficiency gains and economic opportunities, and others by an increasing sense of urgency of the need for action to mitigate or adapt to climate change. As in many developed countries, climate change law and policy in developing countries are often characterized by contradictions. China is a good example, where government policy on the one hand encourages solar and wind power, and pursues world leadership in the design and production of electric vehicles, while on the other hand mandating construction of coal-fired power plants and natural gas pipelines at a breakneck pace. Similarly, in Brazil, governmental policies to develop a national biofuels industry that has been promoted for its economic and environmental benefits have also indirectly encouraged deforestation of the Amazon in order to grow soybeans and other biofuel feedstocks.

This book's attention to the many obstacles to and tensions within climate change law and policy in developing countries should not be taken as suggesting that developed countries are shining examples to be followed. Many of the same problems that afflict climate law in developing countries inhibit effective domestic responses to climate change in developed countries, albeit in different ways and to different degrees. Policy incoherence, insufficient resources, lack of political will, regulatory capture by vested interests, bureaucratic inertia, ideological opposition to action on climate change, incompetence and even outright corruption are all found in developed countries. Developed and developing countries experience these obstacles differently, however, given their deep disparities in wealth, technological capacities, position in the global economic system, and influence in international affairs.

This brings us to the realm of international climate change negotiations. The primary focus of these negotiations, as the following section explains, tends to be on the overall institutional architecture for intergovernmental cooperation rather than the challenges of domestic policy design and implementation.

4. FUTURE DIRECTIONS IN INTERNATIONAL CLIMATE LAW FOR DEVELOPING COUNTRIES

Achieving Climate Justice Beyond Kyoto

What roles and responsibilities may developing countries assume in future international climate change agreements? Continuation of the current regime is surely unlikely. The Kyoto Protocol has been widely criticized for its unambitious emission reduction targets, insufficient incentives to encourage developing nations to move to a low carbon economy, overly complicated policy tools, lack of measures to support adaptation to climate change, and ineffectual enforcement mechanisms (Barrett, 2003; Nordhaus, 2006; Pardy, 2004; Stewart and Wiener, 2003). Comparable weaknesses, however, inhere in many other international environmental law regimes owing to the complex political compromises that are invariably necessary to secure agreement (Boyle, Redgwell and Birnie, 2009). Nonetheless, in the opinion of one of the IPCC's Working Groups, the Kyoto Protocol's '[n]otable achievements' include 'stimulation of an array of national policies, the creation of an international carbon market and the establishment of new institutional mechanisms that provide the foundation for future mitigation efforts' (IPCC, 2007c, p. 21). Moreover, the Kyoto Protocol was arguably only ever intended to be a modest step in an ongoing, long-term effort to mitigate climate change and adapt to its impacts. Thus, the successor to the Kyoto Protocol is expected to provide a more credible response to an environmental threat that is understood much more clearly and perceived much more gravely today than when Kyoto was negotiated.

The negotiations underway to draft a new commitment period under the Kyoto Protocol, or an entirely new protocol under the UNFCCC, hinge greatly on the possible roles, responsibilities and entitlements of developing countries. A consensus has yet to emerge on such issues. At one extreme, some argue that the North should continue (or, in a more cynical view, finally start) to 'take the lead in adaptation and mitigation efforts on the grounds of historical responsibility, distributive justice, economic capacity and technical capability' (Boston, 2008, p. 52). From this stance,

Southern countries should continue to enjoy concessions without any specific obligations to reduce their GHG emissions while receiving technology transfers and financial assistance to enable them to develop sustainably. On the other hand, some maintain that in order to reduce GHG emissions by up to 85 per cent by 2050, as the IPCC (2007c) recommends in order to avoid catastrophic climate change, all nations including developing countries must shoulder responsibilities to act. The South is now possibly the source of just over half of the world's GHG emissions (Boston, 2008, p. 52). In particular, developing countries with high emissions from rapid industrialization (e.g., China and India) or deforestation (e.g., Brazil and Indonesia) could agree to undertake absolute emission reductions as most Annex 1 Parties presently do. Just as Kyoto differentiates the GHG emission targets of Annex I Parties, so too there is growing clamour that the time has come to take a more nuanced view of developing countries and assign individual emission targets to some countries.

While the expectations of how developing countries should act under a future climate law agreement are evolving, the underlying normative principles of international environmental law for reconciling the positions of the North and South have hardly altered. The ideals of 'intragenerational equity' and 'common but differentiated responsibility' remain starting points in international climate law negotiations (with the major exception of the United States (US), whose outspoken opposition to the idea that developed countries should act first, without the major-emitting developing countries, has softened only a little under the Obama administration). Yet, their interpretation in the context of climate change policy is shifting.

The criteria for differentiating national obligations in order to achieve 'climate justice', as many commentators increasingly describe this policy goal (Hayward, 2007; Page, 2006), are debateable. Some commentators advocate equality of result, whereby each country is allowed the same level of emissions, such as on a per capita basis (Berk and Elzen, 2001). The attraction of such a standard for the South is that per capita emissions in developing countries are generally much lower than in the developed world. Yet, because equalizing per capita emissions ignores who is responsible for historical GHG emissions (primarily the North), it may still be an unjust standard. It does not imply any obligation on developed countries to compensate or assist the South. Alternatively, climate justice could be implemented on the basis of equality of effort, whereby each nation is expected to adopt the same policies and measures (Paterson, 2001). This approach is even more objectionable given that for developing countries the additional effort required may be grossly disproportionate to their responsibility for past emissions and capacity to reduce future emissions.

Other commentators have thus suggested a more granular approach to climate justice that apportions obligations on the basis of historical responsibility for GHG emissions, the opportunity to curtail emissions today, and the ability to pay for climate change mitigation (Claussen and McNeilly, 1998).

Such are the monumental normative challenges that climate change law poses to a world community deeply divided, not just between the North and South, but also among countries and regions within these groups, and internally within countries. The foregoing discussion is a drastically simplified picture of the complex policy and ethical debates about climate justice. Chapters in this book by Sumudu Atapattu, Angela Williams and Eric Kwa in particular canvass various aspects of the climate justice debate in greater detail. While it will continue to unfold as a backdrop to the international negotiations for a future agreement on climate change, the nature of such negotiations is that they are conducted mostly with a more pragmatic outlook, focusing on specific tangible issues, such as the future of the CDM, reducing deforestation, and technology transfer, as the next section explains.

The Bali Roadmap and Beyond

In December 2007, the Indonesian island of Bali hosted the thirteenth COP to the UNFCCC and the concomitant third Meeting of the Parties to the Kyoto Protocol. The purpose of the meetings was primarily to chart a 'roadmap' to negotiate a new agreement to take effect after 2012. The resulting Bali Action Plan and related decisions (UNFCCC Conference of the Parties, 2008) were characterized variously from a 'real breakthrough' (United Nations News Centre, 2007) to 'even worse than the Kyoto Protocol' (Monbiot, 2007). Bali was a critical juncture for the international community, as it was held in the context of the IPCC's then freshly released Fourth Assessment Report, containing dire warnings about climate change. The conference also faced continuing hostility from the US to serious action on climate change, as well as the opposition of the rapidly industrializing countries (e.g., South Korea, China and India) to having any non-Annex 1 countries assume legally-binding emission targets (Aldy and Stavins, 2008).

While seeking to avoid taking a precise position on the respective responsibilities of developed and developing nations, the Bali conference at least agreed to recognize the gravity of the challenge in collectively addressing global warming. It adopted the following terms of reference to guide the Ad Hoc Working Group (AWG), which was established in 2005 to work on 'Further Commitments for Annex I Parties under the Kyoto Protocol':

> [r]ecognizing that deep cuts in global emissions will be required to achieve the ultimate objective of the Convention and emphasizing the urgency to address climate change as indicated in the Fourth Assessment Report of the Intergovernmental Panel on Climate Change (UNFCCC Conference of the Parties, 2008, p. 3).

The Bali conference also agreed, as part of the terms of reference of the AWG, to the following goals:

> [m]easurable, reportable and verifiable nationally appropriate mitigation commitments or actions, including quantified emission limitation and reduction objectives, by all developed country Parties, while ensuring the comparability of efforts among them, taking into account differences in their national circumstances; . . .
>
> Nationally appropriate mitigation actions by developing country Parties in the context of sustainable development, supported and enabled by technology, financing and capacity-building, in a measurable, reportable and verifiable manner (UNFCCC Conference of the Parties, 2008, p. 3).

By speaking in terms of 'developing' and 'developed' countries, rather than 'Annex I' or 'non-Annex 1' Parties, the Bali roadmap subtly tilted the debate for future negotiations. It may signal that certain non-Annex I countries should accept mitigation obligations (phrased as 'actions'), although fairly differentiated to reflect the relative prosperity, resources, technical capacity and historical emissions of individual states (Earth Negotiations Bulletin, 2007, p. 19). Such a move would, of course, depend ultimately on developed countries' commitment to make proportionately deep emission cuts. Encouragingly, the European Union (EU) made a unilateral commitment in early 2007 to slash its GHG emissions by 20 per cent by 2020 (compared to a 1990 baseline) and possibly to curtail emissions by as much as 30 per cent if other industrialized countries follow suit (Traynor and Gow, 2007) (although it is also important to note that to be effective at minimizing catastrophic climate change, much deeper global cuts would have to be realized by mid-century). The fact that the new Obama Administration has pledged to take vigorous measures to reduce US emissions makes international cooperation in this area more likely.

Another, more immediately useful, outcome of the Bali meeting was the agreement reached on the implementation of the Adaptation Fund established under the Kyoto Protocol. Sustained by a 2 per cent levy on the CO_2 credits generated through CDM projects, the Fund serves to assist developing countries to adapt to the impacts of climate change. The Fund had well over US$300 million in accumulated reserves as of early 2009.[13] The Bali conference recognized adaptation as a much more serious policy concern than previous COPs, and the Bali Action Plan calls for 'enhanced action on adaptation' including more 'vulnerability assessments', 'capacity-building and response strategies', and 'integration of adaptation actions into sectoral and national planning' for the benefit of 'developing countries that are particularly vulnerable to the adverse effects of climate change' (UNFCCC Conference of the Parties, 2008, p. 4). New and additional financial resources to support such activities have yet to be fully realized, and if history is any guide we should not get our hopes too high. As Christina Voigt, Damilola S. Olawuyi and some other contributors to this book contend, the CDM is a flawed instrument the administrative complexity of which hampers its ability to attract investment and thereby to generate the levy income necessary to support the Adaptation Fund. The CDM itself did not receive much attention at Bali, apart from the discussions about the need to simplify its cumbersome operational procedures, and its potential role in reducing deforestation in developing countries.

The Bali conference discussed how to reduce deforestation by providing incentives in a future agreement for countries to protect significant carbon sinks, as Claire Stockwell and her co-authors explain further in their chapter in this book. At the Montreal COP in 2005, the governments of Costa Rica and Papua New Guinea advocated reforms to the Kyoto regime to address this challenge, such as by making forest conservation activities eligible under the CDM. Later, Brazil proposed that an international fund be established to provide compensatory payments to countries that prevent deforestation (Karousakis and Corfee-Morlot, 2007, p. 39). The Bali COP agreed to launch a work programme called 'Reducing Emissions from Deforestation and Degradation (REDD)', which aims to develop new policy mechanisms including resolving the methodologies to enable REDD projects to be CDM eligible.[14] Reforestation projects are already eligible, but not measures to conserve existing forests. Also at the Bali conference, the World Bank announced a new US$250 million Forest Carbon Partnership Facility, which the Bank's Board had approved in September 2007. These initiatives have however incurred criticism from some indigenous peoples who are concerned about the potential for increased foreign intrusion into the management of their traditional lands (Forest Peoples Programme, 2008). Many forested parts of the Amazon and other regions remain occupied by indigenous tribes who do not necessarily welcome conservation projects that may lead to the loss of their control over resource management decisions.

Also on the agenda of Bali and other recent international negotiations has been the question of how to enhance environmental technology transfers to the South to facilitate their shift to a low carbon economy. The Bali meeting focused on possible institutional arrangements, performance indicators, and financing arrangements for a new technology transfer facility. On financing for technology transfer, developing countries sought a new funding mechanism under the auspices of the UNFCCC. The previously established Expert Group on Technology Transfer (EGTT) was given at Bali an additional five-year mandate, with the further task of designing performance indicators to measure progress on clean technology transfer to the South.

Beyond Bali and other UNFCCC-related meetings, a vast array of policy ideas have been advanced and debated in the academic and scientific community, as well as among nongovernmental organizations (NGOs), the business sector and intergovernmental groups. Many of these initiatives deal directly with developing country concerns. One example is the work of the G8+5 legislators group (UK, Germany, US, Japan, France, Canada, Russia, Italy, as well as Brazil, China, India, Mexico and South Africa), whose 2008 report *Combating Climate Change: An International Cooperation Framework Beyond 2012* contains innovative proposals on issues of adaptation measures and technology transfer funding, and options for mitigation targets for some developing countries (Jay, 2008). Another example is the Global Leadership for Climate Action (GLCA), consisting of government, business and civil society representatives from over 20 nations.[15] The GLCA published a *Framework for a Post-2012 Agreement on Climate Change* in 2007, which focuses on the poorest countries and the disproportionate impact of climate change on such vulnerable nations. The proposed framework includes measures to overcome barriers to dissemination of clean energy technologies to the South, such as intellectual property rights and competition rules, sectoral agreement mitigation targets covering rapidly industrializing countries, policies to protect carbon sinks, and a universal carbon tax. These are merely a few of the many ideas being canvassed to enable the North and South to protect collaboratively the global climate while ensuring social and economic justice to the world's most disadvantaged.

Yet, since these and other initiatives were proposed, the conditions for taking action on climate change have shifted dramatically. The economic recession that swept the world in late 2008 and early 2009, which is forecast to be the most severe since the Great Depression, will likely disrupt progress in various ways. It has stirred governments, in developed and developing countries alike, to concentrate on restoring economic growth. Concomitantly, climate change and other environmental concerns have become a much lower priority for most policy-makers (New Zealand Herald, 2008). While economic stimulus programmes in both developed and developing countries contain some support for renewable energy and low-carbon technologies, they also feature massive investments in coal, oil and natural gas infrastructure and GHG-intensive manufacturing industries. The American and Canadian governments have spent billions of dollars bailing out the American auto makers that have fought almost every effort to regulate fuel efficiency and continue to bet their futures on large gas-guzzling vehicles. With hardly any exceptions, the world's governments have not taken the current financial and economic crisis as an opportunity to reorient their economies fundamentally toward a post-carbon future. Secondly, the economic malaise will surely reduce the resources that governments and businesses can mobilize to commit to climate change actions, such as investments in CDM projects, funding technology transfer and supporting adaptation measures. Unless such commitments are viewed as compatible with the imperatives of economic revival, they are likely to be marginalized.

On the other hand, the current crisis may cause the rate of increase of global GHG emissions to slow. In times of economic contraction, the ecological footprint of the economy usually diminishes. Already, for instance, recent reductions in deforestation rates in the Amazon are being linked to the economic downturn (Brooks, 2009). Yet, the poverty and hardship that accompany periods of economic distress may also intensify other myopic practices that undermine the long-term commitments needed to safeguard the planet's climate.

5. SCHOLARSHIP ON CLIMATE LAW AND DEVELOPING COUNTRIES

This book provides the first sustained scholarly analysis of climate law and policy issues related to developing countries. Most climate law research to date has concentrated on OECD nations and international cooperation. While some of this research has considered developing countries, for example in relation to deforestation, biofuels or the Kyoto Protocol's CDM, the focus is usually on their implications for global or Northern climate law. Relatively little research has addressed climate law challenges specific to the South. Yet the UN climate change conference in Bali in December 2007 focused particularly on the possible role of developing countries in a future climate treaty including the controversial question of whether they should accept limits on their GHG emissions. This book is therefore timely in that it includes several chapters by developing country scholars who speak from a position of direct experience about the interests and needs of their nations. For example, measures to address climate change in the developing world must place even higher emphasis on poverty alleviation and promotion of social justice than is the case in North American or EU climate law.

This book also builds on the growing body of scholarship on environmental law in developing countries. Rajendra Ramlogan's (2004) book, *The Developing World and the Environment,* focuses on the role of international law in relation to climate change in the developing world. The IUCN Environmental Law Program's (2001) *Environmental Law in Developing Countries* is a collection of work written by the Program's visiting fellows from various regions of the world. Roberts' and Parks' (2006) work, *A Climate Change of Injustice: Global Inequality, North-South Politics, and Climate Policy*, and Joyeeta Gupta's (1997) *The Climate Change Convention and Developing Countries: From Conflict to Consensus?*, approach the issue from the perspective of international politics and negotiation. Lavanya Rajamani's (2006) essential work *Differential Treatment in International Environmental Law* addresses a specific sub-issue in this important debate. Two Canadian professors of political science, Jordi Diez and O.P. Dwivendi, in 2008 edited a collection of papers on *Global Environmental Challenges: Perspectives from the South* which includes many case studies on the environmental effects of globalization at various places in the Global South, but from the perspective of politics and economics, not law. Finally, another recent work, edited by Ernesto Zedillo (2008), Director of the Yale Center for the Study of Globalization, is *Global Warming: Looking Beyond Kyoto*. It also combines scholars from the North and South and addresses climate change, but from economic, social and political perspectives, omitting the legal angle. The Working Group on Climate Change and Development (2004) also published a brief report *Up in Smoke? Threats from, and Responses to, the Impact of Global Warming on Human Development,* outlining the connection between climate change and development, and providing specific case studies of the various particular effects of climate change on development in specific parts of the world, but again the focus was not legal. Finally, a book on *Climate Change and Africa* edited by Pak Sum Low (2005), contains many analyses of the various elements of climate change in Africa, including mitigation, adaptation and capacity–building, from a variety of disciplines, yet its primary focus again is not on legal aspects of the problem.

There have also been several special editions of academic legal journals devoted to climate change, and a few specifically to climate law in developing countries. Leading examples include the 2008 twentieth anniversary issue of the *Georgetown International Environmental Law Review*, the focus of which was 'Beyond Kyoto: The Developing World and Climate Change'.[16] The journal *Sustainable Development Law and Policy* also issued a special volume devoted to Climate Law in 2008. This issue included discussion of a post-Kyoto agreement, a paper by Christina Voigt on the CDM (a topic addressed in her contribution to this volume), carbon markets, climate litigation, and other related topics. Most contributors were however from the developed world. A 2007 issue of the *UCLA Journal of Environmental Law and Policy* was dedicated to 'Coping with Global Warming', providing an array of useful analyses of climate justice, Indigenous peoples and other topics also considered in this book.

In terms of other legal literature, there is a growing body of analysis of climate change law, often from the perspective of particular jurisdictions, especially in the North. These include Tim Bonyhady and Peter Christoff (2007), *Climate Law in Australia*; Wybe Douma, Leornardo Massai and Massimiliano Montini (2007), *The Kyoto Protocol and Beyond: Legal and Policy Challenges of Climate Change*; Michael Gerrard (2007), *Global Climate Change and US Law*; Marjan Peeters and Kurt Deketelaere (2006), *EU Climate Change Policy*; and Catherine Redgwell and others (2008), *Beyond the Carbon Economy*.

The present book therefore fills a gap in the literature on the role of law and the unique situation of developing countries as we move toward the post-2012 world.

6. PLAN OF THE BOOK

The remaining 15 chapters in this volume are divided into four thematic sections: (i) Climate Justice; (ii) Expanding the Scope of the Climate Change Legal Regimes; (iii) The Clean Development Mechanism and Mitigation Strategies; and (iv) Climate Policy Bridging the North and South.

Climate Justice

The international legal principle of 'common but differentiated responsibility' is at the core of the UNFCCC and the Kyoto Protocol. How this principle applies to state responsibility in the context of climate change is the question that **Sumudu Atapattu** examines in 'Climate Change, Differentiated Responsibilities and State Responsibility: Devising Novel Legal Strategies for Damage Caused by Climate Change'. She rejects a simplistic classification between developed and developing countries, as some developing countries are major contributors to GHG emissions while some communities in developed countries are as vulnerable to climate change as some of the least developing countries that contribute negligible GHGs. Atapattu thus advocates a nuanced differentiation of responsibilities among developing nations in the negotiations of a post-Kyoto regime. The second part of her chapter demonstrates, in part through a study of the Inuit Petition before the Inter-American Commission on Human Rights, that some communities in developed countries are also vulnerable to climate change. She then proceeds to explore possible new avenues for legal principles of liability and compensation that take into account the complexity of issues attached to climate change claims. Atapattu also stresses the need for adaptation strategies linked to poverty reduction, concluding that liability regimes cannot alone redress the harm caused by climate change.

While Atapattu's chapter highlights differences among developing countries as GHG emitters, **Deepa Badrinarayana** focuses on the disparities within a developing country. India is fast becoming one of the world's largest emitters of GHGs. While it has taken a number of domestic measures to mitigate these emissions, India has nevertheless resisted the call from some other countries to accept international binding obligations to reduce these emissions within a set time frame on the basis of India's need for economic development, historical equity and its low per capita share of emissions. In 'India's Constitutional Challenge: A Less Visible Climate Change Catastrophe', Badrinarayana argues that this stance in the post-Kyoto negotiations could be at the expense of the poorest within India who will suffer most from the consequences of climate change. Given that past litigation based on constitutionally recognized human rights has been successful in environmental cases in India, Badrinarayana examines whether a similar strategy could lead to more meaningful actions by the Indian government, both at the domestic and international level, to respond to the challenges brought by climate change.

Another issue of climate justice tackled in this part of the book is that of climate refugees. The displacement and migration of an increasing number of persons because of climate change has led many to ask whether we should expand the current legal protection for refugees and internally displaced persons to offer some protection to climate change refugees. In 'Promoting Justice within the International Legal System: Prospects for Climate Refugees', **Angela Williams** takes a novel approach, exploring how discourses regarding justice can provide a foundation on which developing countries can build to address the dire consequences of global warming. In turn, she examines remedial justice claims and the promises and difficulties of domestic and international litigation, and distributional justice claims based on the concept of intra-generational equity. Williams concludes that both of these theories of justice are necessary to empower groups, communities and states particularly vulnerable to the effects of climate change.

Indigenous peoples all over the world are also vulnerable to climate change, and some are already climate refugees. In their chapters, both Williams and Attapattu discuss climate change litigation involving indigenous communities. **Eric Kwa** also explores the impact of climate change on these communities in 'Climate Change and Indigenous Peoples in the South Pacific: The Need for Regional and Local Strategies'. Many of the island states located in this region are among the most likely to be seriously harmed by the adverse effects of global warming. As Kwa explains, this threat is particularly serious for small low-lying island states and small atolls in Papua New Guinea and Vanuatu. He stresses that the response to this unprecedented threat should be tailored to this region's needs, and in particular to integrate local traditional knowledge, law and practices from indigenous communities at risk. As Kwa observes, these customs and practices play a seminal role in the promotion of biological conservation and sustainable use. They have legal status in many national laws in South Pacific countries and, in some cases, already help communities to adapt to climate change. In his view, national laws could provide a framework by which customary laws, knowledge and practices might contribute more to climate change mitigation and adaptation. In this regard, recent regional and national initiatives are promising but much remains to be done, including devising regulatory frameworks for CDM projects that are sensitive to indigenous communities' concerns.

Expanding the Scope of the Climate Change Legal Regimes

As our understanding of climate change's causes and consequences evolves, new proposals to respond to this challenge are put forward. In this section of the book, scholars explore various means to expand the scope of climate change legal regimes internationally, regionally and domestically.

Early developments in international climate law focused heavily on mitigating GHG emissions. However, as the adverse effects of climate change became more concrete and frequent, concerns about *adaptation* to these changes have come to the fore and will no doubt figure prominently in the legal regime that will result from the post-Kyoto negotiations. **Jolene Lin**, in 'Supporting Adaptation in Developing Countries at the National and Global Levels', reminds us that the UNFCCC and the Kyoto Protocol do contain provisions on adaptation, even if these provisions are not very explicit or extensive. This deficiency has led some scholars to call for the adoption of a special Adaptation Protocol. Lin believes however that adaptation is now accepted as an essential element in the UNFCCC agenda and, therefore, there is no longer a need to call for the adoption of a separate instrument to rally the international community around this concept. Progress on the adaptation front requires more expertise and financial commitments. International organizations already play a key role, having 'mainstreamed adaptation' in their agenda, as Lin explains. In the second part of her paper, Lin draws on three case studies (Vietnam, China and Laos) to conclude that the domestic institutional and normative framework for adaptation work in developing countries needs to be further strengthened to allow, among other things, for proper input from local communities and NGOs. Imposing legal obligations on the Parties to the UNFCCC to establish a focal point for adaptation issues and to provide for the involvement of local governments would be further steps in the right direction.

The Kyoto Protocol lacks specific means to deal with some of the major causes of global GHG emissions. For instance, deforestation, which accounts for approximately 20 per cent of global emissions, was not included as a component of the CDM. The post-2012 negotiations provide an opportunity to revisit this issue, with some countries supporting the creation of a REDD programme. Given the diversity of causes of deforestation, designing this programme will no doubt be complicated. Key choices will have to be made on several issues that **Claire Stockwell**, **Bill Hare** and **Kirsten Macey** canvass in 'Designing a REDD Mechanism: The TDERM Triptych'. They firstly highlight the host of choices facing the international community when addressing elements that are inherent to this kind of programme. For instance, what will be the nature of commitments (incentive or obligation-based) and will they be confined to developing countries or extend to developed countries? Will this programme be funded by market or non-market means? Will sub-national actors have access to these international mechanisms? What will the nature of institutional arrangements be? As these issues overlap with those that have arisen for the CDM, the authors point to some of the lessons and warnings one can draw from the establishment and implementation of this mechanism. Finally, they propose a Tropical Deforestation Emission Reduction Mechanism (TDERM) to address REDD. One of the proposal's distinctive features is to include some asymmetry in most of its key components in order to ensure participation by developing countries with limited economic and technological capacities.

In 'The Role of Marine "Forests" and Soils as Carbon Sinks: Enhanced Bio-Sequestration as a Mitigation Strategy to Help Avoid Dangerous Climate Change', **Robert Fowler** explores another way in which the scope of an international agreement could be enlarged as part of the post-Kyoto negotiations in order to stabilize GHG emissions at a safe level. Recent scientific studies suggest that the stabilization target could be much higher than what was projected a few years ago. This prompts Fowler to assess the potential of enhanced bio-sequestration. In particular, he focuses on two carbon sinks which, until now, have been neglected: sea-grass meadows and soils. Developing countries could play a major role in the implementation of this new strategy. More generally, Fowler suggests that states should build flexible mechanisms into the agreement that will emerge from post-Kyoto negotiations to be able to react promptly to new scientific developments.

Climate change has surged on the lists of threats to biodiversity since the late 1990s. The legal means that may help the North and the South meet this complex challenge are the object of the last chapter in this part. In 'Adaptation to Climate Change to Save Biodiversity: Lessons Learned from African and European Experiences', **Saja Erens**, **Jonathan Verschuuren** and **Kees Bastmeijer** canvass the extent to which regional instruments related to protected areas in Europe and Africa implement an 'ecological network' approach as a means to respond to the fragmentation and isolation of species caused by climate change. They highlight initiatives in two regions (the Danube River Basin in Europe and the Southern African Development Community), from which they draw useful lessons for climate change adaptation.

The Clean Development Mechanism and Mitigation Strategies

At its inception the CDM was seen as a win-win proposal for the North and the South, as it allowed flexibility for developed states to comply with their reduction commitments under the Kyoto Protocol and enabled the funding of new sustainable projects and transfers of technology to developing countries. Since then many scholars have raised a number of concerns with regard to the CDM and see the post-Kyoto negotiations as an opportunity to improve on it. Likewise, governments in developing countries need to modify their laws to take full advantage of the benefits that the CDM can offer.

As **Christina Voigt** explains in 'The Deadlock of the Clean Development Mechanism: Caught between Sustainability, Environmental Integrity and Economic Efficiency', the mechanism will only be able to deliver long-term benefits once existing tensions within it are properly addressed. In her view, the contribution of the CDM to sustainable development and environmental integrity is counterbalanced by its demands for procedural efficiency and economic feasibility. Currently, the latter considerations too often outweigh the former. Voigt argues that additional legal rules are needed to right this balance, including measures to (i) ensure the additionality of projects; (ii) prevent a CDM project leading to increased emissions in another industry or neighbouring region (known as 'carbon leakage') or a reduction of energy through a project being neutralized by a surge of energy elsewhere (the 'rebound effect'); (iii) require environmental and sustainability impact assessments and an increase in the scope of these assessments beyond GHG emissions; and (iv) decrease the probability that CDM projects will act as a disincentive for developing countries to accept binding obligations to reduce emissions or for developing and developed countries to pursue emission reductions within their territory. Finally, Voigt addresses the ethical dimensions of the CDM, and proposes some political and legal reforms that, in her view, are necessary if the CDM is to survive as an innovative tool for North–South collaboration.

A specific example of the potential of the CDM is the forestry sector. Forestry management can help mitigate GHG emissions and enhance adaptation to climate change while contributing to sustainable development. In his chapter, 'Policy and Legal Dimensions of CDM Projects in the Forestry Sector: Implications for Climate Change Mitigation and Adaptation in Uganda', **Emmanuel B. Kasimbazi** details some of the CDM projects in Uganda's forestry sector. He also explains the legal framework and the key policies necessary for the implementation of these projects, detailing the desired relationships between private entities, government and local communities. Kasimbazi identifies several obstacles to effective implementation of CDM forestry projects, such as gaps in the regulatory framework and conflicts that pit government and private entities against local communities in which environmental regulatory agencies appear to be powerless. Other factors, such as the lack of financial resources and expertise, hamper the proper implementation and verification of CDM projects. Kasimbazi formulates a series of recommendations aimed at the Ugandan government and the international community to improve on a mechanism which still has the potential to play a key role in Uganda's efforts to mitigate and adapt to climate change.

Uganda is not alone in facing challenges in relation to the CDM. To date, Africa has attracted only 2 per cent of all CDM projects, with the lion's share going to so called 'CDM giants', namely large developing countries such as Brazil, China, India and Indonesia. In 'Beautifying Africa for the Clean Development Mechanism: Legal and Institutional Issues Considered', **Damilola S. Olawuyi** explores the main reasons why developed countries have not invested more in this continent. These reasons include: a limited potential to deliver high mitigation at low cost; prevailing insecurity in the political and investment climates; and the absence of sound institutional and legal frameworks governing CDM investments. It is to this last factor that Olawuyi devotes most of the chapter. Obvious barriers are, for many countries, the absence of a legislative framework governing CDM and, as a result, the lack of a legally constituted Designated National Authority. Other obstacles canvassed by Olawuyi include: (i) conflicting environmental regulations in federal states such as Nigeria; (ii) archaic laws on technology transfer as well as lack of capacity to integrate properly this technology into the economy; (iii) gaps in contract laws and limited legal expertise in this area; (iv) the absence of a CDM national master plan; and (v) a lack of intergovernmental linkages and coordination. The chapter concludes with a list of legal and policy initiatives that African countries should take to remove, or at least limit, the negative impact of these barriers.

The fight against climate change has led to investment in sources of energy that were largely untapped until recently. Prominent among these, especially in developing countries, is the rise of the biofuel industry which is promoted by many as a clean and renewable source of energy. In 'Brazilian Policy on Biodiesels: A Sound Means of Mitigating Climate Change?' **Solange Teles da Silva** and **Carolina Dutra** examine the development and regulation of this industry in the country that is the leading producer and consumer of biofuel energy. They explain why, in their view, the current Brazilian regulatory framework for biofuels, in terms both of standards and implementation, does not adequately address some of the main concerns that have been voiced with regard to this industry, from deforestation and its impact on biodiversity to the impact on food prices. According to da Silva and Dutra, the promise of biofuels as a clean energy source will only be realized if they can contribute to sustainable development, notably through the CDM.

Climate Policy Bridging the North and South

The CDM is one of myriad means by which the North and South interact with each other in the climate change context. In the last section of this volume, the authors consider other forums in which the interests of the North and the South are at play: voluntary offset markets, the World Trade Organization and development cooperation policies.

Citizens and businesses from developed countries have access to a growing environmental market in which they can purchase 'carbon credits' to offset their GHG emissions. As **Marjan Peeters** explains in 'Improving Citizen Responsibility in the North and Its Consequences for the South: Voluntary Carbon Offsets and Government Involvement', the reductions in emissions represented by these credits originate in great part from developing countries, including through the CDM. However, an increasing portion of these credits is not linked to the Kyoto mechanisms. The carbon credit market raises for Peeters a question of integrity: is this market, driven by the North, beneficial for the populations in the

South? As well, there is a question of credibility to consider: can the consumer from the North get assurances that the carbon credits he or she is purchasing will indeed translate into emission reductions? At the heart of these questions is the issue of governance. Should this market be regulated and, if so, at what level and by whom? Should it be through the UNFCCC, through 'private' regulation by NGOs, as exemplified by the Gold Standard and the Voluntary Carbon Standard, or through official state regulation? To examine these questions, Peeters draws on the practices of the United Kingdom and the Netherlands, two states with divergent approaches to the regulation of the voluntary private carbon offset market. She concludes with remarks on the role governments can play as important buyers of these carbon offsets.

Reconciling economic development with measures to mitigate and adapt to climate change is a daunting challenge for developing countries. One factor that could affect how they address this problem is the climate change policy conditions that developed nations could impose on these countries when importing their products. In 'Climate and Trade in a Divided World: Can Measures Adopted in the North End Up Shaping Climate Legislative Frameworks in the South?', **Francesco Sindico** explores this issue by examining recent US legislative initiatives and their compatibility with international trade rules. Questions related to the coverage of goods, the criteria according to which a developing country would be targeted and the implementation of climate related trade measures are canvassed. Sindico concludes that real change will not come from looking at climate change as an environmental and trade problem, but rather as an issue rooted in consumption and production patterns.

Effective climate change strategy by developed countries will necessitate the adoption of development cooperation policies toward the South that address the latter's need for adaptation and mitigation. Over the last few years the EU has developed a policy that is examined in detail by **Marie-Pierrre Lanfranchi** and **Sandrine Maljean-Dubois**, in terms both of its legality and legitimacy. In 'Climate Change in the European Union Development Cooperation Policy', they take a critical approach, highlighting differences between discourse and implementation. Maljean-Dubois and Lanfranchi consider, in particular, various funding schemes that have been considered or set up recently, such as the Global Energy Efficiency and Renewable Energy Fund. While acknowledging a number of important measures taken by the EU to clarify and improve its development cooperation policy, they also highlight the elements that hamper the effectiveness of an aid package that the EU itself must deliver in tandem with initiatives taken by some of its Member States. Maljean-Dubois and

Lanfranchi conclude by asking whether the EU has the means to deliver its ambitious programme on climate change.

NOTES

* Professors Richardson and Wood are with the Osgoode Hall Law School, York University, and Professors McLeod-Kilmurray and Le Bouthillier are at the University of Ottawa Faculty of Law.

1. For more information on the conference, see www.iucnael.org/content/view/17/30/lang,english (visited 10 April 2009).
2. See www.iucnael.org.
3. This measure reflects the fact that CO_2 is not the only GHG. In order to create a single measure, other GHGs such as methane and nitrous oxide are converted into a CO_2 equivalent.
4. Sierra Club, *Canada's Kyoto Report Card 2008*, pp. 7–8, www.sierraclub.ca/national/kyoto/kyoto-report-card-2008.pdf (visited 10 April 2009).
5. The target of 350 ppm has attracted a growing grassroots movement as well: see www.350.org (visited 10 April 2009).
6. G. Monbiot (2007), speech at the Camp for Climate Change, London, 18 August 2007, www.indymedia.org.uk/en/2007/08/378866.html (visited 10 April 2009).
7. See Alliance of Small Island States, www.sidsnet.org/aosis (visited 5 April 2009).
8. ILM 31 (1992) 849 (UNFCCC).
9. ILM 37 (1998) 22 (Kyoto Protocol).
10. Ibid., Art. 11(2)(a).
11. Ibid., Art. 11(2)(b).
12. Conference of the Parties to the UNFCCC, 'Bonn Agreements on the Implementation of the Buenos Aires Plan of Action', Decision 5/CP.6, in *Report of the Conference of the Parties on the Second Part of its Sixth Session*, UN Doc. FCCC/CP/2001/5 (Bonn, 16–17 July 2001), p. 36.
13. See Adaptation Fund, www.adaptation-fund.org (visited 5 April 2009).
14. For subsequent developments see REDD-Monitor, an information clearing-house for news and critical analysis related to REDD, www.redd-monitor.org (visited 5 April 2009).
15. See Global Leadership for Climate Action, www.globalclimateaction.com (visited 5 April 2009).
16. This journal has had other contributions on this theme as well, namely K.A. Baumert (2006), 'Participation of Developing Countries in the International Climate Change Regime: Lessons for the Future', *George Washington International Law Review* 38, 365; S.M. Neal (1998), 'Bringing Developing Nations on Board the Climate Change Protocol: Using Debt-for-Nature Swaps to Implement the Clean Development Mechanism', *Georgetown International Environmental Law Review* 11, 163.

REFERENCES

Agarwal, A., S. Narain, and A. Sharma (eds) (1999), *Green Politics: Global Environmental Negotiation* (Centre for Science and Environment).

Aldy, J. and R. Stavins (2008), 'Climate Policy Architectures for the Post Kyoto World', *Environment* 50(3), 6.

Amin, S. (1976), *Unequal Development: An Essay on the Social Formations of Peripheral Capitalism* (Monthly Review Press).

Anand, R. (2004), *International Environmental Justice: A North–South Dimension* (Ashgate).
Andreen, W.L. (2000), 'Environmental Law and International Assistance: The Challenge of Strengthening Environmental Law in the Developing World', *Columbia Journal of Environmental Law* 25, 17.
Barrett, S. (2003), *Environment and Statecraft: The Strategy of Environmental Treaty-Making* (Oxford University Press).
Berk, M.M. and M. den Elzen (2001), 'Options for Differentiation of Future Commitments in Climate Policy: How to Realise Timely Participation to Meet Stringent Climate Goals?', *Climate Policy* 1, 465.
Boer, B. (1999), 'The Rise of Environmental Law in the Asian Region', *University of Richmond Law Review* 32, 1503.
Boer, B., R. Ramsay and D.R. Rothwell (1998), *International Environmental Law in the Asia Pacific* (Kluwer).
Bonyhady, T. and P. Christoff (2007), *Climate Law in Australia* (Federation Press).
Boston, J. (2008), 'Global Climate Change Policies: From Bali to Copenhagen and Beyond', *Policy Quarterly* 4(1), 50.
Boyle, A., C. Redgwell and P. Birnie (2009), *International Environmental Law* (Oxford University Press).
Brooks, B. (2009), 'Amazon Deforestation Drops Sharply', *International Herald Tribune*, 23 January.
Bruch, C., W. Coker and C. VanArsdale (2001), 'Constitutional Environmental Law: Giving Force to Fundamental Principles in Africa', *Columbia Journal of Environmental Law* 26, 131.
Canadell, J., et al. (2007), 'Contributions to Accelerating Atmospheric CO_2 Growth from Economic Activity, Carbon Intensity, and Efficiency of Natural Sinks', *Proceedings of the National Academy of Sciences.*
Carrell, S. (2003), 'Arab States Claim CO_2 Targets Could Cause Slump', *The Independent*, 9 February.
Chen, G. (2009), *Problems and Progress: Politics of China's Environmental Protection* (World Scientific Publishing).
Claussen, E. and L. McNeilly (1998), *Equity and Global Climate Change* (Pew Centre on Global Climate Change).
Diez, J. and O.P. Dwivendi (2008), *Global Environmental Challenges: Perspectives from the South* (Broadview Press).
Donner, S.D. (2007), *Occasional Paper – Canada Country Study. Human Development Report 2007–08. Fighting Climate Change: Human Solidarity in a Divided World* (United Nations Development Program).
Douma, W., L. Massai and M. Montini (eds) (2007), *The Kyoto Protocol and Beyond: Legal and Policy Challenges of Climate Change* (Asser Press).
Earth Negotiations Bulletin (2007), 'Summary of the Thirteenth Conference of Parties to the UN Framework Convention on Climate Change and Third Meeting of Parties to the Kyoto Protocol: 3–15 December 2007', *Earth Negotiations Bulletin* 12(354), 18 December.
Epstein, P.R., et al. (2008), *Healthy Solutions for the Low Carbon Economy – Guidelines for Investors, Insurers and Policy Makers* (Centre for Health and the Global Environment, and Harvard Medical School).
Escobar, A. (1995), *Encountering Development. The Making and Unmaking of the Third World* (Princeton University Press).

Flannery, T. (2005), *The Weather Makers: How Man is Changing the Climate and What it Means for Life on Earth* (Atlantic Monthly Press).
Forest Peoples Programme (FPP) (2008), *The Forest Carbon Partnership Facility: Facilitating the Weakening of Indigenous Peoples' Rights to Lands and Resources* (FPP).
Gathii, J. (1998–99), 'Representations of Africa in Good Governance Discourse: Policing and Containing Dissidence to Neo-Liberalism', *Third World Legal Studies* 18, 65.
Gathii, J. (1999), 'The Limits of the New International Rule of Law on Good Governance', in E.K. Quashigah and O.C. Okafor (eds), *Legitimate Governance in Africa* (Kluwer).
Gathii, J. (2000), 'Retelling Good Governance Narratives About Africa's Economic and Political Predicaments: Continuities and Discontinuities in Legal Outcomes Between Markets and States', *Villanova Law Review* 45, 971.
Gerrard, M. (ed.) (2007), *Global Climate Change and US Law* (American Bar Association).
Gupta, J. (1997), *The Climate Change Convention and Developing Countries: From Conflict to Consensus?* (Kluwer).
Hansen, J., et al. (2008), 'Target Atmospheric CO_2: Where Should Humanity Aim?', *Open Atmospheric Science Journal* 2, 217.
Harashima, Y. (2000), 'Environmental Governance in Selected Asian Developing Countries', *International Review for Environmental Strategies* 1(1), 193.
Hayter, T. and C. Watson (1994), *Aid: Rhetoric and Reality* (Pluto Press).
Hayward, T. (2007), 'Human Rights Versus Emissions Rights: Climate Justice and the Equitable Distribution of Ecological Space', *Ethics and International Affairs* 21(4), 431.
Hughes, H. (2009), *Achieving Industrialization in East Asia* (Cambridge University Press).
Intergovernmental Panel on Climate Change (IPCC) (2000), *Special Report on Emissions Scenarios* (IPCC).
Intergovernmental Panel on Climate Change (IPCC) (2007a), *Fourth Assessment Report: Impacts, Adaptation and Vulnerability*, Working Group II (Cambridge University Press).
Intergovernmental Panel on Climate Change (IPCC) (2007b), *Climate Change 2007: The Physical Science Basis: Summary for Policymakers*, Working Group II (Cambridge University Press).
Intergovernmental Panel on Climate Change (IPCC) (2007c), *Climate Change 2007: Mitigation of Climate Change – Summary for Policymakers*, Working Group III (Cambridge University Press).
Intergovernmental Panel on Climate Change (IPCC) (2007d), *2007 Synthesis Report: Summary for Policy Makers* (Cambridge University Press).
International Energy Agency (IEA) (2008), *CO_2 Emissions from Fuel Combustion 2008* (IEA).
IUCN Environmental Law Programme (ELP) (2004), *Environmental Law in Developing Countries* (IUCN Environmental Policy and Law Paper No. 43), (IUCN ELP).
Jay, M. (2008), *Combating Climate Change: An International Cooperation Framework Beyond 2012. A Consensus Paper by Legislators from the G8 and +5 Countries* (Globe International).
Kameri-Mbote, P. and P. Cullet (1995), 'Law, Colonialism and Environmental

Management in Africa', *Review of European Community and International Environmental Law* 6, 23.
Karousakis, K. and J. Corfee-Morlot (2007), *Financing Mechanisms to Reduce Emissions from Deforestation: Issues in Design and Implementation* (OECD).
Kim, E.M. (1998), *The Four Asian Tigers: Economic Development and the Global Political Economy* (Academic Press).
Klein, N. (2000), *No Logo: Taking Aim at Brand Bullies* (Vintage Canada).
Low, P.S. (2005), *Climate Change and Africa* (Cambridge University Press).
Makhijani, A. (2007), *Carbon-Free and Nuclear-Free* (Institute for Energy and Environmental Research).
Martens, T.K. (1989), 'Ending Tropical Deforestation: What is the Proper Role of the World Bank', *Harvard Environmental Law Review* 13, 485.
McClintock, A. (1995), *Imperial Leather* (Routledge).
McElroy, M., C. Nielsen and P. Lydon (1998), *Energizing China: Reconciling Environmental Protection and Economic Growth* (Harvard University Center for the Environment).
Millennium Ecosystem Assessment (MEA) (2005), *Ecosystems and Human Well-being: Biodiversity Synthesis* (MEA).
Monbiot, G. (2007), 'We've been Suckered Again by the US', *Guardian*, 17 December.
Najam, A., S. Huq and Y. Sokona (2003), 'Climate Negotiations Beyond Kyoto: Developing Countries Concerns and Interests', *Climate Policy* 3, 221.
New Zealand Herald (2008), 'Green Politics Seem a Luxury in Recession', *New Zealand Herald* 18 December.
Nordhaus, W.D. (2006), 'After Kyoto: Alternative Mechanisms to Control Global Warming', *American Economic Review* 96, 31.
Ott, H.E. (2007), *Climate Policy Post-2012 – A Roadmap: The Global Governance of Climate Change* (a discussion paper for the 2007 Tallberg Forum), (Tallberg Foundation).
Otto, D. (1996), 'Subalternity and International Law: The Problems of the Global Community and the Incommensurability of Difference', *Society and Legal Studies* 5, 337.
Page, E. (2006), *Climate Change, Justice and Future Generations* (Edward Elgar Publishing).
Pardy, B. (2004). 'The Kyoto Protocol: Bad News for the Global Environment', *Journal of Environmental Law and Practice* 14, 27.
Paterson, M. (2001), 'Principles of Justice in the Context of Global Climate Change', in U. Luterbacher and D.F. Sprinz (eds), *International Relations and Global Climate Change* (MIT Press).
Peerenboom, R. (2002), *China's Long March Toward Rule of Law* (Cambridge University Press).
Peeters, M. and K. Deketelaere (2006), *EU Climate Change Policy* (Edward Elgar Publishing).
Rajamani, L. (2006), *Differential Treatment in International Environmental Law* (Oxford University Press).
Rajan, G. and R. Mohanram (eds) (1995), *Postcolonial Discourse and Changing Cultural Contexts* (Greenwood Publishing).
Ramlogan, R. (2004), *The Developing World and the Environment* (University Press of America).

Redgwell, C., et al. (eds) (2008), *Beyond the Carbon Economy* (Oxford University Press).
Republic of Costa Rica and Papua New Guinea (2005), *Submission by the Governments of Papua New Guinea and Costa Rica: Reducing Emissions from Deforestation in Developing Countries: Approaches to Stimulate Action*, UN Framework Convention on Climate Change, Conference of the Parties: FCCC/CP/2005/MISC.1.
Rich, B. (1994), *Mortgaging the Earth: The World Bank, Environmental Impoverishment and the Crisis of Development* (Beacon Press).
Richardson, B.J. (2000), 'Environmental Law in Postcolonial Societies: Straddling the Local – Global Institutional Spectrum', *Colorado Journal of International Environmental Law and Policy* 11(1), 1.
Richardson, B.J. (2004), 'Uneven Modernisation: Environmental Regulation in China's Special Economic Zones', *Asia Pacific Law Review* 12(2), 115.
Richardson, B.J. (2005), 'Is East Asia Industrializing Too Quickly? Environmental Regulation in Its Special Economic Zones', *UCLA Pacific Basin Law Journal* 22, 150.
Richardson, B.J. (2009), 'Climate Finance and its Governance: Moving to a Low Carbon Economy through Socially Responsible Financing?', *International and Comparative Law Quarterly* 58, 597.
Roberts, J.T. and B.C. Parks (2007), *A Climate of Injustice: Global Inequality, North-South Politics, and Climate Policy* (MIT Press).
Sharma, K. (2006), *Trade, Growth and Inequality in Developing Countries* (Routledge).
Stern, N. (2007), *The Economics of Climate Change* (Cambridge University Press).
Stewart, S.B. and J.B. Wiener (2003), *Reconstructing Climate Policy* (Aei Press).
Thomas, C. (1992), *The Environment in International Relations* (Royal Institute of International Affairs).
Traynor, I. and D. Gow (2007), 'EU Promises 20% Reduction in Carbon Emissions by 2020', *Guardian*, 21 February.
Trubek, D.M. and A. Santos (eds) (2006), *The New Law and Economic Development* (Cambridge University Press).
United Nations Economic Commission for Africa (2006), *Economic Report on Africa: Meeting the Challenges of Unemployment and Poverty in Africa* (United Nations Publications).
United Nations Framework Convention on Climate Change (UNFCCC) Conference of the Parties (2008), *Report of the Conference of the Parties on its Thirteenth Session, held in Bali from 3 to 15 December 2007. Part Two: Action taken by the Conference of the Parties at its Thirteenth Session*, FCCC/CP/2007/6/Add.1, 14 March.
United Nations News Centre (2007), 'Secretary-General Ban Welcomes Climate Change Breakthrough in Bali', 15 December.
Wang, Y. (2002), *Chinese Legal Reform: The Case of Foreign Investment* (Routledge).
Watson, F. (2008), 'A View from the Forest Floor: The Impact of Logging on Indigenous Peoples in Brazil', *Botanical Journal of the Linnean Society* 122(1), 75.
Weiss, E. (1989), *In Fairness to Future Generations: International Law, Common Patrimony, and Intergenerational Equity* (Transnational Publishers).

Weizsäcker, E.U., von, et al. (1999), *The Kyoto Protocol: International Climate Policy for the 21st Century* (Springer-Verlag Berlin and Heidelberg GmbH & Co.K).
Working Group on Climate Change and Development (2004), *Up in Smoke? Threats from, and responses to, the impact of global warming on human development* (New Economics Foundation).
Wunsch, J. and D. Olowu (eds) (1990), *The Failure of the Centralized State: Institutions and Self-Governance in Africa* (Westview Press).
Zedillo, E. (ed.) (2008), *Global Warming: Looking Beyond Kyoto* (Brookings Institution Press).

PART 1

Climate justice

2. Climate change, differentiated responsibilities and state responsibility: devising novel legal strategies for damage caused by climate change

Sumudu Atapattu*

> Climate change is the defining human development issue of our generation. All development is ultimately about expanding human potential and enlarging human freedom . . . Climate change threatens to erode human freedoms and limit choice. It calls into question the Enlightenment principle that human progress will make the future look better than the past.
>
> *Human Development Report 2007/2008, UNDP, Overview*

1. INTRODUCTION

While global climate change originated as an environmental problem, it now impinges on every aspect of human life with implications for international economy, public health, social issues such as migration and loss of livelihood, and, ultimately, threatening international peace and security (United Nations Department of Public Information, 2007). Climate change is considered a problem created mainly by 'rich countries', but the burden will be disproportionately borne by 'poor countries' – however, the issues are not this simple. For example, deforestation in developing countries, particularly Indonesia and Brazil, is considered a major contributor to greenhouse gases (GHG) (Moutinho and Schwartzman, 2005), while rapidly industrializing countries such as India and China are becoming major contributors themselves. Conversely, some poor communities in developed countries are beginning to be disproportionately affected by the impacts of climate change. Indigenous peoples in North America are one example.

International environmental law and the legal regime governing climate change (UN Framework Convention on Climate Change[1] (UNFCCC) and the Kyoto Protocol[2]) have sought to take the vast disparity between developing countries and developed countries into account, as well as their historic contribution to climate and other environmental problems through the 'common but differentiated responsibility (CBDR) principle'.

This chapter analyses issues of equity surrounding the CBDR principle and the challenges that climate change poses for traditional principles of international law. It examines the relationship between state responsibility and the CBDR principle, and how the latter has been applied to climate change. In particular, it questions the validity of the traditional distinction made between so-called 'developing' and 'developed' countries. Vast disparities exist in these broad categories that are particularly pertinent to how countries approach various environmental issues.[3] The chapter will use the Inuit Petition before the Inter-American Commission on Human Rights as a case study to highlight the inequalities that also exist within developed countries, as well as to examine the challenges of achieving justice for such peoples under current legal rules such as state responsibility for climate change-related impacts. Their claim raises complex issues relating to standing, causation and remedies for environmental harms. The Inuit case is an example of an innovative trend that has been taking shape at the domestic level which has now been extended to the international plane – that of using human rights machinery and remedies to seek redress for environmental wrongs. The value of these human rights approaches to environmental problems will be discussed briefly. This chapter argues that such innovative legal strategies must be adopted if claims for climate change-related impacts are to be successful at the international level.

While states remain the primary subjects of international law, other entities are playing a seminal role in international affairs (Sands, 2003, p. 124). Given the all-pervasive nature of climate change and its effects on humankind and the enjoyment of human rights, its effects on these non-state actors need to be considered. Vulnerable non-state actors such as indigenous people often are paying a disproportionate price as a result of climate change. Any discussion of equity in relation to climate change must accommodate the interests of such groups. The CBDR principle has played an important role in international regime creation in the environmental field. However, the time has surely come to use it to differentiate *within* the broad categories of developing or developed countries. This chapter concludes by proposing some reforms in the post-2012 legal regime to give effect to these circumstances and challenges.

2. CLIMATE CHANGE, EQUITY AND THE NORTH–SOUTH DIALOGUE

The science of climate change and its occurrence are no longer doubted (Intergovernmental Panel on Climate Change (IPCC), 2007). While global warming is a problem created mainly by the global North, some developing countries through land use changes such as deforestation have also contributed to the phenomenon. Moreover, some highly populated, rapidly industrializing nations are catching up with their industrialized counterparts. While no state in the international community would escape the consequences of climate change, those who would be severely affected by it are likely not to be those who caused the problem in the first place. As the United Nations Development Program (UNDP) noted in its 2007/08 Human Development Report, poor and vulnerable communities will disproportionately bear the consequences of climate change. Moreover, these communities have the least capacity to adapt to the potential consequences of climate change including increased risk of disease. As Patz and others explain, '[h]erein lies the ethical dilemma of climate change and health: those most vulnerable to the health risks are also those least responsible for creating the problem' (Patz, 2007, p. 398).

Climate change originated as a typical North–South issue of political contention, involving challenges of meeting intra- and inter-generational equity. Similar to the debate on addressing ozone depletion (Hunter, Salzman and Zaelke, 2007, chapter 9), the debate on climate change has had to address issues of equity not only in relation to the present generation but also in relation to posterity. The main dilemma facing states was how to devise an international legal regime which accommodates the vast disparity between developing and developed countries, not only in terms of wealth (and, therefore, the ability to meet the challenges posed by climate change), but also in terms of their historic and projected GHG emissions. To address this challenge, the international community devised the CBDR principle, which is incorporated into the UNFCCC as one of its guiding principles (Article 3). Other principles featured in the UNFCCC, and other international instruments such as the Rio Declaration on Environment and Development[4] (hereinafter, the 'Rio Declaration'), include inter- and intra-generational equity and the precautionary principle (Atapattu, 2006).

3. COMMON BUT DIFFERENTIATED RESPONSIBILITY PRINCIPLE

The first time the CBDR principle was formally applied was in the Vienna Convention for the Protection of the Ozone Layer of 1985, although the treaty itself does not mention the principle specifically. It merely refers to the need to take into account 'the circumstances and particular requirements of developing countries' (preamble). The Montreal Protocol, while also not specifically referring to the CBDR principle, contains an elaborate set of provisions giving effect to it. Acknowledging that 'special provision is required to meet the needs of developing countries for these substances', the Montreal Protocol adopts, for the first time in international environmental law history, three mechanisms that take into account the special situation of developing countries in formulating their obligations: a different phase out period for such countries with a grace period of 10 years to eliminate the use of controlled substances; the establishment of a fund to help developing countries meet their obligations under the Protocol; and transfer of technology provisions including measures to facilitate their access to environmentally safe alternative substances and technology.[5]

The Rio Declaration was the first international instrument specifically to incorporate the CBDR principle. Principle 7 states, in part:

> [i]n view of the different contributions to global environmental degradation, States have common but differentiated responsibilities. The developed countries acknowledge the responsibility that they bear in the international pursuit of sustainable development in view of the pressures their societies place on the global environment and of the technologies and financial resources they command.

This principle became a very contentious issue at the 1992 Rio Conference. The earlier drafted version of Principle 7 that referred to the legal responsibility of developed countries for their historic contribution to environmental problems was not adopted (Hunter, Salzman and Zaelke, 2007, p. 495). A watered-down version with no reference to legal responsibility was finally adopted. While differential obligations are not new in international law (Atapattu, 2006, p. 400), they constitute an exception to the principles of sovereign equality and reciprocity.

The CBDR principle draws on the principle of equity, an important norm of international law. As I have noted elsewhere:

> [d]espite the rather vague language adopted in Principle 7, there is little doubt that it has broken new ground in international environmental law. While it is open to debate whether it is a legal principle or a guiding principle, it could lead

> to significant legal implications, as more and more environmental treaties adopt differential obligations for states (Atapattu, 2006, p. 387).

The first binding environmental instrument explicitly to incorporate the CBDR principle is the UNFCCC. Both the preamble and Article 3 refer to it. According to Article 3:

> The Parties should protect the climate system for the benefit of present and future generations of humankind, on the basis of equity and in accordance with their common but differentiated responsibilities and respective capabilities. Accordingly, developed country Parties should take the lead in combating climate change and the adverse effects thereof.

Thus, Article 3 places a special responsibility on the global North. It also requires in Article 3.2 that: '[t]he specific needs and special circumstances of developing country Parties, especially those that are particularly vulnerable to the adverse effects of climate change, and of those Parties, especially developing country Parties, that would have to bear a disproportionate or abnormal burden under the Convention, should be given full consideration'.[6] Thus, it is clear that the Parties to the UNFCCC have recognized the need to categorize states within the broad class of developing countries according to whether they are particularly vulnerable to climate change or whether they will bear a disproportionate burden as a result of climate change. Small island states and states with low-lying inhabited areas would fall into the first category, while states with very poor communities could fall into the second.

The Kyoto Protocol, which embodies substantive obligations in relation to climate change, contains no obligations for developing country parties apart from some general obligations that embellish commitments contained in the UNFCCC. These obligations range from developing national inventories of anthropocentric emissions, formulating and implementing national programs to mitigating climate change, promoting and cooperating in the transfer of technology, promoting sustainable development, cooperating in preparing adaptation plans; promoting and cooperating in scientific research; and promoting education, training and public awareness. As can be seen, these are very broad and vague commitments. Only 'Annex I countries' (i.e., developed countries and countries with economies-in-transition) have specific obligations for GHG emission reduction (Article 2).

Thus, unlike the Montreal Protocol which contained specific obligations for developing countries, albeit with a grace period for compliance, the Kyoto Protocol does not contain any emission reduction obligations for developing countries. The United States (US) government cited this lack of binding obligations for developing countries as one of the reasons for not ratifying the Kyoto Protocol. In fact, it passed the Byrd-Hagle resolution[7] (105th Congress, 1997) as early as 1997 which tied US acceptance of binding emission reduction targets, *inter alia*, to the imposition of similar targets for developing countries.

Does this mean that the present climate regime unreasonably favours the South? Certainly, the very purpose of the CBDR principle is to 'favour' one group over another by taking into account certain disparities that exist in international society. In this context, it is useful to recall the wording of Principle 7 of the Rio Declaration: it refers to the *common* obligation of states to protect the environment. This obligation is common to all states, whether they are developing, developed or have economies-in-transition. Moreover, Principle 7 embodies the general obligation of states *to cooperate in a spirit of global partnership* (emphasis added). This general obligation may be qualified by differentiated responsibilities based on different contributions states make to global environmental problems. This position does not, however, mean that the general obligation to cooperate to conserve and protect the environment becomes nullified as a result of differential obligations. Accordingly, the one sided nature of obligations in the Kyoto Protocol becomes problematic as it does not contain any differentiated responsibilities for developing countries – on the contrary, it contains *no* obligations for developing countries for the first commitment period.

Of course, the counter-argument can be made that the emission reduction targets in the Kyoto Protocol apply only during the first commitment period and, therefore, it was not necessary to adopt obligations for the South as their contribution to global GHG emissions was relatively negligible at the time the Kyoto Protocol was negotiated. This argument would have been more plausible if all developing countries were made equal. Countries with large populations and rapidly industrializing economies, such as India, China and Brazil, are anything but equal to other developing countries. It would appear increasingly justified to differentiate such nations from other Southern states when negotiating the post-Kyoto regime. China is now the world's largest carbon dioxide (CO_2) emitter, having surpassed the US in 2007 (Rosenthal, 2008). India's emissions are increasing rapidly as well. Moreover, when taking into account the vast quantities of CO_2 released into the atmosphere by massive deforestation, several developing countries also can be considered major contributors to climate change (Food and Agricultural Organization, 2006). For example, Indonesia ranks as the third highest CO_2 emitter in the world when its deforestation activities are counted (Greenpeace; Houghton, 2005, p. 13). Thus, it is no longer equitable to exclude these countries from substantive commitments in a successor to the Kyoto Protocol. It would be unjust to treat them in the same manner as those countries that are particularly vulnerable to climate change, such as the members of the Alliance of Small Island States.

Unlike the Montreal Protocol, the climate change regime is based on commitment periods that presently do not include developing countries. It is quite likely that in the second commitment period some developing countries will have emission reduction obligations. However, it is not clear what specific obligations they would incur; should they be commensurate with their recent, past GHG emissions or projected emissions as rapidly industrializing economies or countries with high deforestation rates? The challenge facing the international community is to find a workable compromise to address these various circumstances. Certainly, relying on the historic distinction between developed and developing countries made in the UNFCCC and Kyoto Protocol has become redundant in many ways.

While both the climate change and ozone regimes apply the CBDR principle, they approach the challenge of differential treatment differently. While developing countries as a whole do not have any binding obligations under Kyoto during the first commitment period, this did not stop some countries from voluntarily adopting emission reduction targets. Argentina, for example, announced in 1999 that it would reduce its GHG emissions within a range of 2–10 per cent below the baseline level during 2008–2012, thus becoming the first developing country to establish a voluntary target (Pew Center, 2000). South Korea, to its credit, has also expressed its desire voluntarily to reduce GHG emissions from 2018 (Asian Economic News, 1998).

While GHG emission reduction targets have been a sensitive issue for many developing countries, it is encouraging that they indicated their willingness (albeit rather cautiously) for the first time to be subjected to binding commitments under the post-2012 regime. Thus, the Bali Action Plan[8] conveys the need for enhanced national and international action on climate mitigation, including consideration of:

> [n]ationally appropriate mitigation actions by developing country Parties in the context of sustainable development, supported and enabled by technology, financing and capacity-building, in a measurable, reportable and verifiable manner.

Although this wording is rather vague and cautious, it represents a significant shift from developing countries' pre-Kyoto position that climate change was a rich country-made phenomenon and it would be unfair to require the South to pay for mitigation when it faced other pressing challenges such as poverty alleviation (Hunter, Salzman and Zaelke, 2007 p. 664). This position is reflected in the UNFCCC as well. Its preamble affirms that 'responses to climate change should be coordinated with social and economic development in an integrated manner with a view to avoiding adverse impacts on the latter, taking into full account the legitimate priority needs of developing countries for the achievement of sustained economic growth and the eradication of poverty'. The Convention further calls upon parties to integrate policies and measures to protect the climate with national development programs, 'taking into account that economic development is essential for adopting measures to address climate change' (Article 3.4).

While differential treatment is often necessary to redress past imbalances, at some point it becomes necessary to discard differentiation when states have either achieved a certain economic status or become high polluters. Affirmative action is based on the same premise. In other words, differentiation should be a temporary measure, utilized mainly to level the playing field (Atapattu, 2006, chapter 5). If these international norms are temporary in nature, can they become generally applicable, outside particular treaty regimes? It is submitted that differential norms have to be specifically adopted in a treaty regime and cannot be assumed to exist in the absence of such specific inclusion, such as by having the status of a *jus cogens* norm or even a norm of customary international law.

4. SUSTAINABLE DEVELOPMENT AS AN OVERARCHING FRAMEWORK?

It must be stressed that the whole debate on climate change should be placed within the context of sustainable development, which requires humankind to integrate environmental protection into the development process, not consider it *ex post facto* or in isolation (Atapattu, 2006, chapter 2). It also requires humankind to integrate social justice into the process, as well as ensure related procedural rights. Much has been written on sustainable development. It is one of the few concepts in international law and policy that has attracted so much literature (McCloskey, 1999; Pallemaerts, 1996; Dernbach, 1998; Atapattu, 2006, chapter 2). From a rather vague concept at the time of its enunciation in the Brundtland report (World Commission on Environment and Development, 1987), sustainable development has matured into a set of substantive and procedural norms that have achieved wide-ranging legal expression (Richardson and Wood, 2006). The UNFCCC is the first, and so far the only, international instrument to refer to the parties' *right* to sustainable development (Article 3).

Sustainable development is an umbrella term encompassing substantive components such as the principles of integration and equity (both intra- and inter-generational), and procedural components such as public access to information, justice and participation in the decision-making process (the Convention on Access to Information, Public Participation in Decision-making and Access to Justice in Environmental Matters[9] (1999) elaborates on the procedural components of sustainable development). The procedural components overlap with principles of good governance, further reinforcing the importance of sustainable development as a process. Some of these components have now achieved binding normative status: for example, the principle of integration has been given effect primarily through the environmental impact assessment process and the procedural rights in human rights regimes. Thus, it would be difficult for developing countries to raise their age-old argument that economic development must come first and environmental protection can be addressed only once living standards have been raised.

In the *Case Concerning the Gabcikovo Nagymaros Project*,[10] the International Court of Justice (ICJ) explained that '[t]his need to reconcile economic development with protection of the environment is aptly expressed in the concept of sustainable development' (ICJ, 1997). It is interesting to note that the Rio Declaration calls upon states to further develop 'international law in the field of sustainable development' (Principle 27), not international environmental law. The influence of sustainable development philosophy in the field of international environmental law has been such that some writers contend that a separate branch of international law called international sustainable development law has now evolved (Cordonier Segger and Khalfan, 2004).

At the local community level, particularly in poor and disadvantaged communities most vulnerable to environmental changes, global warming poses severe threats to the challenge of sustainable development. At stake is often not just the need to make development sustainable, but also the very existence and future of a community's way of life. The following section, which examines how the Inuit people are responding to climate change, illustrates this challenge and the role of international law in providing redress. This case study also highlights the need to look beyond the broad dichotomy of developing versus developed countries in understanding the threat of climate change and the legal solutions necessary.

5. INUIT LITIGATION: USING HUMAN RIGHTS MACHINERY TO SEEK REDRESS FOR GLOBAL ENVIRONMENTAL PROBLEMS

The Complaint

In December 2005, the Inuit of the US and Canada together with the Inuit Circumpolar Conference filed a petition before the Inter-American Commission on Human Rights. They alleged the violation of their human rights on the basis that the US government, by failing to reduce its GHG emissions, is contributing to global climate change and thus shares responsibility for the consequential environmental changes in the Arctic and the concomitant impacts on the lives and livelihoods of its Inuit inhabitants (Inuit Petition, 2005). They alleged that climate change is causing the Arctic region to melt at an alarming rate, and thereby 'destroying the habitat of polar bears, seals and caribou upon which the Inuit depend for subsistence and cultural identity' (Abate, 2007). The Inuit argued that as a result of such changes, their traditional way of life, hunting practices and even their lives are being jeopardized, which they contend violates their international human rights.

Moreover, the petitioners argued that because their way of life and culture are intrinsically linked to their physical surroundings, climate change is violating their right to practise their culture. They allege that the US, currently the largest contributor to GHG emissions in the world, has consistently refused to take meaningful steps to reduce GHG emissions, despite having ratified the UNFCCC. The specific rights that are allegedly being violated include: the right to use and enjoy traditional lands; the right to enjoy personal property; the rights to health and life; rights to residence and movement, and inviolability of the home; the right to their own means of subsistence; and the right to culture, to the extent that it is recognized under international law. The relief requested includes (i) preparing a report, with facts and applicable law, declaring that the US is internationally responsible for the violation of rights embodied in the American Declaration on Rights and Duties of Man (Organization of American States, 1948); (ii) holding a hearing; (iii) adopting and implementing a plan to protect the Inuit land and resources; and (iv) providing assistance to the Inuit to adapt to the impacts caused by climate change where they cannot be avoided. It is noteworthy that no compensatory damages were sought by the petitioners.

There are several reasons for using human rights machinery for environmental wrongs. International environmental law regimes, particularly in relation to climate change, lack the sophisticated mechanisms sometimes found in human rights regimes with their institutions, relief mechanisms and direct standing for individuals. These features are most evident at the regional level.[11] Environmental law, by contrast, focuses on regulation of activities and prevention of environmental damage, rather than providing institutional mechanisms to seek relief for damage. As long as this situation remains so, it is likely that victims of environmental harm will continue to seek relief under human rights regimes where the causal link between environmental degradation and protected rights can be established.

While this strategy is not new, using it for damage caused by climate change is a recent development. The Inuit case gave a human face to climate change, a problem that was hitherto considered largely as an environmental problem. It also highlighted that the consequences of climate change are taking place now; in other words, it shattered the myth that climate change is an abstract issue that will give rise to undetermined consequences for 'future generations'.

The Inter-American Commission on Human Rights initially declined to entertain the petition. However, after a renewed request in January 2007, the Commission invited Sheila Watt-Cloutier, the then Chairperson of the Inuit Circumpolar Conference, Martin Wagner of Earthjustice and Daniel Magraw of the Center for International Environmental Law to a hearing on global warming and human rights on March 1, 2007 (Spicer, 2007).

In another interesting twist, Tuvalu, a small island developing state in the South Pacific, threatened to file action in the ICJ against the US and Australia for their contribution to global climate change and their failure to accept Kyoto obligations to reduce GHG emissions which will cause sea levels to rise and thereby submerge their country.[12] While this threat has not materialized to date, it seems inevitable that similar law suits will be initiated one day against the world's major polluters if they do not act more responsibly now.

State Responsibility and the CBDR Principle

Introduction

If the Inter-American Commission were to find the US responsible under international law for the alleged damages to the Inuit people, it would have to do so under the CBDR principle. In other words, responsibility would be for the portion of climate change that the US was responsible for causing. This could be determined by reference to GHG emissions data collected in recent decades. However, for purposes of state responsibility, one has to prove that the US was responsible for the damage caused to the Inuit – i.e., it was the emission of GHGs by the US that caused the environmental damage to the Arctic which, in turn, resulted in the violation of the protected rights of the Inuit people. In other words, the causal link between US GHG emissions and the damage to the Inuit must be established.

Linking environmental harm to human rights violations is not difficult. Much has been written on the relationship between the enjoyment of human rights and environmental damage (Hancock, 2003; Zarsky, 2002).[13] For example, in the *Yanomani Case*, the Inter-American Commission on Human Rights held that the construction of a highway through indigenous territory violated the rights to life, liberty and personal security.[14] The European Court of Human Rights held in *Lopez Ostra v. Spain* that 'the consequences of environmental degradation may so affect an individual's well-being as to deprive her of the enjoyment of her private and family life'.[15] More recently in *Social and Economic Rights Action Center and the Center for Economic and Social Rights v. Nigeria*, the African Commission on Human and Peoples' Rights held that the disposal of toxic wastes into the environment and the release of oil into waterways resulted in the violation of peoples' rights to life, property, health, family life, disposal of natural resources and the right to environment.[16] In his separate opinion in the *Case Concerning the Gabcikovo Nagymaros Project* (1997), Judge Weeramantry noted that '[t]he protection of the environment is likewise a vital part of contemporary human rights doctrine, for it is a *sine qua non* for numerous human rights such as the right to health and the right to life itself'. However, whether state responsibility can be based on the CBDR principle must be resolved.

Establishing responsibility of states under international law

Under principles of international law, an internationally wrongful act gives rise to the responsibility of that state. This is codified in Article 1 of the ILC draft articles on State Responsibility adopted in 2001.[17] There is an internationally wrongful act of a state when conduct consisting of an action or omission: (a) is attributable to the state under international law; and (b) constitutes a breach of an international obligation of the state (Article 2). Thus, in order to establish responsibility of the US, one has to establish that: (a) the US breached one of its international obligations; and (b) such act/omission resulting in the violation is attributable to the US under international law. In addition, of course, the link of causation between the alleged damage and the act/omission attributed to the US must be established.

The origins of the application of state responsibility to transboundary air pollution issues can be traced to the *Trail Smelter Arbitration* between the US and Canada in the late 1930s (see generally Okowa, 2000: Sands, 2003, chapter 18).[18] This case involved injury caused by fumes emanating from a smelter situated in Canada. The arbitral tribunal noted:

> [u]nder principles of international law, as well as the law of the United States, no state has the right to use or permit the use of territory in such a manner as to cause injury by fumes in or to the territory of another or the properties or persons therein, when the case is of serious consequence and the injury is established by clear and convincing evidence.

Quite apart from the general obligations in the UNFCCC, which the US has ratified, a rule of customary international law obliges states to ensure that activities within their jurisdiction or control do not cause damage to the environment of other states or to the global commons (Birnie and Boyle, 2002, p. 109). This rule is reflected in Principle 21 of the Stockholm Declaration on the Human Environment of 1972,[19] which was no doubt influenced by the ruling in the *Trail Smelter* Arbitration:

> Under principles of international law and the UN Charter, every state has the right to exploit their natural resources according to their environmental policies, and the responsibility to ensure that activities within their jurisdiction or control do not cause damage to the environment of other states or of areas beyond the limits of national jurisdiction (Principle 21).

Therefore, it is clear that the US has violated at least this norm by refusing to take action to reduce its GHG emissions. Principle 21 has now developed into a general obligation to prevent environmental harm (Birnie and Boyle, 2002, p. 109), and states may incur international responsibility if this damage extends beyond their national borders.

This general obligation does not cover every kind of environmental damage. In the absence of specific duties (such as those established under a treaty), this obligation extends only to significant or serious damage (Okowa, 2000, p. 176; Sands, 2003, p 878). The petition filed by the Inuit before the Inter-American Commission indicates that the damage caused by climate change in the Arctic region is indeed significant. Quoting from the Arctic Climate Impact Assessment, the petition noted:

> The Arctic is extremely vulnerable to observed and projected climate change and its impacts. The Arctic is now experiencing some of the most rapid and severe climate change on Earth. Over the next 100 years, climate change is expected to accelerate, contributing to major physical, ecological, social, and economic changes, many of which have already begun (Inuit Petition, 2005, p. 10).

However, is the US internationally responsible for the damage caused to the Arctic region by its emission of GHGs? Climate change is a global phenomenon – virtually every state in the international community is a perpetrator as well as a victim of it. The *accumulation* of these gases in the atmosphere over a period of time is causing the climate to change. Thus, can one state in a community of some 200 states be held responsible for the damage caused by global warming? Clearly, under existing principles of state responsibility, where the doctrine of joint and several liability (Faure and Nolkaemper, 2007, p 165) does not generally apply, the answer would be no.

However, of the top emitters of GHGs in the world in recent years, the US accounted for approximately 20 per cent of all such emissions, followed by China at nearly 15 per cent of the global total (Hunter, Salzman and Zaelke, 2007, p. 663). It has since been reported that China has overtaken the US as the highest CO_2 emitter (International Herald Tribune, 2007). The European Union (EU) countries collectively accounted for 14 per cent (Hunter, Salzman and Zaelke, 2007, p. 663). With regard to *per capita* emissions, the US is also high on the list with annual *per capita* emissions of 24.5 tons of CO_2, compared to only 3.9 tons per person in China (Hunter, Salzman and Zaelke 2007, p. 663). Thus, by both accounts, the US contribution to climate change is significant, accounting for about one fifth of global GHG emissions. Moreover, during 1990–2006, GHG emissions associated with land use changes and forestry reportedly *increased* in the US by 14 per cent (by contrast, during the same period emissions from such sources *decreased* in Great Britain by 15.6 per cent) (UNFCCC, 2008). While data are available that enable some quantification of individual states' contribution to climate change, it is not possible to establish precisely that a particular damage is due to the activities of a particular state. Thus, the traditional rules of state responsibility for global environmental problems such as climate change must be reevaluated (Okowa, 2000, p. 185).

While the recent trend in international environmental law has been to adopt a preventive and a precautionary approach to environmental problems, particularly for complex global environmental issues involving a multitude of actors (Sands 2003, p. 26 and chapter 6; Birnie and Boyle, 2002, pp. 104, 112–25; Hunter, Salzman and Zaelke, 2007, pp. 505–16), what happens when past emissions cause environmental problems such as for the Inuit, despite precautionary measures being taken to mitigate *future* damage? While the Inuit case will not provide us with precise answers, as no ruling to date has been made on the merits of the case, it did indicate that the legal community will resort to novel legal strategies to seek redress. Okowa's belief that the '[a]pportionment of responsibility between many contributing factors will depend on the factors presented before the tribunal' (Okowa 2000, p.. 188) is not a very useful suggestion for the situation presented by the Inuit petition as she does not provide examples of these factors.

Collective theories of liability

At the domestic level, however, courts have been willing to deviate from the general principle that a specific individual's activities must be proved to have actually caused the damage in question (Hunter, Salzman and Zaelke, 2007, 627). English courts, for example, have adopted 'a material contribution to risk' approach (Okowa, 2000, p. 188). Some US courts have adopted a market-share liability rule, such as in cases involving the drug DES, where liability is allocated according to the percentage of market share held by each company. Thus, in *Sindell* v. *Abbott Laboratories*, the Supreme Court of California held:

> [t]herefore, once plaintiffs joined the manufacturers of a substantial percentage of DES, defendants were required to prove they could not have manufactured the injury-causing product. Absent such proof, liability for damages could be apportioned based on each defendant's share of the appropriate market.[20]

While this theory of liability is not without its critics (Geistfeld, 2006), it may provide a useful model for international law. Could international courts adapt this principle to global climate change and develop a theory of liability based on the GHG emissions of each country? Similar to the theory of market share liability, could international courts reverse the burden of proof and require states to show that their emissions *did not* result in the damage in question? This approach is also tied to the precautionary principle. In the *MOX Plant Case* (2001), for example, Ireland argued in its application for provisional measures that '[t]he precautionary principle places the burden on the United Kingdom to demonstrate that no harm would arise from discharges and other consequences of the operation of the MOX plant'[21] (ITLOS, 2001). The Tribunal, however, siding with Britain's contention, declined to prescribe provisional measures on the ground that there was no urgency to do so. Nonetheless, the Tribunal called upon the parties to cooperate in exchanging information and to enter into consultations on the basis of 'prudence and caution' (ibid., para. 84). It further noted that 'the duty to cooperate is a fundamental principle in the prevention of pollution of the marine environment under Part XII of the Convention and general international law' (ibid., para. 82).

The other theories that domestic courts have adopted include the contribution to risk theory, as used by the Wisconsin Supreme Court in *Collins v. Eli Lilly Company*[22] (1984) (an approach similar to that used by British courts, as noted above), a modified market share liability theory adopted by the Washington Supreme Court in *Martin v. Abbott Labs*[23] (1984) in which the court applied the test in the *Sindell* case but required the plaintiff to sue only one defendant while defining 'the market' as narrowly as possible (Daley 1997, p 6) and a liberal market share theory applied by the New York Supreme Court in *Hymowitz v. Eli Lilly*[24] (1989) (Daley, 1997, p 12), in which the court dispensed with the requirement that the defendant must have marketed the type of DES consumed by the plaintiff.

If courts were to adopt a novel theory of liability based on the GHG emissions for each country, the temporal element of the issue would become relevant. Given that it was only recently that it was unequivocally established that anthropocentric emissions of GHG are causing global climate change (IPCC, 2007), can we hold states and companies liable at all for their past emissions? In other words, at what point in time do their emissions become wrongful, given the pervasive, lawful uses of fossil fuels in nearly every economy worldwide? Could we address this issue by applying the precautionary principle and using 1992 as the relevant date – when the UNFCCC was adopted? These are some of the difficult questions that the legal community will have to grapple with if it wishes to use liability as a tool to address climate change.

In advocating a liability regime for climate change, Cullet contends that:

> [s]ome degree of climate change is unavoidable in the coming years and decades because of past and current greenhouse gas emissions. As a result, besides mitigation, adaptation measures have to be taken. Regardless of the adaptation measures taken, damages will occur. It is thus necessary to provide a framework for allocating responsibility for damage that has and will occur (Cullet, 2007, pp. 99–100).

However, Cullet accepts that while a separate liability regime is desirable, it would be complicated to draft, and developed countries in particular would very likely oppose such a regime. Indeed, the international community has always been reluctant to develop an international liability regime for environmental damage; in general, the Protocol on Liability and Compensation for Damage Resulting from the Transboundary Movement of Hazardous Wastes and their Disposal[25] (1999) and Annex VI to the Protocol on Environmental Protection to the Antarctic Treaty[26] (2005) are the only exceptions – conventions on marine pollution generally use civil liability (Sands, 2003, p. 904). While the policy emphasis on prevention is important, it is equally important to ensure that victims of environmental damage have some redress available to them. However, devising a liability regime for climate change would be more complicated compared to other environmental problems given the complexity of the issue and the plethora of actors involved.

A novel approach has been proposed to base liability on the amount of 'natural debt' created by each country. This conceptual framework, proposed by Patz and others, contends that '[j]ust as nations often borrow financial resources from the future, creating a national debt, they also essentially borrow assimilative capacity from the future by emitting pollutants faster than Earth can assimilate, creating a 'natural debt' (Patz and others, 2007, p. 401). The current natural debt of developing countries is relatively small. Patz and others argue that this is a more accurate indicator of responsibility for global warming. Thus, according to this approach, liability would be directly commensurate with the natural debt created by each country. The question, of course, is whether this approach would be sufficient for legal purposes.

Two further issues need to be addressed in designing a liability regime: these are who has standing to bring a claim, and the appropriate remedies in the event that liability of a state is established. Under traditional international law, only states had standing to bring a claim at the international level (Article 34, ICJ Statute). However, international law is gradually providing standing to non-state entities, such as individuals under human rights regimes. For example, Optional Protocol I to the International Covenent on Civil and Political Rights[27] (1966) gives individuals the right to refer a communication directly to the UN Human Rights Committee in Geneva against violations of human rights by their states (Brown Weiss, 2002, p. 809). Similarly, the regional human rights treaties allow individuals to bring claims against their states before the regional human rights bodies. This is how the Inuit were able to bring a claim against the US before the Inter-American Commission on Human Rights.

The issue of remedies is more complicated. The traditional remedies available under international law such as restitution, compensation, and guarantees against non-repetition (Gray, 1990) would not bring much relief to the victims of climate change. Restitution of the environment is often impossible. A declaratory judgment may bring some relief, as would compensation, but several questions remain unresolved. How would courts compute damage to the environment itself, loss of lives, loss of culture and even the loss of an entire country (in the case of a submerged small island state)? While international law has rules on state succession, it has never dealt with the disappearance of an entire state, as could happen with sea level rise associated with climate change.

Teubner argues that '[i]t is crucial to the further development of ecological liability to understand more precisely under what circumstances and in what way courts and legislatures are shifting liability away from the individual responsibility of single actors towards a new collective responsibility of risk networks' (Teubner, 1994, p. 18). He notes that lawyers are relying less on traditional causation, and resorting to novel ways of establishing liability:

> Causation-in law, *prima facie*, enhanced *res ipsa loquitur*, reversing the burden of proof, probabilistic causation, joint and several liability in multiple causation, enterprise liability, market share liability, Superfund liability-all these new forms of 'risk liability' tend to reduce or even eliminate individual causal linkages between acts and damages and to replace them with an overarching cupola of quasi-collective responsibility for ecological damages (Teubner, 1994, p. 19).

While his analysis is mainly in relation to liability at the domestic level, some of his observations can be equally applied to the international level. Thus, the inadequacy of individual responsibility, the trend towards creating new forms of risk pooling, and new forms of collective risk management, are all relevant to global environmental issues. However, given that climate change is tied to pervasive, day-to-day activities of many people worldwide, the question arises how these novel forms of establishing liability can be applied in relation to these activities.

Given the challenges in designing a climate change liability regime, and the likely political resistance to it (Brunnée, 2004), it may be more practical to provide redress through special funding mechanisms, such as the Special Climate Change Fund established under the UNFCCC in 2001 (Cullet, 2007, p 115)[28] or the Adaptation Fund established under the Kyoto Protocol, to allow victims of global warming to obtain compensation. Contributions to the fund should be proportionate to each nation's GHG emissions. This alone would be an incentive for high emitters to reduce their use of fossil fuels. While this would avoid the need to establish the responsibility of a particular state, it would ensure that victims of climate change receive some compensation. Of course, monetary compensation cannot ever compensate adequately for loss of life or loss of culture. Hopefully, the Copenhagen Protocol being negotiated as a successor to Kyoto will provide a framework for a more comprehensive funding mechanism, and provide the necessary guidance on compensable damage and eligible claimants.

UN Declaration on the Rights of Indigenous Peoples, 2007

While the UN Declaration on the Rights of Indigenous Peoples[29] (UNDRIP) was not adopted by the UN General Assembly at the time of the Inuit Petition, it is interesting to see how some of its provisions are directly applicable to the case. Although the UNDRIP is a 'soft law' instrument, and therefore without any binding effect, many of its provisions reaffirm or clarify existing human rights law in relation to indigenous peoples.[30]Adopted in September 2007 after years of tense debate in the UN system, the UNDRIP recognizes that 'respect for indigenous knowledge, cultures and traditional practices contributes to sustainable and equitable development and proper management of the environment' (Preamble). Article 1 of the Declaration affirms that indigenous peoples have the right to the full enjoyment of all human rights recognized under international human rights law (Article 1).

Affirming their right to self-determination, the Declaration provides that indigenous peoples have the right freely to determine their political status and freely to pursue their economic, social and cultural development (Article 1). Article 8 affirms the right not to be subjected to forced assimilation or destruction of their culture, and states should provide effective mechanisms to prevent such forced assimilation or destruction of culture and provide redress in the event that it takes place. Further, indigenous peoples cannot be forcibly removed from their lands or territories (Article 10) and they have the right to practise and revitalize their cultural traditions and customs (Article 11). In addition, they have the right to the dignity and diversity of their cultures, traditions, histories and aspirations. Indigenous peoples also have the right to the conservation and protection of their environment (Article 29) as well as the right to maintain, control, protect and develop their cultural heritage, traditional knowledge and traditional cultural expressions (Article 31).

It is therefore obvious that whether or not the US is found legally responsible for the damage caused to the Inuit, global climate change is having a serious impact on the indigenous rights recognized in the UNDRIP, including the right to health, the right to practise their culture, protection of their environment and maintenance of their cultural diversity, as well as the right to life in extreme instances. The UNDRIP, however, does not provide any compliance machinery that would enable groups such as the Inuit to seek redress. They must use other international human rights regimes, such as the Inter-American system.

6. POST-2012 REGIME: RECOMMENDATIONS

While international law on climate change has thus far focused on mitigating emissions, albeit rather lamely (both the UNFCCC and the Kyoto Protocol are based on this premise), there is increasing recognition that adaptation to climate change as highlighted by the Inuit petition must also be addressed more earnestly. Due to time-lags in the materialization of some climate change impacts, the IPCC warns that atmospheric warming and sea level rise could continue for *centuries* even if GHG concentrations were stabilized today. It advises that more extensive adaptation measures than those that are currently being practised would be necessary in order to reduce vulnerability to climate change, regardless of the mitigation methods adopted. Therefore, for the present generation and the next, adaptation strategies[31] will be at least as crucial as mitigation strategies.

The IPCC report notes that a wide array of adaptation options is available. While most societies have long managed to deal with weather and climate-related events,

> [a]dditional adaptation measures will be required to reduce the adverse impacts of projected climate change and variability, regardless of the scale of mitigation undertaken over the next two to three decades. Moreover, vulnerability to climate change can be exacerbated by other stresses. These arise from, for example, current climate hazards, poverty and unequal access to resources, food insecurity, trends in economic globalisation, conflict and incidence of diseases such as HIV/AIDS (IPCC, 2007, p. 14).

Clearly, climate change risks are exacerbating global inequalities between the North and South, and also between groups within countries. Poor and vulnerable communities are often not only the most prone to climate change, they are less able to adapt because of poverty and unequal access to resources. Human rights standards and remedies should play a role in the adaptation strategies to address such challenges. Otherwise, such strategies themselves run the risk of further violating the human rights of these vulnerable communities, rather than helping them to adapt. In designing such adaptation strategies, it is also essential to ensure the input of those groups for whose benefit these strategies are being prepared. Giving a voice to them is crucial to ensure that the adaptation strategies are suitable for these affected peoples.

The UNDP Human Development Report also recognizes the importance of a two-pronged strategy of adaptation and mitigation, particularly in the context of addressing poverty and promoting social equity:

> [a]daptation priorities must also be addressed. For too long, climate change adaptation has been treated as a peripheral concern, rather than as a core part of the international poverty reduction agenda. Mitigation is an imperative because it will define prospects for avoiding dangerous climate change in the future. But the world's poor cannot be left to sink or swim with their own resources while rich countries protect their citizens behind climate-defence fortifications. *Social justice and respect of human rights demand stronger international commitment to adaptation* (Human Development Report, Summary 2007/08, p. 13, emphasis added).

Thus, the UNDP is advocating that climate change adaptation should be considered part of the international poverty reduction agenda. De-linking the two issues would result in a fragmented approach to both poverty alleviation and climate change management.

Yet, while adaptation strategies must give more consideration to the plight of vulnerable peoples in developing countries, as well as vulnerable groups in developed countries such as the Inuit, some developing countries should begin to shoulder some of the burden of mitigating climate change beyond 2012 under the proposed Copenhagen Protocol. It is hardly possible to rely solely on the simple distinctions made between developing countries (i.e., non-Annex 1 Parties) and developed countries as well as emerging economies (Annex 1 Parties) in the present climate law regime. Emissions of GHGs in some rapidly industrializing economies such as China and India are rising exponentially, while emissions associated with deforestation in Brazil and Indonesia are similarly of concern already. Thus, unless such states also take on substantive obligations, any 'stabilization of emissions by Annex I countries will be more than counterbalanced by an ongoing and strong rise in emissions in non-Annex I countries' (den Elzen and Hohnes, 2008, p. 261). Of course, the present Annex I countries should carry more stringent targets under the Copenhagen Protocol. Some already appear committed to this task. The UK's Climate Change Act of 2008 commits Britain to reduce its emissions by 80 per cent below the 1990 baseline by 2050, with an interim target of reducing emissions by 26 per cent below this level by 2020. Presently, the UK and Germany are the only two Annex I countries to have achieved their Kyoto emission reduction targets, in contrast to the US whose 2008 emissions had grown by 14 per cent since 1990 (UNFCCC, 2008).

7. CONCLUSION

Global climate change poses unprecedented challenges to the international community, and in particular to international law. No longer is it possible to rely on traditional principles of international law such as those relating to standing, given that non-state actors play an increasingly significant role in international affairs and have interests distinct from those of the state. Likewise, the rules and remedies under the existing state responsibility regime increasingly look redundant, given the complexity of climate change as a policy challenge.

It is also imperative that robust adaptation measures be adopted along with mitigatory measures. For much of the present generation and the next, adaptation will likely be as crucial a challenge as mitigation. Greater attention should be paid to particularly vulnerable communities such as indigenous groups whose way of life and culture are intrinsically linked to their land.

In advancing reform, the traditional distinctions made between developing and developed countries must be re-examined. As the case of the Inuit demonstrates, the victims of climate change are not only communities in the South, but can also be found within the North where significant disparities in affluence and ability to cope with climate change exist. Thus, while the CBDR principle has helped to redress imbalances between the North and South, the time has come also to acknowledge differences among groups within these categories, such as indigenous peoples.

This chapter has argued that liability and litigation are not the best approach to climate change. If novel legal strategies of redress, which dispense with the traditional principles of causation and standing, can be adopted liability may be used to complement other approaches taken to mitigate and adapt to climate change. Alternatively, a more comprehensive compensation fund for victims of climate change could be established. While such a mechanism might eschew some of the complexities and delays that commonly afflict a liability regime, the nature of remedies would remain problematic. For example, how can compensation for the loss of one's land or country as a result of higher sea levels be fairly measured? This is not a hypothetical question – for small island states this unfortunately is a grave threat.

Developing countries as a whole are getting a raw deal from the consequences of climate change – they are being disproportionately burdened by the negative consequences of a phenomenon that they neither significantly contributed to nor benefited from. Thus, one may pose the question – 'is this equitable?'. In the final analysis, however, while the issue is one of equity, time may be already running out for adopting precautionary measures. This is no time to split hairs about historic responsibility or whether developing countries must address economic development first. The time has come for both developing and developed countries alike to take measures together in a spirit of global partnership and within a framework of sustainable development to protect our generation and future generations from the negative consequences of climate change.

NOTES

* Sumudu Atapattu is the Associate Director of the Global Legal Studies Center at University of Wisconsin Law School. For a discussion of the link between human rights and climate change see the author's article on 'Global Climate Change: Can Human Rights (and Human Beings) Survive this Onslaught?', *Colorado Journal of International Environmental Law and Policy*, 20(1), 35 (2008).

1. (1992) ILM 31, 849.
2. (1997) ILM 37, 22.
3. The United States and European Union positions in relation to ozone depletion, climate change and the precautionary principle are good examples.
4. (1992) ILM 31, 874.
5. Article 5 of the Protocol is entitled 'Special Situation of Developing Countries'.
6. The Convention on Biological Diversity also refers to the special conditions of least developed countries and small island states and the need for special provision to meet the needs of developing countries including new and additional financial resources and access to technology (preamble). It does not, however, specifically refer to the CBDR principle.
7. Byrd-Hagle Resolution (105th Congress, 1997), www.nationalcenter.org/KyotoSenate.html (verified 15 March 2009).
8. Bali Action Plan, Advanced unedited version, Decision /CP.13, (2007) http://unfccc.int/files/meetings/cop_13/application/pdf/cp_bali_action.pdf (verified 15 March 2009).
9. (1999) ILM 38, 517, www.unece.org/env/pp/documents/cep43e.pdf (verified 15 March 2009).
10. *Case Concerning Gabcikovo Nagymaros Project (Hungary v. Slovakia)*, [1997] ICJ Rep, 7.
11. The two regional human rights systems, the African system and the Inter-American system, recognize a right to environment (African Charter on Human and People's Rights, adopted in 1981, and Additional Protocol to the American Convention on Human Rights in the Area of Economic, Social and Cultural Rights, 1988, 28 ILM 156). The European human rights system, by contrast, has yet to recognize a right to environment specifically, although it has dealt with environmental issues under other protected rights, such as right to health and the right to privacy. See Desgagne (1995) and Atapattu (2002).
12. Australia ratified the Kyoto Protocol just before the Bali Climate Change conference in December 2007, leaving the US the only industrialized country not to have ratified the Kyoto Protocol.
13. While not all human rights violations are a result of environmental damage, most environmental harms have an impact on human beings, and their enjoyment of human rights.
14. Case 7615, Inter-Am.C.H.R., OEA/ser.L/V/II.66,doc.10 rev.1 (1985) reprinted in [1985] *Inter-American Yearbook on Human Rights* 264.
15. App. No. 16798/90, 20 Eur. H.R. Rep. 277 (1995).
16. *Social and Economic Rights Action Center and the Center for Economic and Social Rights v. Nigeria*, African Commission on Human and People's Rights, Comm. No 155/96 (2001), available at: www1.umn.edu/humanrts/africa/comcases/155-96.html (verified 15 March 2009).
17. Draft articles on Responsibility of States for Internationally Wrongful Acts, adopted by the ILC at its 53rd session (2001), www.ilsa.org/jessup/jessup06/basicmats2/DASR.pdf (verified 15 March 2009).
18. *Trail Smelter Arbitration (USA v. Canada)*, 3 RIAA 1905 (1941).
19. Report of the United Nations (UN) Conference on the Human Environment, U.N. Document A/CONF/48/14/Rev. 1 (1972).
20. 26 Cal. 3d 588 at 588; 607 P.2d 924; 163 Cal. Rptr. 132; 1980 Cal. LEXIS 151; 2 A.L.R.4th 1061; CCH Prod. Liab. Rep. P8648.
21. *MOX Plant Case, Ireland* v. *United Kingdom*, 3 December 2001, International Tribunal for the Law of the Sea, No 10 (Request for Provisional Measures).
22. 342 N.W. 37, 53–54 (Wis.), 469 U.S. 826 (1984).
23. 689 P 2d 368 (Wash. 1984).
24. 539 NE 2d 1069 (NY 1989).
25. 10 December 1999, U.N. Doc. UNEP-CHW.1-WG-1-9-2 (1999).
26. Protocol on Environmental Protection to the Antarctic Treaty (2005), at: www.ats.aq/documents/cep/Annex_VI_e.pdf (verified 15 March 2009).

27. (1966) ILM 3, 383.
28. The Special Climate Change Fund (SCCF) under the Convention was established in 2001 to finance projects relating to adaptation; technology transfer and capacity building; energy, transport, industry, agriculture, forestry and waste management; and economic diversification. This fund should complement other funding mechanisms for the implementation of the Convention, Decision 7/CP.7, available at UNFCCC website: http://unfccc.int/cooperation_and_support/financial_mechanism/special_climate_change_fund/items/3657.php (verified 15 March 2009).
29. A/RES/61/295, 2 October 2007.
30. It is of concern that neither the US nor Canada signed the Declaration.
31. Adaptation is defined as 'finding and implementing ways of adjusting to climate change': see UNFCCC website: http://unfccc.int/adaptation/items/4159.php (verified 15 March 2009).

REFERENCES

Abate, R. (2007), 'Climate Change, The United States, and the Impacts of Arctic Melting: A Case Study in the need for Enforceable International Environmental Human Rights', *Stanford Environmental Law Journal* 26, 3.

Asian Economic News (1998), 'S. Korea to Reduce Greenhouse Gas Emissions from 2018', *Asian Economic News*, 15 July.

Atapattu, S. (2002), 'The Right to a Healthy Life or the Right to Die Polluted? The Emergence of a Human Right to a Healthy Environment under International Law', *Tulane Environmental Law Journal* 16, 65.

Atapattu, S. (2006), *Emerging Principles of International Environmental Law,* (Transnational Publishers).

Birnie, P. and A. Boyle (2002), *International Law and the Environment* (Oxford University Press).

Brown Weiss, E. (2002), 'Invoking State Responsibility in the Twenty-First Century', *American Journal of International Law* 96, 798.

Brunnée, J. (2004), 'Of Sense and Sensibility: Reflections on International Liability Regimes as Tools for Environmental Protection', *International and Comparative Law Quarterly* 53, 351.

Climate Change and International Security (2008), Paper from the High Representative and the European Commission to the European Council, (S113/08).

Cordonier Segger, M.C. and A. Khalfan (2004), *Sustainable Development Law: Principles, Practices and Prospects* (Oxford University Press).

Cullet, P. (2007), 'Climate Change Liability and the Allocation of Risk: Liability and Redress for Human-induced Global Warming: Towards an International Regime', *Stanford International Law Journal* 26, 99.

Daley, R. (1997), 'Market Share Liability in DES and Lead Pigment Cases – Bridging the Gap: A Suggested Proposal to Apportion Liability in Lead Pigment Cases', *Duquesne Law Review* 36, 79.

den Elzen, M. and N. Hohne (2008), 'Reductions of Greenhouse Gas Emissions in Annex I and Non-Annex I Countries for Meeting Concentration Stabilization Targets', *Climate Changes* 91, 249.

Dernbach, J. (1998), 'Sustainable Development as a Framework for National Governance', *Case Western Reserve Law Review* 49, 1.

Desgagne, R. (1995), 'Integrating Environmental Values into the European Convention on Human Rights', *American Journal of International Law* 89, 263.

EarthJustice (2007), Global Warming and Human Rights Gets Hearing on the World Stage', Press Release, 1 March.

Faure, M. and A. Nolkaemper (2007), 'International Liability as an Instrument to Prevent and Compensate for Climate Change', *Stanford Environmental Law Journal* 26, 123.

Food and Agricultural Organization (FAO), (2006), 'Deforestation Causes Global Warming', FAO Newsroom, 4 September, at: www.fao.org/newsroom/en/news/2006/1000385/index.html.

Geistfeld, M. (2006), 'The Doctrinal Unity of Alternative Liability and Market-Share Liability', *University of Pennsylvania Law Review* 155, 447.

Gray, C. (1990), *Judicial Remedies in International Law* (Clarendon Press).

Greenpeace (undated), 'Deforestation and Climate Change, Greenpeace', at: www.greenpeace.org.uk/forests/climate-change.

Hancock, J. (2003), *Environmental Human Rights: Power, Ethics and Law* (Ashgate).

Houghton, R. (2005), 'Tropical Deforestation as a Source of Greenhouse Gas Emissions', in P. Moutinho and S. Schwartzman (eds), *Tropical Deforestation and Climate Change* (Environmental Defense).

Hunter, D., J. Salzman and D. Zaelke (2007), *International Environmental Law and Policy* (3rd edn, Foundation Press).

Intergovernmental Panel on Climate Change (IPCC) (2007), *Climate Change 2007: The Physical Science Basis: Summary for Policymakers,* Working Group II (Cambridge University Press).

International Herald Tribune (2007), 'China Overtakes US in Greenhouse Gas Emissions', *International Herald Tribune*, 20 June.

Inuit Petition (2005), *Petition to the Inter American Commission on Human Rights Seeking Relief from Violations Resulting From Global Warming Caused by Acts and Omissions of the United States*, at www.Ciel.Org/Climate/Climate_Inuit.Html.

Knight, A. (2005), 'Global Environmental Threats: Can the Security Council Protect our Earth?', *New York University Law Review* 80, 1549.

McCloskey, M. (1999), 'The Emperor has No Clothes: The Conundrum of Sustainable Development', *Duke Environmental Law and Policy Forum* 9, 153.

Moutinho, P. and S. Schwartzman (2005), *Tropical Deforestation and Climate Change* (Amazon Institute for Environmental Research).

Okowa, P. (2000), *State Responsibility for Transboundary Air Pollution in International Law* (Oxford University Press).

Osofsky, H. (2007), 'Climate Change Litigation as Pluralist Legal Dialogue', *Stanford Environmental Law Journal* 26, 181.

Pallemaerts, M. (1996), 'The Future of Environmental Regulation: International Environmental Law in the Age of Sustainable Development: A Critical Assessment of the UNCED Process', *Journal of Law and Commerce* 15, 623.

Patz, J., et al. (2007), 'Climate Change and Global Health: Quantifying a Growing Ethical Crisis', *EcoHealth* 4, 397.

Penny, C. (2005), 'Greening the Security Council: Climate Change as an Emerging Threat to International Peace and Security', *International Workshop on Human Security and Climate Change*, at: www.gechs.org/downloads/holmen/Penny.pdf.

Pew Center on Global Climate Change (2000), *Developing Countries & Global Climate Change: Electric Power Options in Argentina* (Pew Center on Global Climate Change).
Richardson, B.J. and S. Wood (2006), *Environmental Law for Sustainability* (Hart Publishing).
Richardson, M. (2008), 'Southeast Asia and Climate Change: Deforestation in Indonesia', *OpinionAsia,* 13 October, at: www.opinionasia.org/DeforestationinIndonesia.
Rosenthal, E. (2008), 'China Increases Lead as Biggest Carbon Dioxide Emitter', *New York Times*, 14 June.
Sands, P. (2003), *Principles of International Environmental Law* (2nd edn, Cambridge University Press).
Spicer, J. (2007), 'Hearing to Probe Climate Change and Inuit Rights', (Reuters), 21 February, at: www.unep.org/Indigenous/pdfs/Climatechange-and-Inuit-rights.pdf.
Taylor, P. (1998), *An Ecological Approach to International Law* (Routledge).
Teubner, G., L. Farmer and D. Murphy (eds) (1994), *Environmental Law and Ecological Responsibility: The Concept and Practice of Ecological Self-Organization* (John Wiley & Sons).
United Nations Framework Convention on Climate Change (UNFCCC), *National Greenhouse Gas Inventory Data for the Period 1990–2006*, FCCC/SBI/2008/12.
United Nations Department of Public Information (2007), 'Security Council Holds First-Ever Debate on Impact of Climate Change on Peace, Security, Hearing over 50 Speakers', Security Council, SC/9000, 17 April, at: www.un.org/News/Press/docs/2007/sc9000.doc.htm.
United Nations Development Program (UNDP) (2008), *Human Development Report 2007/2008: Fighting Climate Change: Human Solidarity in a Divided World* (Palgrave Macmillan).
Wang, J. and M. Oppenheimer (2005), *The Latest Myths and Facts on Global Warming* (Environmental Defense).
World Commission on Environment and Development (WCED) (1987), *Our Common Future* (Oxford University Press).
Zarsky, L. (2002), *Human Rights and the Environment: Conflicts and Norms in a Globalizing World* (Earthscan).

3. India's constitutional challenge: a less visible climate change catastrophe

Deepa Badrinarayana*

1. INTRODUCTION

At the heart of dissonance between the United States (US) and developing countries such as India in the climate change debate are issues of economic growth and equity. Particularly during the Bush Administration, the US government has argued that it would be economically disadvantaged if it accepted legally binding responsibility to reduce greenhouse gas (GHG) emissions, despite its historic contribution to the problem. India, for its part, argues that it cannot be equitably required to accept legal obligations, given its economic problems and its historically low contribution to the climate change problem.

India's negotiation stance, however, fails to reflect grave internal economic and equitable problems that will be exacerbated by climate change. Some of the poorest people in India are included in the numbers that the Indian government uses to argue that the country's per capita share of GHG emissions is very low. However, the benefit from relying on their low per capita use of resources in order to justify the government's refusal to accept binding emissions targets will not flow to these poorest people; they will in fact pay most of the price of inaction.

The current position of the Indian government needs to be reviewed, and will, it is hoped, be reviewed in the negotiations to revise the Kyoto Protocol for the period beyond 2012, in light of two potential, critical concerns analysed in this chapter. The first is that climate change induced catastrophes would directly deprive many Indians of their rights to life and livelihood. The second concern is that domestic government mechanisms, *not even* the much heralded (and sometimes maligned) public interest litigation can safeguard the core constitutional rights of 'We, the people of India . . .' (Preamble, the Constitution of India). Because climate change could affect the ability of Indians to pursue their constitutionally guaranteed rights to lead healthy and productive lives, this chapter argues that in defining its international bargaining position, the government of India should concentrate on these domestic human rights challenges, rather than support an international regime based on the creation of equal international rights to emit GHGs.

The first section of the chapter provides a snapshot of the Government of India's position on climate change. The second and third sections briefly discuss the scope of the rights under Article 21 of the Indian Constitution and the reasons why climate change could infringe rights guaranteed under that provision. The fourth and fifth sections assess the scope of public interest litigation as a strategy to evince effective responses to climate change related rights violations, and the alarming legal and pragmatic limits even of this innovative mechanism to provide redress. The final sections consider more broadly the actual and potential utility of climate litigation as a basis for shaping a new international climate regime.

2. INDIA'S MITIGATION FOCUS AND NUMBERS STRATEGY

On August 26, 2002, India acceded to the Kyoto Protocol to the United Nations Convention on Climate Change (UNFCCC). Although the UNFCCC requires India to monitor and report its GHG emissions, neither the framework Convention or the Protocol impose any obligation on India to meet specific time-bound targets in reducing its emissions. However, India's gross domestic product (GDP) has grown steadily by about 8 per cent annually since 2000,[1] and such economic growth has correspondingly resulted in increased GHG emissions. India is now reputed to be the fifth largest emitter of GHGs in the world (World Bank, 2007). As a result of these changing realities, the Government of India is under growing international pressure to accept binding obligations, much like existing Annex I nations, under the new instrument being negotiated to replace the Kyoto Protocol.[2]

The reasons for differentiating between developed and developing countries on grounds of economic differences and historic contributions to the build up of GHGs in the atmosphere are specifically acknowledged in the Preamble to the UNFCCC. The Indian Government resists proposals that it accept binding emission reduction obligations on grounds of economic development and international equity (Havlvorssen, 2007). India argues that, if one were to compare the total GHG emissions of the US and India, for instance, both the past and current emissions of India are significantly smaller. The vast gap is also evident in the per capita emissions. In 2005, the US was ranked fifth in the world for its per capita GHG emissions, whereas India was ranked 121st. The US emitted 19.8 tonnes per person while India emitted 0.8 tonnes per person. Between 1994 and 2004 the cumulative carbon dioxide emissions of the US totaled 128.7 tonnes per person as opposed to India's 5.7 tonnes, each country accounting for 23.5 per cent and 3.8 per cent respectively of the world emissions for the same period.[3] Such vast disparities are not surprising, given that in 2000 the US constituted approximately 22 per cent of the world economy, whereas India comprised 4.34 per cent.[4]

Nonetheless, the Indian administration is undertaking several voluntary actions that may stem the rate of GHG emission increases. These actions include the promotion of renewable energy and investments in clean development technologies. For example, the Ministry of New and Renewable Energy is exploring alternatives such as solar, wind, hydropower and biogas.[5] Also, pursuant to Article 12 of the Kyoto Protocol, there are several Clean Development Mechanism (CDM) projects underway in India. Most CDM projects focus on generating renewable energy, such as biomass/biogas, small scale hydro power, solar, and wind. Other projects include energy efficiency in the cement sector.[6]

Further, the Indian government has adopted energy regulations and established new ministries or administrative agencies. For example, the Energy Conservation Act was passed in 2001 to promote efficient use of energy. The Bureau of Energy Efficiency (BEE), created under this Act, is working with key industries including cement, paper and pulp, and aluminum, to establish voluntary energy efficiency practices. It is also drafting standards for energy labeling, building codes, and certification programs, among other initiatives to promote energy conservation and efficiency.[7] The National Tariff Policy (2006) also encourages State Regulatory Commissions (SERCs) to purchase a mandated percentage of their energy from renewable sources.[8] The government has also set up special committees, including an Expert Committee on Climate Change, headed by the Chairperson of the Intergovernmental Panel on Climate Change (IPCC), Rajendra Pachauri, and other prominent government and non-government representatives, to consider additional venues for action (Gandhi, 2007). Also significant is the establishment of the Prime Minister's Council on Climate Change, which released the National Action Plan on Climate Change in June 2008. The Action Plan primarily focuses on energy efficiency, sustainability, and protection of the Himalayan glaciers.[9] India's eleventh Five Year Plan also specifically refers to the need to address climate change, but reiterates the point that developed countries must bear the burden of mitigation while India should focus on adaptation and stimulate economic growth to achieve this goal.[10]

Competing policies and interests, however, present significant challenges to these efforts. Demand for energy and carbon intensive materials to sustain development activities has risen exponentially. Notably, there is growing demand for electricity for hi-tech industries and for modern amenities such as air conditioners; for materials to build infrastructure for the spiraling land and air traffic; and for cement for commercial and residential constructions (Shukla, et al., 2003; Ministry of Environment and Forests, 2004; Ferrey, 2007; Yergin, Eklof and Edmond, 1998). These energy requirements have driven the government to exploit its coal reserves and increase its dependence on petroleum resources (Ferrey, 2007; Ministry of Environment and Forests, 2004; Yergin, Eklof and Edmond, 1998).[11] Further, the central administration has focused on encouraging internal and foreign direct investments to jumpstart some of these projects, particularly major infrastructure projects such as power plants construction, marine ports, telecommunications and real estate. The Indian Electricity Act of 2003 provides incentives such as a fixed return on investment of 16 per cent, the removal of licensing requirements to operate and maintain power-generation stations, and entitlements to sell power directly without government intervention. The government is also drafting an Open Skies Policy and an Airport Economic Regulatory Authority to expand construction of airports, which will inevitably increase GHG emissions (Riedy and Kurth, 2007; Nelivigi, et al., 2007).

However, the increased carbon footprint from such developments is hardly addressed in India's environmental legislation. For instance, India's environmental impact assessment (EIA) regulation, which requires projects to be reviewed for potentially negative impacts on the environment such as pollution and harm to wildlife, does not currently assess potential impacts of a project on climate change.[12] Indeed, the Govindrajan Committee specifically recommended the removal of hurdles to infrastructure development, including streamlining and expediting EIA requirements (Government of India, 2002; Paliwal, 2006). By contrast, in other jurisdictions EIA legislation is increasingly being interpreted as encompassing climate change impacts. In *Border Power Plant Working Group v. Department of Energy*,[13] the US District Court of Southern California held that the US government was required to conduct an EIA to assess the potential increase in carbon dioxide emissions before constructing utility plants. EIA legislation in Australia is also increasingly being applied to assess and avoid impacts on climate change (Bonyhady and Christoff, 2007).

With a booming economy that is fueling a rapid increase in its GHG emissions, the Indian government thus faces the daunting challenge of designing policies that balance the country's insatiable appetite for rapid economic growth with the need to mitigate GHG emissions and to act in an internationally responsible manner on climate change. While the National Action Plan on Climate Change of June 2008 promises increased attention to alternative and renewable energy sources such as solar energy, it will take significant investment to meet projected energy demands.

Recent projections by the International Energy Agency (2007, pp. 461–67) indicate that India's energy demand could grow at a rate of 3.3 per cent per year, but with higher levels of growth forecast for some fossil fuels, such as a 4.7 per cent annual increase in the demand for coal. Consequently, the Indian economy is likely to continue to rely heavily on fossil fuels, despite the renewable energy emphasis articulated in the Draft Report of the Expert Committee on Integrated Energy Policy. Indeed, the Policy also focuses on privatizing coal production to increase investment and output (Planning Commission, 2005, pp. 110–13).

These developments may not only reflect on India's ability to accept binding international obligations to reduce its GHG emissions beyond 2012; they may also infringe the right to life guaranteed under the Indian Constitution. It is these domestic constitutional implications that are the focus of the remainder of this chapter.

3. FUNDAMENTAL RIGHTS JURISPRUDENCE UNDER THE INDIAN CONSTITUTION: A BRIEF INTRODUCTION

Adopted in 1949, the Indian Constitution establishes a number of basic citizen rights that have acquired implications for climate and other environmental policy issues. Most importantly, Article 21 of the Indian Constitution provides: '[n]o person shall be deprived of his life or personal liberty except according to procedure established by law'.[14] The Indian Supreme Court has interpreted this Article to serve as a conduit for recognizing and enforcing a variety of ancillary rights, such as rights to livelihood,[15] health,[16] basic necessities,[17] travel abroad[18] and privacy.[19] The Supreme Court bundled these rights into Article 21 based on other, non-binding, provisions of the Indian constitution,[20] on decisions of non-Indian courts, and on international law and principles.[21]

Of particular importance to this discussion is the fact that the Indian Supreme Court has included within Article 21 rights that are affected due to government inaction on environmental and related matters. Examples of such matters include protection of the Taj Mahal from coal and coke pollution;[22] cleaning up the Ganga river;[23] relocation of hazardous industries in Delhi;[24] curbing of vehicular pollution;[25] requiring compulsory environmental education;[26] and re-directing an illegally diverted river,[27] to name a few.[28] The Court has also gone so far as to take executive action, in the form of establishing expert committees and issuing directives on several matters.[29]

Thus, the Indian Supreme Court has interpreted the right to life under Article 21 in a manner that extends beyond mere survival, to cover conditions that are necessary for higher standards of living. Climate change, as discussed below, could potentially affect both the survival and standards of living of Indians. Therefore, constitutional litigation may be one avenue to accelerate government responsiveness to climate change, and this possibility may influence the tenor of the Indian Government in negotiations for a post-2012 climate regime. Of course, even if litigation is a feasible option to influence India's stance on addressing global warming, it may not amount to very much internationally, given India's current limited influence in the post-Kyoto negotiations. On the other hand, such litigation may be a more promising avenue to leverage change domestically than any other currently available legal strategy. I therefore now turn to discussing specifically how climate change could infringe constitutional rights in India, before moving on to discussion of domestic constitutional litigation as a tool for combating climate change.

4. CONSTITUTIONAL IMPACT OF CLIMATE CHANGE

Climate change, if unmitigated, will directly and indirectly bear upon Article 21 rights that are guaranteed under the Indian Constitution. This is evident in some of the early predictions made by scientists and some preliminary observations of possible climate-related occurrences. For example, the IPCC predicts that glacial melts in the Himalayan region alone will increase flooding, trigger avalanches and landslides, and cause extinction of species and ecosystems (IPCC Working Group II, 2007; Revkin, 2007). As such, Himalayan glaciers, including the Gangotri which is a source of the perennial and holy river Ganga, have receded by 30 metres, endangering water supply in the dry season (Wax, 2007). Other changes in hydrological cycles are also expected to cause extreme drought or flood conditions in some river basins (Gosain, et al., 2006), shorten crop duration periods to the detriment of agricultural yields (Sathaye, et al., 2006), threaten biological diversity (Ravindranath, et al., 2006), increase the risk of malarial outbreaks by creating conditions favorable to disease carrying vectors (Bhattacharya, et al., 2006), and increase cyclones in coastal regions (Unnikrishnan, et al., 2006).

Some early observations of the effects of climate change are also becoming visible. In early 2007, an Indian farmer was reportedly forced to abandon his ancestral agricultural land because it was part of one of two islands submerged in the Sunderbans region (Harrabin, 2007). Another farmer faced a similar threat temporarily (Dhar, 2007). Absent compensation and support from their government, the former moved to urban areas in search of alternative livelihood, whereas the latter stayed on despite the risk of future flooding and limited access to food. Both incidents, which have been attributed to climate change related sea level rise, portend the fate of some of the nearly 65 per cent of India's population that is dependent on agriculture, forestry and fisheries for a living (Indian Communication to the UNFCCC, 2004).

Thus, climate change will affect not only the economic growth of India but more importantly the wide array of fundamental rights – to life, to livelihood and to health – that have been guaranteed under the Indian Constitution. Pursuing constitutional remedies, however, presents risks as well as opportunities.

5. REASONS FOR PROCEEDING UNDER THE CONSTITUTION

Bringing climate change within the constitutional dialogue may not only be strategically desirable, but may be the only viable option in India to compensate victims for their loss and, more importantly, to deter continuing GHG emissions in India for numerous reasons. Currently, if one were to seek to impose liability for climate change problems in India, those primarily responsible would be nations other than India, such as Australia, Canada, some European nations, Japan, and the US. Few of these countries provide remedies to their own private citizens, particularly on a scale guaranteed and interpreted under the Indian Constitution, and it is almost impossible for foreign victims to sue in the domestic courts of these countries. Furthermore, even the limited window of redress provided by some of these nations may be impossible to open.

For instance, the US Alien Tort Claims Act permits foreign nationals to bring an action in US domestic courts against the US government for acts committed abroad.[30] However, claimants must prove violation of international law,[31] and not of their own domestic law. Thus, in order to sue, Indian citizens would have to demonstrate that the US government caused climate change in India and in violation of international law.[32] Such a claim is wrought with a range of evidentiary and jurisprudential problems[33] which most affected Indians would find insurmountable, both because of their economic circumstances and the extent and nature of scientific data that would have to be garnered to prove causation.

The Indian Constitution, on the other hand, could serve as a potential basis for pursuing climate change litigation within India, not only because of its substantive provisions but also because the Supreme Court of India has facilitated enforcement of fundamental constitutional rights by relaxing several formal procedural rules, which generally impede access to courts. Firstly, the Court has waived 'ripeness' requirements for bringing an action, on the ground that in a country where most people are unaware of their rights, violations should be addressed *before* they occur.[34] Thus, the presence of substantial threats of climate-related violations should be sufficient to invoke the Court's writ jurisdiction under Article 32.[35]

Secondly, the Court has the authority to determine whether an injury has occurred,[36] without relying on statutory enactments.[37] Further, petitioners need not satisfy any additional standing requirements such as causation and redressability[38] (i.e., the remedy), both of which are required in American courts. Further, *any* person with 'sufficient interest'[39] in helping poor and vulnerable sections of the population can seek judicial review.[40] In the alternative, the Court can assume *suo moto* jurisdiction by treating newspaper reports or letters as writ petitions (the latter is also known as 'epistolary jurisprudence').[41] Finally, the Court can provide broad remedies; it can issue a writ of mandamus not only ordering the government to perform non-discretionary functions, or enjoining it from performing statutorily prohibited actions, but also requiring it to perform *discretionary* functions (Sathe, 2001, p. 23). The judiciary could also issue 'continuing mandamus',[42] obliging the government to take specific actions and report progress on a regular basis,[43] as it has in the past.[44]

The unique alignment of procedural flexibilities and substantive rights that will be affected if climate change is not mitigated provide a sound basis for pressing forth a constitutional rights violation argument in the Indian context.

6. POSSIBLE ROADBLOCKS TO CONSTITUTIONAL LITIGATION

These strong advantages of constitutional litigation, however, do not necessarily promise favorable or predictable outcomes, for both legal and practical reasons. For one thing, these constitutional rights are not enforceable against the foreign governments that bear the historic burden of the problem. Even if the Supreme Court were to find an Article 21 violation, it could not exercise its jurisdiction over foreign nations. Article 32 of the Indian Constitution provides for judicial review only in the case of violation of fundamental rights by the Indian government.[45]

Even the option of bringing a constitutional challenge against the Indian government as a strategy to catalyze action on climate change is not without problems. The Constitution vests foreign affairs powers exclusively in the central government.[46] It grants to parliament, the legislative branch,[47] the power to enact laws regarding foreign affairs,[48] including the power to conclude legal arrangements.[49] However in practice, the executive branch of government[50] enters into and implements treaties and international obligations,[51] and the parliament merely has the power to enact executing domestic legislation (Franck and Thiruvengadam, 2003, pp. 483–84). No provision of the Indian Constitution explicitly grants the judiciary the authority to review matters related to foreign affairs. In fact, the judiciary is specifically excluded from adjudicating international disputes,[52] except for advising the President upon request.

Of course, one could argue that the Indian Supreme Court has placed checks on the executive treaty-making power in the past (Franck and Thiruvengadam, 2003). For instance, in *Madhav Rao Scindia* v. *Union of India*,[53] the Court held that the government does not have the authority to use its foreign affairs power unilaterally to withdraw recognition of royalty status to former princes (Neuborne, 2003). Thus, unlike in the US,[54] the Indian judiciary has not decisively concluded that foreign affairs are a prerogative of the executive branch subject only to limited intervention from the legislative branch.

Further, in order to support an argument in favor of judicial intervention in foreign affairs, one could point to the Indian Supreme Court's reasoning for developing epistolary jurisdiction. The judiciary interpreted *locus standi* liberally because the Court believed that the promises of a constitutional democracy were beyond the reach of many Indians due to financial and cultural constraints (Varshney, 2007). Judges who pioneered public interest litigation reasoned that in a society where oppression and poverty were cultural norms, most people did not have the knowledge or the means to claim their constitutional rights (Callahan, 1994; Varshney, 2007). None of these conditions have changed, despite India's economic growth. India remains home to some of the poorest people in the world (United Nations Development Program, 2008). It contains some of the most polluted and hazardous sites on the planet (Harvey, 2007). Administrative systems remain prone to corruption (Mehta, 2003; Patel, 1998),[55] while access to the courts remains abysmally expensive and slow (Prasad, et al., 2007). In other words, a large percentage of Indians remain vulnerable to climate change related violations of their fundamental rights without redress. Thus, the Indian judiciary could be justified in exercising its jurisdiction within Article 32 to intervene in foreign affairs on the grounds that this was necessary to prevent or remedy violations of constitutional rights related to climate change. Of course, to sustain such judicial intervention would be challenging, because it raises a host of complex issues, such as on what specific grounds would the Indian Government be accountable (e.g., its negotiating stance in international climate law negotiations).

Furthermore, the judiciary may be hard pressed to grant appropriate remedies. Short of directing the government on international climate negotiations and policies, there is little that the Indian judiciary can effectively do by way of shielding fundamental rights from climate change related violations, without risking its institutional legitimacy.[56] If and when the violations actually occur, the Supreme Court could order the government to pursue international adjudication, but, as current evidence suggests, international adjudication has limited utility (Romano, 2000, pp. 39–41). International principles under which any such remedy might be claimed, such as the duty to prevent transboundary pollution (even though interpreted as customary international law in the Trail Smelter Arbitration),[57] are often not complied with by states (Bodansky, 1995, p. 116).

The Indian judiciary could perhaps mitigate these concerns by following the lead of the US Ninth Circuit Court, which in a 2006 decision established the liability of foreign entities for acts committed within their territory that led to adverse consequences within the US. In *Pakootas* v. *Teck Cominco Metals Ltd.*,[58] the Court held that Canadian smelter industries that released hazardous waste into the Washington River, with proper permits from Canadian environmental authorities, were liable under a US law on hazardous waste disposal.[59] The Court held that even though the waste was released on the Canadian side, the companies responsible for the pollution were liable under US law to clean up the waste in the US where the effects materialized. It also held that the US Environmental Protection Agency had a non-discretionary duty to enforce domestic environmental legislation against the companies.

In effect, the US Court extended the application of a domestic statute to a foreign entity for acts committed outside the US, on the ground that the effects were felt in the US. In arriving at this conclusion, the Court rejected the companies' argument that they were not responsible for the flow of the river, which essentially carried the waste from Canadian soil, and noted that Washington taxpayers ought not to bear the economic burden for external actions.[60] While this case is situated in the context of a long history of cross-border pollution incidents involving Canada and the US, including failed diplomatic interventions adequately to address the problem, the rationale for the judgment nevertheless provides a salutary lesson. It demonstrates that domestic legislation can be the most accessible safeguard of citizens' rights against acts of foreign entities, in the absence of effective international regulations.

Thus, the Indian Supreme Court could order the Indian government to pass legislation on climate change which includes provisions that hold persons or states responsible for climate change liable to redress violations of Indian constitutional rights, and requires them to compensate those who suffer economic losses as a result. However, such legislation would not only be diplomatically and politically undesirable, but difficult to implement given the transnational limits to judicial power. Moreover, the government has traditionally been reluctant to enact any legislation that potentially hinders economic growth, and, as mentioned earlier, this may affect the separation of powers and raise questions of institutional legitimacy.

The Indian government has also been unsuccessful in litigating on behalf of those people affected by environmental harms, as demonstrated by two critical cases. The Bhopal gas leak incident[61] of 1984 illustrates the weaknesses of the government and the domestic legal system to deliver effective justice to citizens. Following the fatal gas leak at a facility owned by a Union Carbide subsidiary, the government passed the Bhopal Act[62] to consolidate the thousands of civil suits brought before Indian and US courts, and to represent the interests of its citizens against a foreign entity based on the *parens patriae* doctrine. The Act not only allowed the government to act on behalf of its citizens, but also provided mechanisms for distribution of compensation among victims.

However, the litigation remained contentious because the government brought an action before the District Court of New York instead of its own courts, on the ground that a foreign entity was involved and that its own courts lacked the capacity to redress the matter.[63] However, the New York court dismissed the claim on the ground of *forum non conveniens*.[64] Eventually, the case was brought before the Madhya Pradesh High Court, based on the Indian rule of absolute liability.[65] However, after much bargaining on both sides, the Indian government settled the matter in 1989, with Union Carbide agreeing to pay the government US$465 million, of which the Indian subsidiary paid US$45 million. The settlement not only led to criticisms about the government bargaining away the right to justice of thousands of victims, but the actual distribution of compensation was also the subject of critique, especially since a large proportion of the victims and their families have still not received any compensation to this day.[66]

Similarly, *Narmada Bachao Andolan* v. *Union of India and Others*,[67] involving the damming of the Narmada River, popularly known as the

Sardar Sarova or the Narmada Dam project (Ramachandra, 2006), exemplifies the inadequacy of domestic legal protection for Indians affected by natural resource abuses. Most people displaced by its rising water levels remain without adequate compensation, primarily because the issue of land-related displacement, which is governed by the British-era Land Acquisition Act 1894,[68] confers on the government the right to expropriate private property in the public interest, by simply paying compensation at the prevailing market value.[69] Moreover, given the complex caste system and the remnants of a near feudal-like land ownership system in India, especially in rural areas, many farmers either lease or simply work on the land, and may not be entitled to compensation.[70] Rehabilitation is also plagued by administrative gaps and delays, as State governments fail to comply with awards given by the Narmada Tribunal fully and in a timely way, and the judiciary has chosen in this case to give deference to the executive branch (Mehta, 2005).

The Bhopal and Narmada Dam cases are but two well-known instances of administrative failures and judicial inefficacy to prevent, protect against and adequately redress constitutional violations of the environmental rights of thousands of Indians. Rights violations occur routinely in India (Neuborne, 2003). There is no indication at this point that such violations will be prevented or compensated for when climate change related catastrophes unravel, especially if the incidents in the Sunderbans are any indication (Harrabin, 2007). More importantly, the Indian judiciary may be facing the limits of its capacity to deliver justice in this context, because of the complexity of climate science, the nature and breadth of remedies from different branches of government that will be required effectively to mitigate the problem, and the role of international governments.

7. THE ROAD AHEAD

The constitutional challenges that could be triggered by climate change lend themselves to adjudication. This is why judicial intervention has become an important strategic tool to catalyze action on climate change in other jurisdictions, notably the US. In *Massachusetts v. EPA*, a group of states and environmental public interest groups brought an action against the federal Environmental Protection Agency (EPA) for failing to take action to combat climate change.[71] The majority of the US Supreme Court in that decision dismissed the EPA's arguments that it did not have the power under the Clean Air Act to enact legislation to regulate vehicular emissions and that preventive action was better left to international negotiations, which were outside the purview of the EPA's authority. The majority of the Court held that the EPA was legally required to exercise its jurisdiction under the Clean Air Act to come to an opinion as to whether or not vehicular emissions posed a danger, and were not free to shirk this obligation based on the rationale of scientific uncertainty.

Constitutional litigation in India could serve the same purpose – to catalyze action on climate change by the Indian government. This has been the purpose of public interest litigation in India, to fill the void created by incompetent administration. Thus, despite the potential hurdles discussed earlier, constitutional litigation could serve the purpose of forcing the government to think in terms of the rights of its people that will be violated if climate change occurs, rather than thinking in terms of the right of its people to add to the problem.

Such a shift could possibly lead to the Indian government taking a more cooperative approach in the current post-Kyoto negotiations that would focus on both its responsibilities for mitigation and adaptation strategies. The outcome of international meetings could become more meaningful if citizens in several similarly located countries forced their governments to think in terms of the constitutional rights that their administration is abandoning while fighting for an equal right to emit GHGs. In fact, many developing countries such as Bangladesh, Pakistan and South Africa have expanded the bundle of rights, including environmentally-related rights, under their national constitutions.[72] These countries could therefore negotiate under a mandate that their position and demands in international climate talks could fundamentally affect the constitutional rights of their citizens.

For their part, domestic courts must tread cautiously so as not to usurp executive or legislative powers and functions. The Indian Supreme Court's directions may require the government to rethink its negotiating positions, but it would be dangerous for the judiciary to attempt to play a role in the negotiation of a climate deal because any such attempts to overreach would also affect the entire constitutional structure upon which the nation is founded.

8. CONCLUSION

India's commitment to climate mitigation and adaptation is considered by many to be critical to the future of an effective post-2012 international climate regime. This view stems from concerns about the potential effects that a rapidly growing hydrocarbon economy of a billion people can have on the global climate and on efforts by other nations to reduce GHG emissions. This is a legitimate concern and one that requires attention. While playing its part in the negotiation of this international regime, the Indian government must also be spurred to action by considering the threat that the absence of an effective regime presents to the constitutionally guaranteed rights of its citizens, and its own potential domestic liability for climate induced harm.

This emerging domestic constitutional challenge in the case of climate change illustrates that by focusing on limiting their international obligations, states may actually be endangering their own constitutions, even in countries such as India where the right to judicial review of government action or inaction is very expansive. In light of this challenge, India and other nations will have to determine a way to preserve their constitutions in situations where their protections are jeopardized by events beyond their control and when such events can be managed only by means of international law. What is at threat is not just the economy, but the very structural beams of modern civilization.

NOTES

* Assistant Professor, Chapman University School of Law, United States.

1. Reserve Bank of India, 'RBI Increases Cash Reserve Ratio', Press Release (13 February 2007), www.rbi.org.in/scripts/BS_ViewBulletin.aspx?Id=8333 (visited 15 March 2009). This figure may change given the current global economic recession.
2. The Byrd–Hagel Resolution passed by the US Senate unanimously, 95–0, expressly advised the Clinton Administration not to accept binding obligations under the Kyoto Protocol unless developing countries did so as well: Byrd-Hagel Resolution, Letter to Members of the Senate on the Kyoto Protocol on Climate Change) for a statement by President Bush rejecting the Kyoto Protocol.
3. See Cumulative Emissions, Concentrations, and Temperature Increases, Climate Analysis Indicators Tool (CAIT) (World Resources Institute), http://cait.wri.org/cait.php?page=compcoun (visited 15 March 2009).
4. Ibid.
5. Ministry of Renewable Energy, *Annual Report 2005–2006; Administrative Circular Notification*, No. 24/1/1993-Admn.II, Government of India, 20 October 2006).
6. See CDM India, India's Designated National Authority for CDM, http://cdmindia.nic.in (visited 15 March 2009).
7. See www.bee-india.nic.in (visited 15 March 2009).
8. Tariff Policy, Resolution, No. 23/2/2005-R&R (Vol. III), *The Gazette of India* (6 January 2006).
9. See http://pmindia.nic.in/Pg01-52.pdf (visited 15 March 2009). For an update on India's climate mitigation measures as of September 2008 see also Climate Change Mitigation Measures in India, International Brief 2, September 2008, www.pewclimate.org/docUploads/India-FactSheet-09-08.pdf (visited 15 March 2009).
10. Planning Commission (2007, pp. 203–07), http://planningcommission.nic.in/plans/planrel/fiveyr/11th/11_v1/11v1_ch9.pdf (visited 15 March 2009).
11. The Ministry of New and Renewable Energy (2006) estimates that only 5 to 6 per cent of India's energy needs will be produced through renewable resources for the foreseeable future.
12. Environmental Impact Notification, S.O. 1533, 14 September 2006, Ministry of

Environment and Forests, *Gazette of India*, Extraordinary, Part II, and s.3(ii), http://envfor.nic.in/legis/eia/so1533.pdf (visited 15 March 2009).
13. 260 F. Supp. 2d 997 (S.D. Cal. 2003).
14. See http://lawmin.nic.in/legislative/Art1-242%20(1-88).doc (visited 15 March 2009).
15. *Hoskot v. State of Maharstra*, A.I.R. 1978 S.C. 1533.
16. *Bandhua Mukti Morcha v. India*, A.I.R. 1984 S.C. 802.
17. *Olga Tellis v. Bombay Municipal Corporation*, A.I.R. 1985 S.C. 180.
18. See *Satwant Singh v. Asst. Passport Officer*, A.I.R. 1967 S.C. 1836; *Maneka Gandhi v. Union of India*, A.I.R. 1978 S.C. 597, where the petitioner argued that, by impounding her passport, the Government had violated her Art. 21 right to personal liberty.
19. See *Kharak Singh v. State of U.P.*, A.I.R. 1963 S.C. 1295, holding that personal liberty implied the right to privacy. See also Sathe, pp. 51–57, for a detailed discussion of the interpretation of Art. 21 by the Supreme Court.
20. Directive principles contained in Part IV of the Constitution set out non-binding goals, such as free legal aid, gender non-discrimination, compulsory education, and provision for livelihood, for the Government to implement. Article 37 states: '[t]he provisions contained in this Part shall not be enforceable by any court, but the principles therein laid down are nevertheless fundamental in the governance of the country and it shall be the duty of the State to apply these principles in making laws'. See also *M.C. Mehta v. Union of India* (1992) 2 S.C.C. 256; *Sachidananda Pandey v. State of West Bengal*, A.I.R. 1987 S.C. 1109.
21. See *Indian Council for Enviro-Legal Action* (1996) 5 S.C.C. 212, holding that the polluter pays principle was the law of the land; *Vellore Citizens Welfare Forum* v. *Union of India* (1996) 5 S.C.C. 647, holding that the precautionary principle was part of India's environmental laws; *Andhra Pradesh Pollution Control Board v. M. Nayudu* (1999) 2 S.C.C. 718, explained in detail the evolution of the precautionary principle in response to a challenge brought against the A.P. Pollution Control Board by a vegetable oil industry for rejecting a permit to site the industry close to a water reservoir; *M.C. Mehta v. Kamal Nath*, using the Roman law doctrine of public trust, as applied by the Supreme Court of California in *National Audubon Society v. Superior Court*, 33 Cal. 3d 419 (1983). Also, for a study of trends in the Indian judiciary's use of international and foreign decisions see Smith (2006), who provides statistics demonstrating that the Supreme Court's reliance on foreign law, not just British law, has declined since 1990s, although it relied heavily on Privy Council decisions in the 1950s.
22. *M.C. Mehta v. Union of India* (1997) 2 S.C.C. 353.
23. *M.C. Mehta v. Union of India* (1998) 6 S.C.C. 63.
24. *M.C. Mehta v. Union of India*, A.I.R. 1996 5 S.C.C. 281. The Court ordered the closure and relocation of more than 1,300 major polluting hazardous industries from Delhi to sites in neighboring states. See *M.C. Mehta v. Union of India* (1998) 6 S.C.C. 63.
25. *M.C. Mehta v. Union of India* (1999) 6 S.C.C. 12, 14; *M.C. Mehta v. Union of India* (2002) 4 S.C.C. 359. In both cases, the Court ordered the Government to implement Euro I and II emission standards for reducing automobile pollution. Further, in a far reaching case, the Court ordered that all public buses be run on compressed natural gas to reduce pollution. For an overview of the Euro emissions standards see Emission Norms, Society of Indian Automobile Manufacturers, at http://www.siamindia.com/scripts/emission-standards.aspx (visited 15 March 2009).
26. *M.C. Mehta v. Union of India*, A.I.R. 1988 S.C. 103; *M.C. Mehta v. Union of India* (1998) 6 S.C.C. 63.
27. *M.C. Mehta v. Kamal Nath*, A.I.R. 2000 6 S.C.C. 213, in which Mr. Mehta challenged the diversion of the river Beas by a hotel in which close relatives of Kamal Nath, the Minister of Environment and Forests, held shares. The Court held that disturbing the 'basic environment', such as air and water, constituted a violation of the right to life, which encompassed preservation of ecological balance. See G8 + 8 Climate Change Dialogue, Brussels Legislators Forum, GLOBE International, www.globeinternational.org (visited 15 March 2009).

28. An overview of all the cases brought by M.C. Mehta regarding environmental protection can be found at the M.C. Mehta Environmental Foundation website, www.mcmef.org/landmark.htm (visited 15 March 2009).
29. See *M.C. Mehta v. Union of India* (1998) 6 S.C.C. 63 (known as the 'Delhi Pollution Case'), in which the Court set up a Committee to advise it on the implications of shifting from traditional fuel to compressed natural gas for public buses, and ordered all related government agencies to coordinate with each other, and monitored the implementation by requiring periodic requirements. Similarly, in *T.N. Godavarman Thiurmulpad v. Union of India* (2006) 5 SCC 57, the Court issued a series of orders regarding the management of national forests.
30. 28 U.S.C. §1350 (1994).
31. Ibid.
32. For a discussion of the *Alien Tort Claims Act*, its scope and limits see D'Amato (1998) and Dodge (1996). For a detailed analysis of the potential of bringing environmental claims under the Act see Unger (2001).
33. The difficulties in deciding the ongoing suit brought by the Inuit against the US demonstrate the limitations of litigation as a strategy to persuade nations to participate in an international climate regime. For a discussion of the suit see the chapter by Sumudu Atapattu in this book.
34. See *Basheshar Nath* v. *Commissioner of Income Tax*, A.I.R. 1959 S.C. 149, holding that a waiver of fundamental rights could not be upheld in a country where many people were ill-informed about their rights (see also Sathe, 2001, p. 67).
35. Art. 32(1) states: '[t]he right to move the Supreme Court by appropriate proceedings for the enforcement of the rights conferred by this Part is guaranteed'. Art. 32 provides for writ petitions such as mandamus and habeas corpus.
36. The Court has held that it has the authority to 'decide whether proper procedure was prescribed by the legislature and followed by the executive' under Art. 21 (Sathe, 2001, p. 67).
37. This position differs from the US test for standing, where injury-in-fact is determined on the basis of statutory provisions. In fact, in a recent climate-related case, *Massachusetts v. US Environmental Protection Agency*, 127 S. Ct. 1438 (2007), the US Supreme Court held that injury-in-fact involved a substantive test based on statutory rights and not on judicial determination of actual injury. See Environmental Law Institute (2007), discussing briefly the development of standing jurisprudence in the US, leading up to the Supreme Court's recent decision regarding the power of EPA to regulate carbon dioxide emissions.
38. See *Lujan* v. *Defenders of Wildlife*, 504 US 555 (1992), explaining each of the tests. Although the Indian judiciary followed a three prong test similar to that in US law, it relaxed these requirements in *S.P. Gupta* v. *President of India*, A.I.R. 1982 S.C. 149, in which the executive arbitrarily transferred judges for their opinions, threatening the independence of the judiciary. The Court did not lower standing requirements, and also held that the fundamental right to freedom of speech and expression enshrined in Art. 19(1) of the Constitution included the right to information. Further, the Court ruled that the Government was required to consult with the Chief Justice and other judges before appointing judges, even though the final decision remained with the executive (Sathe, 2001, pp. 70, 96 and 102).
39. As a result, non-governmental organizations and public interest lawyers have filed many writ petitions on behalf of those affected, which the Court had balanced by allowing only genuine petitions and not *mala fide* actions (Sathe, 2001, p. 81).
40. *Bandhua Mukti Morcha v. India,* A.I.R. 1984 S.C. 802. The Court observed that judicial review proceedings to enforce fundamental rights are not limited to any person or proceedings under Art. 32(1) of the Constitution. The Court has taken such a broad approach to ensure that rights of Indians who are not in a position to claim fundamental rights protection enjoy full constitutional protection. Judges in the Supreme Court, notably Justice P.N. Bhagawati, view public interest litigation as a means for poor,

under-informed, and underprivileged Indians to access expensive judicial systems, and the judiciary must therefore open up access through flexible rules (Bhagawati, 1985; Cooper, 1993).

41. *Sunil Batra v. Delhi Administration*, A.I.R. 1978 S.C. 1675, in which the Court treated a letter from a prisoner complaining about prison conditions as a writ petition seeking to enforce fundamental rights. See also *P. U.D.R. v. India*, A.I.R. 1982 S.C. 1473.
42. Ibid., 82.
43. Sathe (2001), discussing the orders issued by the Court to the Central Bureau of Investigation in *Vineet Narain v. Union of India*, A.I.R. 1996 S.C. 3386 and in *Union of India v. Sushil Kumar Modi*, A.I.R. 1997 S.C. 314.
44. For instance, in *Kishen v. State of Orissa*, A.I.R. 1989 S.C. 677, the Court ordered the Government to prevent death by poverty and starvation.
45. Art. 12 states: '[i]n this Part, unless the context otherwise requires, 'the State' includes the Government and Parliament of India and the Government and the Legislature of each of the States and all local or other authorities within the territory of India or under the control of Government of India'. 'In this Part' refers to the section on Fundamental Rights.
46. The subject matters with respect to which the state and central governments have jurisdiction are listed in the Seventh Schedule. List I sets out areas for the central government, List II the areas within the jurisdiction of the states, and List III those within the power of both governments, even though the central government has preemptory powers in case of overlap or conflict. The central government also has residuary powers regarding subject matters not covered in Lists II and III. See Arts 246 and 254.
47. The Parliament is composed of the President and two Houses – the Council of States and the House of the People. See Art. 79.
48. However, because the President heads the Executive branch and is also part of the Legislature, the division of powers under the Indian Constitution is considered a mix of both the British and the US systems (Franck and Thiruvengadam, 2003, p. 483).
49. List I, Item 10, lists foreign affairs as: 'all matters which bring the union into relation with any foreign country'. A series of foreign affairs related powers are listed in List I – preventive detention with respect to foreign affairs (Item 9); treaty-making and implementation (Item 14); 'foreign jurisdiction' (Item 16); and foreign exchange and foreign loans (Items 36, 37).
50. The Executive branch is headed by the President. The President is advised by a Council of Ministers, headed by the Prime Minister. See Art. 74.
51. A similar practice has been observed in the US (White, 1999, pp. 4–5), arguing against exclusive federal executive authority to pursue foreign affairs.
52. Art. 363 of the Constitution.
53. A.I.R. 1971 S.C. 530 (2003).
54. In the US, foreign matters are generally excluded from judicial review under the political question doctrine (Frank, 1992).
55. But see *Report on the Transparency International Global Corruption Barometer* (Transparency International, 2006), noting, however, some improvements; Schurer (2005), 158, noting that despite problems of administrative corruption the 'new cabinet in New Delhi has an impeccable anti-corruption record and that India is well on its way to building both domestic and international trust in its legal and regulatory climate'.
56. As such there are criticisms that the judiciary has ignored the separation of powers doctrine. See Rosencranz and Jackson (2003), arguing that although intervention to abate Delhi's pollution was timely, the Supreme Court undermined the development of administrative capacity to address environmental matters by usurping executive functions. See also Waldman (1996). See also Rajamani (2007), arguing the need for cautious judicial intervention based on two case studies involving municipal solid waste management and the Delhi Vehicle Pollution cases.
57. *Trail Smelter Arbitration (US v. Canada), Convention for Settlement of Difficulties Arising from Operation of Smelter at Trail*, B.C.US Treaty Series No. 893, signed at

Ottawa, 15 April 1935, ratifications exchanged 3 Aug 1935, www.lfip.org/laws666/trailsm.htm (visited 15 March 2009).

58. 452 F. 3d 1066 (2006).
59. *Comprehensive Environmental Response, Compensation and Liability Act*, 42 USC §9603.
60. For an analysis of the history of smelter waste-related disputes between the US and Canada and the Ninth Circuit Court decision see Robinson-Dorn (2006).
61. The Bhopal gas leak case involved the leakage of methyl isocyanate from a Union Carbide India plant in India, killing 2,100 people and injuring some 200,000 people, resulting in a tort claim. See Covell (1991).
62. *The Bhopal Gas Leak Disaster (Processing of Claims) Act* of 1985.
63. The Union Carbide Company held nearly half the shares in the Indian subsidiary company.
64. *In re Union Carbide Corp. Gas Plant Disaster*, 634 F. Supp. 842 (S.D.N.Y. 1986); affirmed in 809 F.2d 195 (2d Cir.) (1987).
65. Based on the strict liability rule, under the absolute liability rule any multinational enterprise carrying on hazardous activities is liable for damage resulting from such activity and cannot claim any of the exceptions available under the original rule. See *M.C. Mehta v. Union of India (Oleum Gas Leak case*), A.I.R. 1987 S.C. 1086. For a discussion of the settlement process see Covell (1991).
66. A chronology of the case and its present status has been posted by Union Carbide Company at www.bhopal.com/chrono.htm (visited 15 March 2009).
67. *Narmada Bachao Andolan v. Union of India and Others*, A.I.R. 1994 S.C. 319. In this case, representatives of people who stood to be displaced by the dam's construction brought public interest litigation arguing violation of fundamental rights.
68. A copy of the legislation is available at http://punjablaws.gov.pk/laws/12.html (visited 15 March 2009).
69. For a discussion of the problems with the land acquisition law in India and takings jurisprudence see Mehta (1992).
70. See generally Jacobsohn (2004), referring to India as an 'essentially feudal society'. Property rights were initially part of fundamental rights, but have since then been repealed. Yet, the zamindari system that was recognized by the British colonial government in India perpetuated a feudal approach to land ownership. Amendments in several states to redistribute the land by abolishing this system through a series of constitutional amendments have led to constitutional challenges based on the amount of compensation, which is now generally not subject to challenge. For a discussion of the key cases that shaped the constitutional jurisprudence of property in India see Neuborne (2003), p. 490. See also Allen (1993), discussing the challenges that judiciaries face in interpreting property rights and the role of the influence of property jurisprudence among Commonwealth countries, including India.
71. 127 S. Ct. 1438 (2007).
72. See Jonas Razzaque, 'Human Rights and the Environment: The National Experience in South Asia and Africa', Joint UNEP–OHCHR Expert Seminar on Human Rights and the Environment, Background Paper No. 4, 2002, www.unhchr.ch/environment/bp4.html (visited 15 March 2009).

REFERENCES

Allen, T. (1993), 'Commonwealth Constitutions and Right not to be Deprived of Property', *International and Comparartive Law Quarterly* 42, 523.

Baxi, U. (2000), 'Constitutionalism as a Site of State Formative Practices', *Cardozo Law Review* 21, 1183.

Bhagawati, P.N. (1985), 'Judicial Activism and Public Interest Litigation', *Columbia Journal of Transnational Law* 23, 561.

Bhattacharya, S., et al. (2006), 'Climate Change and Malaria in India', *Current Science* 90(3), 369.

Bodansky, D. (1995), 'Customary (And Not So Customary) International Environmental Law', *Indiana Journal of Global Legal Studies* 3, 105.

Bonyhady, T. and P. Christoff (2007), *Climate Law in Australia* (Federation Press).

Callahan, M.B. (1994), 'Cultural Relativism and the Interpretation of Constitutional Texts', *Willamette Law Review*, 30, 609.

Cooper, J. (1993), 'Poverty and Constitutional Justice', *Mercer Law* Review 44, 611.

Covell, T. (1991), 'The Bhopal Disaster Litigation: Its Not Over Yet', *North Carolina Journal of International Law and Commercial Regulation* 16, 279.

D'Amato, A. (1998), 'The Alien Tort Statute and the Founding of the Constitution', *American Journal of International Law* 82, 62.

Dhar, S. (2007), 'Environment-India: Rising Seas Threaten Bengal's Deltaic People', *Inter Press Service News Agency* (21 October).

Dodge, W.S. (1996), 'The Historical Origins of the Alien Tort Statute: A Response to the "Originalists"', *Hastings International and Comparative Law Review* 19, 221.

Environmental Law Institute (2007), 'Access to Courts After Massachusetts v. EPA: Who Has Been Left Standing?', *Environmental Law Reporter* 37(September), 10692.

Ferrey, S. (2007), 'Why Electricity Matters, Developing Nations Matter, and Asia Matters Most of All', *New York University Environmental Law Journal* 15, 113.

Franck, T. (1992), *Political Questions/Judicial Answers: Does the Rule of Law Apply to Foreign Affairs?* (Princeton University Press).

Franck, T.M. and A.K. Thiruvengadam (2003), 'International Law and Constitution-Making', *Chinese Journal of International Law* 2, 468.

Gandhi, D. (2007), 'Committee on Climate Change Set Up', *The Hindu* (11 May).

Gosain, A.K., S. Rao and D. Basuray (2006), 'Climate Change Impact Assessment on Hydrology of Indian River Basins', *Current Science* 90(3), 346.

Government of India (2002), *Report on Reforming Investment Approval & Implementation Procedures, Part I (Investment approval procedures – Government and Public Sector Projects)*, http://dipp.nic.in/implrepo/implrepo1.pdf.

Harrabin, R. (2007), 'How Climate Change Hits India's Poor', *BBC News* (1 February).

Harvey, F. (2007), 'Planets Most Polluted Sites Unveiled', *Financial Times* (13 September), 4.

Havlvorssen, A.M. (2007), 'Common, But Differentiated Commitments in the Future Climate Change Regime – Amending the Kyoto Protocol to Include Annex C and the Annex C Mitigation Fund', *Colorado Journal of International Environmental Law and Policy* 18, 247.

Intergovernmental Panel on Climate Change (IPCC) (2007), *Climate Change Impacts, Adaptability and Vulnerability*, Working Group II (Cambridge University Press).

International Energy Agency (IEA) (2007), *World Energy Outlook 2007, China and India Insights* (IEA).

Jacobsohn, G.J. (2004), 'The Permeability of Constitutional Borders', *Texas Law Review* 82, 1763.
Mehta, P. (2005), 'Internally-Displaced Persons and the Sardar Sarovar Project: A Case for Rehabilitative Reform in Rural India', *American University International Law Review* 20, 613.
Ministry of Environment and Forests (2004), *India's Initial National Communications to the United Nations Framework Convention on Climate Change* (Ministry of Environment and Forests).
Ministry of New and Renewable Energy (2006), *Annual Report (2005–2006)* (Ministry of New and Renewable Energy).
Nelivigi, N., D. Peel and C. Krishnamoorthy (2007), 'Infrastructure Project Finance in India: Recent Developments', *Corporate Handbook Series* 1587, 133.
Neuborne, B. (2003), 'The Supreme Court of India', *International Journal of Constitutional* Law 1, 476.
Nzelibe, J. (2004), 'The Uniqueness of Foreign Affairs', *Iowa Law Review* 89, 941.
Osofsky, H.M. (2006–07), 'The Inuit Petition as a Bridge? Beyond Dialectics of Climate Change and Indigenous People's Rights', *American Indian Law Review* 31, 675.
Paliwal, R. (2006), 'EIA Practice in India and its Evaluation using SWOT Analysis', *Environmental Impact Assessment Review* 25(5), 492.
Parasuraman, S. (1999), *The Development Dilemma: Displacement in India* (Macmillan Press).
Patel, T. (1998), 'Corrupt Practices in India: No Payoff', *Loyola of Los Angeles International and Comparative Law Journal* 20, 389.
Planning Commission (2007), *Eleventh Five Year Plan* (Government of India).
Prasad, A.S. and V.I. Balan (2007), 'Strategies for US Companies to Mitigate Legal Risks from Doing Business in India', *Corporate Law and Practice Course Handbook Series*, PLI Order No. 11926 (February–March).
Pratap B.M. (2003), *The Burden of Democracy* (Penguin India).
President G.W. Bush (2001), 'Letter to Members of the Senate on the Kyoto Protocol on Climate Change', *Weekly Compilation of Presidential Documents* 37(444).
Rajamani, L. (2007), 'Public Interest Environmental Litigation in India: Exploring Issues of Access, Participation, Equity, Effectiveness and Sustainability', *Journal of Environmental Law* 19(3), 293.
Ramachandra, K. (2006), 'Sardar Sarovar: An Experience Retained?', *Harvard Human Rights Journal* 19, 275.
Ravindranath, N.H., et al. (2006), 'Impact of Climate Change on Forests in India', *Current Science* 90(3), 354.
Revkin, A.C. (2007), 'Poor Nations to Bear the Brunt as World Warms', *New York Times* (1 April).
Riedy, M.J. and A. Kurth (2007), 'Project Finance India 2007 – Overcoming Hurdles to Growth: Current Trends and Innovative Transactional Structures in India', *Practicing Law Institute* 1587, 143.
Robinson-Dorn, M.J. (2006), 'The Trial Smelter: Is What's Past Prologue? EPA Blazes a New Trial for CERCLA', *New York University Environmental Law Journal* 14, 233.
Romano, C.P. (2000), *The Peaceful Settlement of International Environmental Disputes* (Kluwer Law International).
Rosencranz, A. and M. Jackson (2003), 'The Delhi Pollution Case: The Supreme

Court of India and the Limits of Judicial Power', *Columbia Journal of Environmental Law* 28, 223.
Sandalow, D. (2008), *Freedom from Oil* (McGraw Hill).
Sathe, S.P. (2001), 'Judicial Activism: The Indian Experience', *Washington University Journal of Law and Policy* 6, 29.
Sathaye, J., P.R. Shukla and N.H. Ravindranath (2006), 'Climate Change, Sustainable Development and India: Global and National Concerns', *Current Science* 90(3), 314.
Schurer, W. (2005), 'A Geopolitical and Geo-Economic Overview: On the Rise of China and India as Two Asian Giants', *Fletcher Forum World Affairs* 29, 145.
Shukla, P.R., et al. (2003), *Development and Climate: An Assessment of India* (Indian Institute of Management) (on file with author).
Smith, A.M. (2006), 'Making Itself at Home Understanding Foreign Law in Domestic Jurisprudence: The Indian Case', *Berkeley Journal of International Law* 24, 218.
Transparency International (2006), *Report on the Transparency International Global Corruption Barometer* (Transparency International).
Unger, R. (2001), 'Brandishing the Precautionary Principle Through the Alien Tort Claims Act', *New York University Environmental Law Journal* 9, 638.
United Nations Development Program (UNDP) (2008), *Statistics of Human Development Report* 2007/2008 (UNDP).
Unnikrishnan, A.S., et al. (2006), 'Sea Level Changes Along the Indian Coast: Observations and Projections', *Current Science* 90(3), 362.
Varshney, A. (2007), 'Democracy vs. Growth in India', *Foreign Affairs* 93 (March/April).
Waldman, P. (1996), 'Jurist's Prudence India's Supreme Court Makes Rule of Law A Way of Governing: Splintered Political System Leaves Judges to Battle Corruption and Pollution', *The Wall Street Journal* (6 May).
Wax, E. (2007), 'A Sacred River Endangered by Global Warming', *Washington Post* (17 June), A14.
White, G.E. (1999), 'The Transformation of the Constitutional Regime of Foreign Relations', *Virginia Law Review* 85, 1.
World Bank (2007), *The Little Green Data Book* (World Bank).
World Resources Institute (n.d), *Climate Analysis Indicators Tool*, Version 6.0 (WRI).
Yergin, D., D. Eklof and J. Edmond (1998), 'Fueling Asia's Recovery', *Foreign Affairs* 77, 34.

4. Promoting justice within the international legal system: prospects for climate refugees

Angela Williams*

1. THE PROBLEM OF CLIMATE CHANGE DISPLACEMENT

Climate change creates and contributes to numerous problems worldwide cutting across economic, social, cultural, technological and environmental dimensions. One of the many consequences of our rapidly changing climate is a significant increase in forced migration and displacement as anthropogenic greenhouse gas emissions alter the atmosphere and the global environment within which we all live. The emergence of so-called 'climate change refugees' and the subsequent recognition of this rapidly escalating phenomenon present an important challenge to be addressed by all states. Whilst discussion continues regarding how such displaced individuals, communities or (in some situations) entire nation states should be legally recognised, the objective of this chapter is to examine how the problem might be best managed through justice discourse. Climate change displacement creates enormous injustice and inequity around the world and moreover is facilitated by unfair and inequitable international legal and economic systems. So how might notions of justice be promoted for those affected by climate change displacement? And in what way might the international legal system be able to contribute to a fairer and more just outcome for all concerned?

The many and varied consequences of climate change are now widely recognised and comprehensively documented (for example Stern, 2007; IPCC, 2008; and more generally Monbiot, 2006; Lynas, 2007). In particular, the link between climate change and environmental vulnerability has been the focus of much attention in light of the increased prevalence of droughts, desertification, rising sea levels and extreme weather patterns (UNDP, 2007, pp. 73–107). The potential for environmental damage is on its own cause for considerable global concern. The impact of such environmental change on individuals, communities and in some cases entire countries demonstrates however an unprecedented challenge for which the global community must offer an urgent response. Climate-induced environmental change continues to pose challenges for many people and requires increasingly numerous and complex mitigation and adaptation techniques. One specific challenge to have emerged from the plethora of climate related problems is that of human displacement, whereby environmental change attributed to the climate triggers forced migration.

Displacement can occur as a result of numerous different environmental factors and combinations thereof, but certain types of environmental change are commonly identified as leading to climate change displacement. The most commonly identified cause of climate change displacement is rising sea levels which threaten small island states and low-lying coastal communities. Thermal expansion, along with the melting of glaciers and polar ice caps, has led scientists to estimate conservatively that a global sea level rise of between 28 and 43 centimetres is likely by the end of the century (IPCC, 2007). However, more recent projections put the figure closer to 150 centimetres within the same timeframe.[1] Secondly, displacement attributed to food security occurs where storm surges lead to coastal erosion, salt contamination and crop degradation, whilst coral bleaching destroys stocks of natural marine resources. Similarly, the availability of clean water supplies is threatened by changing rainfall and unpredictable weather patterns. Finally, an increased prevalence in cyclones and extreme weather patterns are having disastrous effects in many areas around the world, as illustrated by Hurricane Katrina which decimated much of New Orleans in 2005.[2]

The impacts of such environmental change on displacement are stark. There have been a number of attempts to predict displacement numbers attributable to climate change. While specific calculations vary (see for example Brown, 2008; Baird, et al., 2007; Nicholls, 2004), current estimates indicate that up to 200 million people will be displaced by 2050 as a direct result of climate change (Myers, 2005). The difficulties in agreeing on future estimates for climate change displacement must be acknowledged. Ongoing developments in climate science and modelling techniques, incomplete data on migration flows, and the complexity of identifying causal links for displacement (for example to what extent should economic factors be take into account) make it difficult to predict with certainty the number of people likely to be affected. Despite such uncertainties current research indicates the numbers are going to be significant. To put the estimated figure of 200 million people into context, the United Nations Refugee Agency (UNHCR) currently recognises approximately 16 million refugees and 51 million Internally Displaced Persons (IDPs) (UNHCR, 2008, p. 2). So whilst currently there may be up to 67 million refugees and IDPs (notably not all of which fall under the protection of the UNHCR), by the year 2050 there could be 200 million people displaced *by climate change alone*. Moreover, the international legal system currently fails to attribute any significant legal protection to such people (see for example Williams, 2008). There remain a number of key challenges in terms of recognising those affected by climate change displacement and predicting the extent of the problem in global terms. Nevertheless, whilst discussion continues and governments carry on debating the extent of the problem and possible responses, the reality for millions of people is loss of livelihood, home, family, cultural heritage, and possibly, country.

2. IDENTIFYING THE EFFECTS OF CLIMATE CHANGE DISPLACEMENT

A major problem with climate change and with the international legal framework that has been created as a response to it is the global inequity and enhanced vulnerability of developing states (Gordon, 2007). Such inequity and unfairness are exemplified by climate change displacement whereby the most vulnerable members of global society are commonly the ones who suffer the greatest. Indeed, economically, socially or environmentally vulnerable communities are much less likely to have the capacity to adapt to a rapidly changing climate. Such limitations in respect of adaptation frequently translate to subsistence problems regarding the availability of clean water supplies and the depletion or degradation of natural resources (UNDP, 2007). Yet the same vulnerable communities usually make a comparatively small contribution – if indeed any recognisable contribution at all – to the global climate change problem. Accordingly, vulnerable members of our global community suffer disproportionally as they contribute very little by way of greenhouse gases, yet experience the greatest loss due to an inability to adapt to climate change:

> People living in the Ganges Delta and lower Manhattan share the flood risks associated with rising sea levels. They do not share the same vulnerabilities. The reason: the Ganges Delta is marked by high levels of poverty and low levels of infrastructural protection (UNDP, 2007, p. 78).

Similar observations can be made at the state level where the biggest contributors to the climate change problem are frequently the same states that demonstrate the greatest capacity to mitigate or adapt to the changing climate, with the result that they are comparatively much less affected. As a country, the United States may be responsible for one of the largest contributions of global greenhouse gas emissions but it demonstrates comparatively enormous capacity to respond and adapt to the changing climate. Conversely, those states that make a very small contribution to greenhouse gas emissions frequently suffer disproportionately as they lack the capacity to mitigate and adapt. Two examples illustrate the extent to which climate change enhances existing global inequity and vulnerability in terms of displacement.

Kiribati

The small island state of Kiribati is located in the South Pacific, approximately half-way between Australia and Hawaii. Made up of 33 island and coral atolls, it currently has a population of more than 90,000 people. However, the very existence of the nation state of Kiribati is currently under threat due to climate change, as rising sea levels, increased storm surges, salt contamination and drought all have a debilitating effect on the islands' natural resources and inhabitants (Pearce, 2000; Loughry and McAdam, 2008). Because most of the people of Kiribati (referred to as I-Kiribati) rely on subsistence activities, changes to the natural environment have a direct and immediate impact on the survival of the population. Much internal migration has already taken place, resulting in the main island atoll of Tarawa now being heavily populated. Moreover, the World Bank has advised that in the absence of adaptation, Kiribati could face a combined annual damage bill from climate change and sea level rise equivalent to 17–34 per cent of Kiribati's 1998 GDP (Bettencourt, et al., 2006, p. 7).

In light of these developments, Kiribati's President, Anote Tong, has called for international assistance to evacuate the country before it completely disappears. He used the 2008 World Environment Day to highlight the plight of his nation, commenting 'we may be beyond redemption . . . we may be at the point of no return, where the emissions in the atmosphere will carry on contributing to climate change, to produce a sea level change so in time our small, low-lying islands will be submerged'.[3]

Emphasising that the entire global community has a responsibility to address the challenge of climate change, Tong suggested that a worse case scenario would see Kiribati uninhabitable within 50 years.[4] The impacts of climate change on the I-Kiribati and their land are now clearly evident. Failing a significant and immediate response from the international community, it appears likely the entire population will be displaced and the land submerged in the Pacific Ocean. Some calculate that the year 2020 represents a tipping point where atmospheric concentrations of greenhouse gases will reach the point at which we will likely no longer be able to avert potentially irreversible climate change.[5] In light of this, Kiribati's already incredibly small window of opportunity to gain support from the international community to develop sufficient adaptation mechanisms and practices diminishes considerably. Moreover, as relocation appears to be an increasingly likely option for the I-Kiribati, other issues are raised regarding *inter alia* the loss of social identity, cultural heritage, nationality, and the possible right of the I-Kiribati to self-determination following relocation.

As a small island developing state Kiribati, like many others in the South Pacific, is responsible for a tiny proportion of global greenhouse gas emissions, yet stands to suffer the most severe of consequences. Kiribati does not only face limited capacity regarding adaptation, however. The ability of Kiribati to respond to what is inherently a global problem is also restricted due to its status as a small island developing state and its comparatively weak global negotiating position. As a member of the Alliance of Small Island States (AOSIS), Kiribati belongs to 'a coalition of small island and low-lying coastal countries that share similar development challenges and concerns about the environment, especially their vulnerability to the adverse effects of global climate change'.[6] AOSIS has worked hard to negotiate on behalf of small island states and provide a strong collective voice within the international climate change framework and negotiating processes. Nevertheless, whilst highlighting the plight of small island states and bringing issues such as displacement to the attention of the international community, it has enjoyed little success in achieving any meaningful commitments from other states (Gillespie, 2003–2004, p. 119; Gordon, 2007, pp. 1603, 1621). Thus not only does Kiribati face problems in terms of suffering disproportionately from climate change, but the international legal process also presents difficulties regarding global inequity.

Mali

With two-thirds of the country physically located within the Sahara Desert, the vast majority of Mali's population live in rural areas throughout the south of the country where they are heavily dependent on the country's natural resources for subsistence living. Over recent years climate change has accentuated the harsh environmental conditions endured by Malians. Ominously, these trends appear set to continue as predicted increases in temperature and decreases in precipitation threaten water and food security (Baird, et al., 2006). One consequence of unpredictable weather patterns, declining soil productivity and reduced crop production is forced migration of many young people from their local communities to urban centres to find work to support their families (Baird, et al., 2007, p. 43). Although local charities offer funding to help rural communities respond to such challenging circumstances, climate change displacement increasingly jeopardizes such aid efforts. During 2006 funds were made available to the village of Toulabé for the construction of a small-scale dam that, by catching water throughout the wet months, would allow villagers to grow food during the dry season. However, because many of the young able-bodied men had migrated elsewhere in search of work, construction could not begin until their return late in the season. Consequently, construction of the dam was not completed until later than planned, coinciding with the first heavy rains of the season. As a result more than 60 per cent of the dam was destroyed and washed away (Baird, et al., 2007, p. 43).

Whilst farmers in Mali have contributed very little to the build-up of greenhouse gases in the atmosphere, they are nonetheless living in a region highly vulnerable to the impact of climate change and suffer the most drastic consequences (ibid., p. 44). Indeed this pattern can be seen throughout many low-income countries worldwide (World Bank, 2008, p. 124). The inevitable displacement that occurs in such situations is cause for serious concern. Whilst economic factors no doubt contribute to some migration within and from Mali, whereby people relocate primarily to avail themselves of better economic and employment opportunities, the link between climate induced environmental change and displacement is clearly evident and increasingly problematic.

The consequences of climate change displacement in Mali highlight the injustice and inequity of the situation. As many villagers have been forced to leave their native rural communities to secure a living elsewhere, a migration path has developed from Mali, through neighbouring Mauritania, and onwards to the Atlantic Ocean, where migrants attempt to travel by boat to Europe.[7] Since Spain enhanced surveillance and control of the Strait of Gibraltar to prevent migrants entering the Spanish mainland, traffickers now appear to be targeting the longer and more dangerous route between Mauritania and the Canary Islands. Recent reports from the Red Cross and the UN Refugee Agency indicate a sharp increase in the number of people attempting to make the sea crossing from North Africa to Southern Europe (Wynter, 2006). Whilst comparatively few migrant deaths at sea are reported each year, the actual mortality rate is estimated to be much higher as the bodies of many refugees are never recovered. The Red Cross estimates that between 2,000 and 3,000 people die trying to reach Spain every year.[8] Such alarming trends highlight and reaffirm the injustice of climate change displacement whereby those responsible for the least greenhouse gas emissions suffer the most dire and extreme consequences.

Kiribati and Mali provide two examples of the way in which climate change displacement fosters injustice and creates disproportionate levels of suffering for those most vulnerable. There are however many more such examples, prompting the questions of how justice might be better achieved within the international legal system in respect of climate change displacement.

3. JUSTICE AND CLIMATE CHANGE

In its broadest sense, the idea of justice is usually associated with notions of fairness and the type of behaviour society generally deems morally right or wrong. Furthermore, justice is intrinsically linked with the law as indicated by its Latin root *jus*, which translates to 'law' or 'right'. One problem with employing justice as a means to promote equity for those affected by climate change displacement however is the inherently subjective nature of the concept. While it may appear straightforward to identify the type of activities or behaviour that embody ideas of justice, in fact any such determination is ultimately based largely on social and moral values. Most people believe they can determine when justice has been achieved or if an end-result is 'just'. Whilst generally reflective of society's values, this approach is inevitably subjective. Within criminal law for example some people might deem acceptable a punishment involving restorative justice when loss has been suffered, whereby an offender is brought face to face with the victim to learn how the offender's actions have affected others and created suffering or loss. In the very same situation however, other people in the same society might consider that the crime necessitated imprisonment in order to achieve the same level of justice or fairness. Accordingly in this situation, the question whether or not justice is achieved for the victim depends largely on a subjective judgement based on moral values.

The universal popularity of the term and its associated social and political importance provide another challenge for the employment of justice discourse, since there is much discussion as to its exact interpretation and application within both law and political theory. There is an enormous volume of literature examining the notion of justice and the complexities of its interpretation and application (see for example Campbell, 2001; Pogge, 2001). In some of the earliest theorising, Aristotle considered that justice consists of treating 'equals' equally and 'unequals' unequally, in proportion to their inequality (see for example von Leyden, 1985). More recently, building on John Rawls' (1971) seminal thesis of 'justice as fairness', justice has subsequently been recognised as the 'normative ideal which serves to distinguish between good and bad outcomes or states of affairs, to inform present and future acts and choices, and to evaluate proposed and past actions' (Vanderheiden, 2008, p. 49). Developing this idea further, a 'virtuous social institution' may be recognised as one that uses its power justly and in the pursuit of justice, tantamount to what moral goodness is to individual persons (ibid., pp. 48–49).

Justice discourse has subsequently evolved to encompass different forms or subdivisions that have, in their own right, been recognised and developed within political theory (Campbell, 2001, p. 15). Two such examples are commonly identified as representing the fundamental components of justice discourse. First, justice can be distinguished as being corrective or remedial in character, whereby its primary function is to correct wrongdoings by way of punishment or compensation. This is known as 'remedial justice'. Secondly, justice can be identified as performing a social or distributive function to ensure a fair distribution of benefits and burdens throughout groups identified by social, racial, class or gender characteristics. This is known as 'distributive justice'. Both of these forms of justice are integral to any consideration of the problem of climate change displacement and are accordingly dealt with in turn.

Remedial Justice

The theory of remedial justice can be traced to Aristotle's early description of ideas of 'compensatory justice' where he considered that the role of the judge was

> to make the parties equal by the penalty he imposes, whereby he takes from the aggressor any gain he may have secured . . . [t]his explains why the disputants have recourse to a judge; for to go to a judge is to do justice . . . [w]hat the judge does is to restore equality (Aristotle, 1956, pp. 148–49).

In this way, Aristotle recognises 'compensatory justice' as the mean between loss and gain. This thesis has been further elaborated and developed so that the primary function of remedial justice is now commonly considered to be rectifying the wrong done to a victim or, to put it simply, correcting the injustice (Shelton, 2000, p. 38). Accordingly, there are three essential requirements for remedial justice: first, the parties are treated as equal; next, there is a damage inflicted by one party on another; and finally, the remedy seeks to restore the victim to the condition he or she was in before the unjust activity occurred (ibid.). Much of the theory underlying remedial justice is applicable to the problem of climate change displacement. Notably, questions arise regarding equity (and equality of options especially in relation to adaptation), the damage suffered by one party as a direct result of the actions of others, and the need for an appropriate remedy in order to compensate and restore victims displaced by climate change. It is important to note that much of the discourse concerning remedial justice relates to actions between individuals, rather than acts committed by a government or a state against an individual. Given that remedial justice provides a basis for public law remedies however, the same theory could similarly be applied to those suffering climate change displacement (ibid., p. 39).

An example of remedial justice being sought in the context of climate change displacement is the *Kivalina v. ExxonMobil Corp et al.* dispute, lodged in a United States District Court in February 2008. The plaintiffs are residents of the native Alaskan Inupiat village of Kivalina comprised of approximately 400 people whose ancestors have occupied the area since 'time immemorial'.[9] Kivalina is located on the tip of a six-mile barrier reef located between the Chukchi Sea and the Kivalina and Wulik Rivers on the Northwest coast of Alaska, some 70 miles north of the Arctic Circle. The plaintiffs claim that global warming is destroying their village and as a result they must be relocated soon, or alternatively the village must be abandoned and cease to exist. The United States Army Corps of Engineers and the United States Government Accountability Office have undertaken studies of the viability of a number of native villages in Alaska, concluding in the case of Kivalina that '[r]emaining on the island . . . is no longer a viable option for the community'.[10] Both bodies conclude that Kivalina must be relocated due to the effects of climate change and estimate the cost at somewhere between $95 and $400 million.[11]

The Plaintiffs in *Kivalina* complain that climate change has reduced the sea ice, which acts as a protective barrier to coastal storms that batter the coast. Specifically, they claim that '[d]ue to global warming, the sea ice forms later in the year, attaches to the coast later, breaks up earlier, and is less extensive and thinner, thus subjecting Kivalina to coastal storm waves and surges. These storms and waves are destroying the land upon which Kivalina is located'.[12] Moreover, the damage suffered by Kivalina as a result of the impacts of climate change is so grave that the village is becoming uninhabitable and the entire community must now be relocated.[13] Based on these facts, Kivalina has filed a complaint for damages against a group of 19 oil, coal and power companies. The Plaintiffs allege three main causes of action: nuisance, conspiracy and concert of action.[14] First, Kivalina seeks monetary damages for the Defendants' contribution to global warming through emissions of large quantities of greenhouse gases. In respect of this cause of action, the Plaintiffs claim public nuisance under federal common law, and in the alternative private and public nuisance under state law. Secondly, Kivalina seeks monetary damages for civil conspiracy against eight of the Defendants for allegedly participating in an agreement with each other to mislead the public about the science of global warming and to delay public awareness of the issue. Finally, Kivalina alleges that the Defendants have engaged in tortious acts in concert with each other relating to the creation, contribution to, and/or maintenance of the public nuisance of global warming.

The Defendants responded in June 2008 by seeking to have Kivalina's action dismissed due to what they allege to be the inordinately difficult problem of factual proof in tracing property losses suffered by the Plaintiffs to human-induced changes in the global climate.[15] Moreover, the Defendants claim that there is no precedent for holding a collection of Defendants liable for such global atmospheric damage. Specifically, the Defendants assert that the claims raised by Kivalina must be dismissed due to problems with causation and jurisdictional limitations regarding federal law. They also argue that the conspiracy and concert of action assertions are entirely derivative of the nuisance claims and therefore unable to survive the dismissal of those claims.. The motion had not been heard when this book went to press.

Based on existing precedent, the prospects for the *Kivalina* case do not look very good. US courts have dismissed other common law climate change-based claims because, among other things, they raised non-justiciable political questions that are beyond the competence of the federal courts.[16] Accordingly, there remain significant questions regarding the possible success of the complaint lodged by Kivalina. Even in the event that the court accepts the complaint as justiciable, challenges persist in the form of establishing standing and causation, which could well hamper its success. As noted by Hsu, '[c]ourts have erected a number of jurisprudential gates that policy-making plaintiffs would have to pass through in order to win, and the courts have historically made use of them to avoid dealing with pressing social issues in a judicial forum' (Hsu, 2008, p. 765).

Nevertheless, the complaint lodged by Kivalina presents a clear example of the theory of remedial justice being applied to the problem of climate change displacement whereby the community of Kivalina is attempting to seek redress and compensation to correct the injustice it has suffered as a result of the actions of others, namely the oil, power and coal companies.

Whilst Kivalina may not ultimately be successful in respect of this claim, such complaints raise interesting questions about the broader role of litigation in achieving justice for the victims of climate change displacement. Public interest litigation is increasingly popular in environmental and human rights law, whereby a case may be brought for the benefit of the public in general (see generally Schall, 2008). In many such cases, the action of bringing a litigious claim may not necessarily be carried out with the primary objective of achieving a legal result but instead to raise public awareness of the problem (possibly with a view to reaching an out of court settlement). Conceivably, the consideration of climate change displacement through justice discourse is especially well suited to such claims which rely on negotiated settlements rather than legal judgments. Indeed, such an approach may well offer a more effective technique of achieving justice by employing increasingly flexible and tailored solutions, as opposed to relying on the (perhaps) limited remedies available to a court, which ultimately must be guided by judicial precedent. As such, the exertion of public pressure or an out of court settlement could allow parties to reach a more 'just' result, whilst the threat of legal proceedings remains as an incentive to reaching agreement.

A similar approach to remedial justice for those displaced by climate change exists at the state level. There remain logistical and jurisdictional challenges in pursuing such a claim and, again, there currently exists no legal precedent. However, it is conceivable that at some not too distant point one state could bring an action against another state to remedy an injustice in relation to climate change displacement. Tuvalu, along with a number of other small island states, currently faces being completely submerged and the entire nation displaced due to climate change.[17] Indeed Tuvalu threatened to bring an international action against Australia in 2002 as one of the major greenhouse gas emitters contributing to the drastic environmental consequences suffered by the remaining population of approximately 90,000 Tuvaluans.[18] States responsible for significant quantities of greenhouse gas emissions are obvious targets for such legal claims, and it could be alleged that the United States (and Australia, which only recently agreed to participate in the Kyoto Protocol) should be held responsible for the damage suffered by the nation of Tuvalu. In this way, as the victim, Tuvalu may seek to achieve remedial justice by securing compensation for the wrong done almost entirely by other states. To achieve justice for those displaced by climate change, such a remedy might include financial compensation by way of damages (in a similar vein to the *Kivalina* claim above), or compulsion of (for example) Australia to accept the population of Tuvalu as climate refugees. Nevertheless in either scenario there remains serious concern regarding the ability of Tuvalu to maintain its social and cultural values, identity and heritage.

In terms of jurisdiction, the International Court of Justice (ICJ) appears the obvious forum for such a dispute, but the Court's lack of compulsory jurisdiction over the United States renders this option problematic. Another possible avenue might be to request an Advisory Opinion from the ICJ on the matter, although an organ of the UN must initiate this process where legal questions arise within the scope of the specific UN organ's activities (see for example Strauss, 2003; Koivurova, 2007). Other possible forums in which remedial justice could be sought at the state level for climate change displacement include, perhaps most notably, the United Nations Law of the Sea Tribunal. The Law of the Sea Convention arguably extends to marine environmental damage caused by greenhouse gas emissions, but it is similarly limited in its applicability due to the United States refusing to adhere to the Tribunal's binding adjudicatory system. Nonetheless, scope remains for other states to be held to account by this Tribunal (Strauss, 2003, p. 10188).

Remedial justice therefore has an important role to play in achieving justice for those affected by climate change displacement, and demonstrates scope for much further development. There may be problems with quantifying justice when an entire country is submerged and ceases to exist, or damages are sought to compensate communities suffering displacement from climate change. How does one determine what level of compensation is adequate when entire communities or nations are forced to leave their native lands? Similarly, logistical and jurisdictional challenges must be met in locating an appropriate forum for such disputes in domestic or international legal systems. The theory of remedial justice nonetheless presents an important legal framework for addressing the suffering occasioned by climate change displacement, and its further development will be integral to addressing current global inequity.

Distributive Justice

Focusing on fair distribution of benefits and burdens for all participants, the second subdivision of the theory of justice is similarly applicable to climate change. Distributive justice is concerned more generally with welfare and fairness than with compensation or remedy. Rawls considers distributive justice as a function of 'the basic structure of society', that is, primarily although not exclusively, a virtue of the basic social institutions of society, such as the political constitution, tax, education and health systems (Rawls, 1971, p. 7; Page, 2006, p. 13).

While the link between climate change and distributive justice has been well established in political theory (for example Page, 2006; Vanderheiden, 2008), evidence of this relationship can also be found in the international legal system. The theory of distributive justice is articulated mainly through the principle of generational equity, a concept now commonly employed throughout international law (for example Redgwell, 1999). The principle of intergenerational equity is now well established as a mechanism for ensuring that the needs of future generations are balanced with those of the current generation. This principle has been examined in academic discourse (for example, Brown Weiss, 1988; Redgwell, 1991 and 1999), and incorporated into many international agreements including the UN Climate Change Convention,[19] the International Convention for the Regulation of Whaling,[20] the Convention on Biological Diversity,[21] and the Convention Concerning the Protection of the World Cultural and Natural Heritage.[22] Moreover, recognition of the principle in ICJ jurisprudence, including Judge Weeramantry's dissenting opinion in the 1995 *Request for an Examination* decision, confirms the relevance and importance of intergenerational equity, whereby the interests of future generations are endorsed within the international legal system.[23] Indeed, much of the current discussion regarding distributive justice and climate change has focused on the distribution of benefits and burdens across generations in line with this principle (see for example Page, 2006).

Whilst the distribution of justice across generations is unquestionably a valid and important claim, it is the principle of intra-generational equity that holds greater significance for those affected by climate change displacement. Intra-generational equity is concerned with the distribution of benefits *within the current generation* (see for example French, 2001). Intergenerational equity may be an important objective, but it seems somewhat futile if there continue to be such vast degrees of inequity within the current generation (Williams, 2006). Priority must therefore be accorded to the more immediately relevant principle of intra-generational equity when pursuing justice in respect of climate change displacement. By its nature, intra-generational equity implies some notion of minimising the North–South divide and rectifying the present imbalance found within the international economic and legal systems to achieve justice and equity within the current generation. Accordingly, such an approach appears well suited as a response to the vast inequity faced by those affected by climate change displacement, in an attempt to rebalance existing benefits and burdens within the current generation. Given that the theory of distributive justice shares this focus on realigning benefits and burdens, the principle of intra-generational equity is in effect the practical implementation of distributive justice, or at least evidence of the theory in practice within the international legal system.

In terms of the legal significance of the intra-generational equity principle, there has, perhaps unsurprisingly, been less support and enthusiasm for its adoption compared to its intergenerational counterpart, primarily due to the commitment required by states to address the global poverty gap. The establishment in 1974 of the New International Economic Order (NIEO) recognised that 'the gap between the developed and the developing countries continues to widen in a system which was established at a time when most of the developing countries did not even exist as independent States and which perpetuates inequality'.[24]

Whilst the intention of the NIEO to banish prevailing disparities and secure prosperity for all[25] can hardly be criticised, many of the principles included to guide the NIEO toward this objective were met with strong resistance by developed states which objected to any agreement, non-binding or otherwise, that implied the existence of an obligation to provide development assistance (French, 2001). Whilst today an increasing number of international agreements provide in some way for aid or support for developing countries, there remains a great deal of reluctance on the part of developed states to recognise or commit to any legal obligation which implies redistributive justice. Recent attempts to recognise 'a collective responsibility to uphold the principles of human dignity, equality and equity at the global level' have resulted in a universal pledge by states to meet the UN Millennium Goals by 2015, but such commitment remains of moral significance only.[26]

Accordingly, it has become apparent that any explicit reference to creating a legal obligation that necessitates redistributive justice is generally met with resistance by developed states. A number of principles and mechanisms can however be identified as aiming to promote distributive justice by focusing on discrete obligations, rather than calling for a general move towards social redistribution. The principle of 'common but differentiated responsibility' is one example that is fairly widespread within international environmental law (see generally French, 2000; Cordonier Segger and Khalfan, 2004, pp. 132–43). Both the UN Climate Change Convention[27] and Kyoto Protocol[28] recognise the principle that efforts to protect the climate system should be done on the basis of equity and in accordance with states' common but differentiated responsibilities and respective capacities. In a similar way distributive justice is evident via the requirement for the wealthiest states to provide new and additional financial resources to support developing states' efforts to adapt to and mitigate climate change.[29] The requirement of technology transfer to developing country parties is another example.[30] Accordingly, whilst there is no overarching international obligation for states to redistribute benefits and burdens per se, a number of principles within the international legal system aim to achieve distributive justice and equity among states by concentrating on specific mechanisms. In this way, much can be achieved toward the objective of minimising inequity and thereby promoting justice for those affected by climate change displacement. Attention can be focused on these more specific mechanisms which aim to recognise the special circumstances of developing states, and the transfer of resources, finance and technology can be fostered as common and regular practice.

4. PROMOTING JUSTICE IN RESPECT OF CLIMATE CHANGE DISPLACEMENT

Promoting justice for those affected by climate change displacement encompasses both remedial and distributive forms of justice. Justice can be seen to operate on a dual level in this context. First, a requirement for a fair distribution of benefits and burdens must be implemented into relevant law and policy to allow participants to mitigate and adapt to problems of climate change displacement. This initial level focuses on creating a legal framework within which principles of distributive justice can be developed so that the legal system promotes and supports a fair distribution of justice among all participants, be they states, communities or individuals. Next, remedial justice provides a second layer so that, should injustice be suffered as a result of climate change displacement despite a system based on distributive justice already being in place, there exists an opportunity for corrective action to be taken to compensate the victim and move towards a just outcome. The recognition of both types of justice is imperative to address the problem of climate change displacement.

Ultimately however both forms of justice require some element of redistribution, whether a distribution of benefits and burdens to realign social welfare or a redistribution of responsibility to provide compensation where, for example, a state contributes to an injustice. Indeed, an approach encapsulating both remedial and distributive justice attempts to provide a more holistic solution: distributive justice is employed by way of specific legal mechanisms and principles to create an infrastructure that properly reflects the value and importance of justice, whilst remedial justice is employed when the system fails and corrective action is required. Working together, both forms of justice are instrumental in promoting and aiming to achieve justice in respect of climate change displacement.

NOTES

* Dr Angela Williams is a Lecturer in Law at the University of Sussex, United Kingdom.

1. R. Black (2008), 'Forecast for Big Sea Level Rise', *BBC News,* 15 April, http://news.bbc.co.uk/1/hi/sci/tech/7349236.stm (visited 2 February 2009).
2. For a more detailed discussion of these issues see Williams (2008).
3. K. Marks (2008), 'Paradise Lost: Climate Change Forces South Sea Islanders to Seek Sanctuary Abroad', *The Independent,* at www.independent.co.uk/news/world/australasia/paradise-lost-climate-change-forces-south-sea-islanders-to-seek-sanctuary-abroad-841409.html, 6 June (visited 2 February 2009).
4. Ibid.
5. New Economics Foundation (2008), *100 Months: Technical Note*, www.neweconomics.

org/gen/uploads/sbfxot55p5k3kd454n14zvyy01082008141045.pdf, 1 August (visited 2 February 2009).

6. Alliance of Small Island States (AOSIS), 'The Alliance', at www.sidsnet.org/aosis/ (visited 2 February 2009).
7. J. Orr (2008), '35 Africa Migrants Die on Boat Crossing to Spain', *The Guardian*, www.guardian.co.uk/world/2008/aug/26/spain (visited 2 February 2009).
8. G. Keeley (2008), 'Grim Toll of African Refugees Mounts on Spanish Beaches', *The Observer*, www.guardian.co.uk/world/2008/jul/13/spain, 13 July (visited 2 February 2009).
9. *Native Village of Kivalina v. ExxonMobil Corp*, No. CV-08-01138 SBA (N.D. Cal.) (Complaint, filed 26 February 2008), para. 1, www.pawalaw.com/assets/docs/kivalina-vs-exxon-08-1138-sba.pdf (visited 2 February 2009) ('Kivalina complaint').
10. Ibid., para. 185, quoting United States GAO 2003, p. 32.
11. Kivalina complaint, para. 186.
12. Ibid., para. 16.
13. Ibid., para. 17.
14. Ibid., paras 249–82.
15. *Native Village of Kivalina v. ExxonMobil Corp*, No. CV-08-01138 SBA (N.D. Cal.) (Oil Companies' Motion to Dismiss for Lack of Jurisdiction, filed 30 June 2008), www.pawalaw.com/assets/docs/motion-to-dismiss-12b1-by-oil-co.pdf (visited 2 February 2009) ('Oil Companies' Motion to Dismiss for Lack of Jurisdiction'). Three other motions to dismiss were filed, one by the oil company defendants for failure to state a claim, one by the electric utility defendants, and one by defendant Peabody Coal. All four were consolidated for hearing. The motions and the plaintiffs' consolidated opposition to them are available at the website of plaintiffs' counsel, Law Offices of Matthew F. Pawa, P.C., at www.pawalaw.com/cases/ (visited 2 February 2009).
16. See for example *California v. General Motors Corp.*, 2007 WL 2726871 (N.D. Cal. 2007), *Comer v. Murphy Oil, USA, Inc.*, No. 1:05-cv-436, (S.D. Miss., 30 Aug. 2007), *appeal docketed*, No. 07-60756 (5th Cir. 2007), *Connecticut v. American Electric Power Company*, 406 F. Supp. 2d 265 (S.D.N.Y. 2005).
17. Statement by His Excellency the Rt. Hon. Bikenibeu Paeniu, Prime Ministerial Special Envoy on Climate Change, Tuvalu, to the Third Conference of the Parties to the United Nations Framework Convention on Climate Change (Kyoto, 8 December 1997), www.tuvaluislands.com/kyoto-panieu.htm (visited 2 February 2009).
18. *BBC News* (2002), 'Tiny Pacific Nation Takes on Australia', http://news.bbc.co.uk/1/hi/world/asia-pacific/1854118.stm, 4 March (visited 2 February 2009).
19. United Nations Framework Convention on Climate Change, 9 May 1992, ILM 31, 849, Preamble and Art. 3(1) ('UNFCCC').
20. 2 December 1946, 161 UNTS 143, Preamble.
21. 22 June 1992, 31 I.L.M. 818, Preamble and art. 2 (definition of 'Sustainable Use').
22. 16 November 1972, ILM 11, 1358, Art. 4.
23. *Request for an Examination of the Situation in Accordance with Paragraph 63 of the Court's Judgment of 20 December 1974 in the Nuclear Tests case* (*New Zealand v. France*) [1995] ICJ Reps 341.
24. United Nations General Assembly, Declaration on the Establishment of a New International Economic Order (1 May 1974), UN Doc. No. A/RES/3201 (S-VI), para. 1.
25. Ibid., para. 4(b).
26. United Nations General Assembly, United Nations Millennium Declaration (18 September 2000), UN Doc. No. A/Res/55/2 (18 September 2000).
27. Arts 3(1) and 4(1).
28. Kyoto Protocol to the United Nations Framework Convention on Climate Change, 11 December 1997, ILM 37, 22, Art. 10 ('Kyoto Protocol').
29. Art. 4(3).
30. Art. 4(3) and (7).

REFERENCES

Aristotle (1956), *The Ethics of Aristotle: The Nicomachean Ethics* (trans. J.A.H. Thompson) (Penguin).

Baird, R., et al. (2007), *Human Tide: The Real Migration Crisis* (Christian Aid).

Bettencourt, S., et al. (2006), *Not If But When: Adapting to Natural Hazards in the Pacific Islands Region: A Policy Note* (World Bank).

Brown, O. (2008), 'The Numbers Game', *Forced Migration Review* 31, 8.

Brown Weiss, E. (1988), *In Fairness to Future Generations: International Law, Common Patrimony, and Intergenerational Equity* (UN University and Transnational Publishers).

Butt, T., B. McCarl and A. Kergna (2006), 'Policies for Reducing Agricultural Sector Vulnerability to Climate Change in Mali', *Climate Policy* 5, 583.

Campbell, T. (2001), *Justice: Issues on Political Theory* (2nd edn, Palgrave Macmillan).

Cordonier Segger, M.-C. and A. Khalfan (2004), *Sustainable Development Law: Principles, Prospects and Practices* (Oxford University Press).

French, D. (2001), 'International Environmental Law and the Achievement of Intragenerational Equity', *Environmental Law Reporter* 31, 10469.

French, D. (2000), 'Developing States and International Environmental Law: The Importance of Differentiated Responsibilities', *International and Comparative Law Quarterly* 49, 35.

Gillespie, A. (2003–2004), 'Small Island States in the Face of Climatic Change: The End of the Line in International Environmental Responsibility', *UCLA Journal of Environmental Law and Policy* 22, 107.

Gordon, R. (2007), 'Climate Change and the Poorest Nations: Further Reflections on Global Inequality', *University of Colorado Law Review* 78, 1559.

Hsu, S. (2008), 'A Realistic Evaluation of Climate Change Litigation through the Lens of a Hypothetical Lawsuit', *University of Colorado Law Review,* 79, 701.

Intergovernmental Panel on Climate Change (IPCC) (2007), *Climate Change 2007: The Physical Science Basis* (Cambridge University Press).

Intergovernmental Panel on Climate Change (IPCC) (2008), *Fourth Assessment Report: Climate Change 2007* (Cambridge University Press).

Koivurova, T. (2007), 'International Legal Avenues to Address the Plight of Victims of Climate Change: Problems and Prospects', *Journal of Environmental Law and Litigation* 22, 267.

Loughry, M. and J. McAdam (2008), 'Kiribati: Relocation and Adaptation', *Forced Migration Review* 31, 51.

Lynas, M. (2007), *Six Degrees: Our Future on a Hotter Planet* (Harper).

Monbiot, G. (2006), *How to Stop the Planet Burning* (Penguin).

Myers, N. (2005), 'Environmental Refugees: An Emergent Security Issue', paper presented to Organization for Security and Cooperation in Europe, 13th Economic Forum, Prague, 23–27 May 2005, at www.osce.org/documents/eea/2005/05/14488_en.pdf (visited 2 February 2009).

Nicholls, R. (2004), 'Coastal Flooding and Wetland Loss in the 21st Century: Changes Under the SREA Climate and Socio-economic Scenarios', *Global Environmental Change* 14, 69.

Page, E. (2006), *Climate Change, Justice and Future Generations* (Edward Elgar).

Pearce, F. (2000), 'Turning Back the Tide', *New Scientist* 2225, 44.

Pogge, T. (ed.) (2001), *Global Justice* (Blackwell).
Rawls, J. (1971), *A Theory of Justice* (Belknap).
Redgwell, C. (1999), *Intergenerational Trusts and Environmental Protection* (Manchester University Press).
Redgwell, C. (1991), 'Intergenerational Equity and Global Warming', in R. Churchill and D. Freestone (eds), *International Law and Global Climate Change* (Graham and Trotman).
Schall, C. (2008), 'Public Interest Litigation Concerning Environmental Matters Before Human Rights Courts: A Promising Future Concept?', *Journal of Environmental Law* 20(3), 417.
Shelton, D. (2000), *Remedies in International Human Rights Law* (Oxford University Press).
Stern, N. (2007), *The Economics of Climate Change: The Stern Review* (Cambridge University Press).
Strauss, A. (2003), 'The Legal Option: Suing the United States in International Forums for Global Warming Emissions', *Environmental Law Reporter* 33, 10185.
United Nations Development Programme (UNDP) (2007), *Human Development Report 2007/2008: Fighting Climate Change: Human Solidarity in a Divided World* (Palgrave Macmillan).
United Nations High Commissioner for Refugees (UNHCR) (2008), *2007 Global Trends: Refugees, Asylum-seekers, Returnees, Internally Displaced and Stateless Persons* (UNHCR).
United States General Accounting Office (GAO) (2003), *Alaska Native Villages: Most are Affected by Flooding and Erosion, But Few Qualify for Federal Assistance,* Report to Senate and House Committees on Appropriations (GAO).
Vanderheiden, S. (2008), *Atmospheric Justice: A Political Theory of Climate Change* (Oxford University Press).
Von Leyden, W. (1985), *Aristotle on Equality and Justice: His Political Argument* (Macmillan).
Williams, A. (2008), 'Turning the Tide: Recognising Climate Change Refugees in International Law', *Law and Policy* 30(4), 502.
Williams, A. (2006), 'Re-assessing Traditional Notions of Sustainable Development', *Environmental Law and Management* 18, 181.
Wynter, A. (2006), 'Deadly Passage', *Red Cross Red Crescent* 2, 12.
World Bank (2008), *World Development Indicators 2008* (World Bank).

5. Climate change and indigenous peoples in the South Pacific: the need for regional and local strategies

Eric Kwa*

1. INTRODUCTION

This chapter examines what challenges climate change poses to the indigenous peoples of the South Pacific, and the policies and laws being adopted in the region to address this threat. It is important to consider these environmental challenges at a regional and local level because the forecast apocalyptic impacts of climate change will materialise in diverse and often locally-specific ways across the globe. The South Pacific, consisting of 16 island states including Australia and New Zealand,[1] is in some ways the planet's canary in the coal mine on this issue, as it is likely to bear the brunt of some of the first impacts, including inundation of low-lying islands by rising sea levels.

Apart from Australia, Papua New Guinea (PNG) and New Zealand, the majority of the Pacific island countries and territories (PICT) are atolls and low-lying islands. The developing countries with bigger land mass and higher terrain with extensive forest cover are located within the sub-region of Melanesia, comprised of PNG, the Solomon Islands, Fiji, Vanuatu and New Caledonia. In 2008, the South Pacific (excluding Australia and New Zealand) had some 9.5 million people, with PNG alone having a population of approximately 6.1 million. The overwhelming majority of the inhabitants of the South Pacific are indigenous to the region. Many live on customarily-owned land, using customary rules and practices to govern their affairs. They depend heavily on their land, its natural resources and the health of the environment for their daily sustenance.

Their way of life will likely be seriously challenged by climate change (Burns, 2000). Its effects are already being felt in the region through sea level rise, emerging health problems and the prospect of an uncertain future (Brookfield, 1989; Connell, 1993; Hay, 1996; Hoegh-Guldberg, 2000). The PICT are particularly vulnerable to the adverse effects of climate change, as they consist mostly of coral and atoll islands, whose small size and low elevation provide few options to adjust to unfavourable environmental changes. Indigenous peoples, living closest to nature, are perhaps the most vulnerable of all. A well known maxim in the South Pacific is: 'land is life, without land there is no life'. While indigenous peoples of the South Pacific have traditionally guarded their land and natural resources fearlessly against foes since time immemorial, climate change is a new, seemingly invisible 'enemy' that requires entirely novel and untested strategies to resist.

This chapter canvasses the threats posed by climate change in this region and examines the key issues that governments and communities in the South Pacific must consider as they search for political, legal and economic solutions to protect their peoples and their environments. Some of the answers to climate change in the South Pacific may lie in the traditional knowledge, customs and practices of its indigenous peoples. In devising local, national or regional strategies to mitigate, or adapt to, climate change, governments of the South Pacific should promote and strengthen indigenous knowledge, law and practices that will best suit the indigenous people who will be the most affected.

While international climate law, including the proposed Copenhagen Protocol, will likely be a seminal part of the future governance framework for the South Pacific, legal strategies to cope with climate change must also be tailored to this region. The South Pacific has had, for some time, a reasonably extensive set of regional environmental conventions and organisations, and therefore potentially has the institutional framework to forge further regional cooperation specifically on climate issues (Boer, Ramsay and Rothwell, 1999, pp. 243–64). At the same time, however, even well-intended regional approaches to environmental management in the Pacific have sometimes encountered trenchant local opposition from indigenous communities who perceive that their social or economic interests have been compromised (Alley, 1999). Local customary laws, including long-standing cultural norms that restrict use of natural resources, have traditionally been poorly recognised in regional and international environmental regimes (Giraud-Kinley, 1999). A regional approach, therefore, is not always isomorphic with a local approach. This chapter examines the roles that *both* regional and local policy approaches can play in helping the indigenous peoples of the South Pacific deal with climate change.

2. INDIGENOUS PEOPLE AND ENVIRONMENT GOVERNANCE IN THE SOUTH PACIFIC

The islands of the South Pacific were first inhabited 50,000 years ago by humans who arrived from what are now East Indonesia and the Southern

Philippines. Later, some 3,500 to 4,000 years ago, most of what is now Polynesia and Micronesia was settled. Only from about 800 years ago were New Zealand and the rest of the inhabitable lands in Polynesia, Micronesia and Melanesia occupied (Crocombe, 2001, p. 9). A reference to the South Pacific for most people would usually mean the countries located south of the Equator. For the people of the South Pacific, it means the countries of the region which are inhabited by indigenous peoples. They are traditionally categorised into three major cultural groups: Polynesia, Micronesia and Melanesia. Polynesia is made up of Samoa, Tonga, Cook Islands, Tokelau, Tuvalu and Niue. Micronesia is comprised of Kiribati, the Marshall Islands, Palau and the Federated States of Micronesia. Melanesia has the largest islands, including PNG, and just over two-thirds of the region's population.

Lacking the commercial and industrial infrastructure of major economies, the peoples of the South Pacific depend largely on their immediate lands and the ocean for their daily sustenance. Throughout most of the PICT, indigenous peoples own 90 per cent of the land while the state owns most of the remainder. Land is intrinsic to their way of life. It is not a fungible economic commodity, as in Western economic systems. Without land, the indigenous peoples have no place and no cultural identity. It is therefore an intensely guarded resource (Haynes, 1981; Fingleton, 1982; AusAID, 2008a and 2008b). Indigenous peoples utilise their land in accordance with traditional knowledge and practices which have been transmitted culturally from time immemorial (Johannes, 1989; Veitayaki, 1998). They have developed highly sophisticated techniques of gardening and fishing which have proved to be generally environmentally sustainable for eons. In fact, archaeological evidence suggests that some of the world's first agriculturalists were from the South Pacific 10,000 years ago, particularly in present-day PNG (Ketan and Muke, 2001). While environmental historians such as Flannery (1994) suggest that the first inhabitants of the South Pacific sometimes wrought significant ecological changes, including the extinction of some animals, the more recent pattern suggests that indigenous peoples in the region have cultivated an intricate and mostly sustainable relationship with nature and its biodiversity. It is a relationship of interdependence, expressed through various customary rites and norms that embody appreciation of the environment (Kwa, 2005).

Many PICT recognise the importance of customary law and the traditional cultures of their peoples, having taken policy and legislative measures to protect and strengthen them, particularly in the context of environmental governance. Generally, the protection of traditional knowledge and practices is entrenched in the national constitutions of the countries of the South Pacific (Kwa, 2005, p. 121). These documents typically adopt the customary law, traditions and cultures of the indigenous people as guiding legal principles and also a source of law for each state (Sack, 1982; Ntumy, 1993; Ghai, 1988; Boer, 1996; Dang, 1999).[2] This constitutional mandate has been translated into various national laws that entrench the position of custom. Examples include the Village Fono Act 1990 and the Land and Titles Act 1981 of Samoa; the Laws of Tuvalu Act 1987; the Laws of Kiribati Act 1989, which is based on the Customs Recognition Act 1969 of PNG; the Custom and Adopted Act 1971 of Nauru; and the Tokelau Village Incorporation Regulations 1986 of Tokelau (Ntumy, 1993).

Many PICT have also sought to promote biological conservation and sustainable use by recognising customary and traditional practices and innovations in legislation (Boer, 1996). In Vanuatu, the Environment Management and Conservation Act 2002 adopts as one of its key goals the protection, promotion and strengthening of fundamental traditional values and principles pertinent to biological conservation and sustainable use. A similar legislative arrangement can be found in PNG's Conservation Areas Act 1978, Fisheries Management Act 1998, and Forestry Act 1991; the Conservation Act 1986 of the Cook Islands; the Coast Conservation Act 1988 of the Marshall Islands; the Parks and Reserves Act 1988 of Tonga; the *Fisheries Act* 1998 of the Solomon Islands; the Marine Protection Act 1994 of Palau, and the Madolenihmw Protected Areas Act 2002 of Pohnpei. These statutes provide strong frameworks to implement relevant regional and international environmental treaties, such as Article 8(j) of the Convention on Biological Diversity 1992,[3] which requires states parties to 'respect, preserve and maintain knowledge, innovations and practices of indigenous and local communities embodying traditional lifestyles relevant for the conservation and sustainable use of biological diversity' (Kwa, 2005, p. 123).[4] Although these national laws do not relate directly to climate change, they could provide the legal basis by which relevant customary laws, knowledge and practices may contribute to climate change mitigation and adaptation, as discussed later in this chapter.

These legislative measures to promote and protect the interests of indigenous peoples in local environmental governance are also consistent with evolving international standards for such peoples (Charters, 2009; Richardson, 2009). The ILO Convention Concerning Indigenous and Tribal Peoples in Independent Countries[5] of 1989 and the United Nations (UN) Declaration on the Rights of Indigenous Peoples[6] of 2007 recognise indigenous rights to territory and natural resources. The Declaration proclaims indigenous peoples' rights to own, develop and control the use of their traditional lands (Article 26), as well as the need for indigenous consent for the approval of any development project affecting their lands (Article 30). The ILO Convention contains similar standards, including an obligation on states parties to 'respect the special importance for the cultures and spiritual values of the peoples concerned of their relationship with the lands or territories' (Article 13). To secure these rights and values, the Convention declares that indigenous peoples have the right to 'participate in the use, management and conservation' of their natural resources. Of course, the authority of such lofty standards is debatable. Few states have ratified the ILO Convention; Fiji is the only state from the South Pacific to have done so. Although the UN Declaration fares better, with over 140 signatories including several from the South Pacific, it is a 'soft law' standard that is not necessarily as legal binding as the ILO Convention is to its state parties (Charters, Malezer and Tauli-Corpuz, 2009). Given that many governments in the South Pacific have already legislated to protect customary land rights and environmental practices, perhaps questions about the formal legal status of such international instruments in the region should not matter.

3. CLIMATE CHANGE ISSUES FOR THE SOUTH PACIFIC

Vulnerable Peoples and Environments

What, then, are the climate change issues for the South Pacific? Given their geographical location, topography and mostly small size, island countries are acutely vulnerable to the adverse effects of climate change. Even larger countries such as those in Melanesia cannot afford to be indifferent to global warming (Nunn, 2001, pp. 59–61). The South Pacific Regional Environment Programme (SPREP), the leading environmental organisation for regional cooperation in this part of the world, advises:[7]

> [m]any Pacific islands are extremely vulnerable to climate change, climate variability, and sea level rise and will be among the first to suffer the impacts of climate change and among the first to be forced to adapt or abandon or relocate from their environment. The islands are low lying or have coastal features and characteristics that make them particularly vulnerable to climate change, variability and sea level change. In addition to significant coastal impacts climate change will affect biodiversity, soils and the water supplies of small islands. Most small island states will find it extremely difficult to adapt to these changing conditions. The impacts will be felt for many generations because of the small island states' low adaptive capacity, high sensitivity to external shocks and high vulnerability to natural disasters.

Climate change will also affect agriculture production and food security of the island communities, and their water resources, land use practices,[8] public health, and ultimately perhaps the very culture of some indigenous groups (UNDP, 2008). To address these issues, the PICT must: (i) recognise the impacts of climate change both at the national and local levels; (ii) secure relevant technical and financial support to initiate adaptation programmes; (iii) support indigenous livelihoods and their environment; and (iv) address the uncertain future of low lying atoll states.

Already, most PICT suffer considerably in their human development. Apart from Tonga and other countries in the Polynesian region, the human development indicators of the countries in Melanesia (apart from Fiji) and most in Micronesia are poor. Most PICT endure poverty, high mortality and low literacy rates, limited access to education and other basic services. Many such states are also unlikely to achieve the eight Millennium Development Goals by 2015 (World Vision, 2006, p.1). Climate change threatens to make attainment of these basic human development goals even more difficult over the long term. For example, a changing climate will likely affect food production through increased destructive cyclones and droughts, thereby affecting rural livelihoods (Barnett, 2007).

Most vulnerable of all are the indigenous peoples of the small, low-lying island countries such as Kiribati, Niue, the Marshall Islands, Tokelau, Tuvalu, and Nauru, as well as those on the small atolls within PNG and Vanuatu. These people may not only lose their land, natural resources and their livelihoods, but also, in some cases, their countries. Their status as nation states would presumably be lost forever, causing their citizens to become climate refugees forced to adopt new and unfamiliar customs and practices, as Angela Williams explains in another chapter in this book. Three examples that raise such issues of climate justice are: (i) Tuvalu and Kiribati; (ii) Lateu Village in Vanuatu; and (iii) Carteret Islands and the Labutali people in PNG (Vidal, 2005).

In the case of Tuvalu, the whole island nation with a tiny population of just 9,500 faces the possibility of extinction as a sovereign nation.[9] In Kiribati, two uninhabited islands, Tebua and Abanuea, already disappeared underwater in 1999. In his address to the 63rd UN General Assembly in 2008, Anote Tong, the President of Kiribati, painted a bleak future of the island nation, predicting that, 'with the projected sea level rise, Kiribati could be looking at 50 to 60 years before the islands are inundated by the rising sea, saltwater intrusion into its freshwater [bodies] rendering the islands uninhabitable'.[10] In fact, Kiribati may need to be evacuated much sooner than this dire projection, as encroaching sea water is contaminating precious fresh water supplies and lands are quickly being rendered uninhabitable (Marks, 2008).

The detrimental effects of climate change on Kiribati and other small island countries in the South Pacific are therefore imminent. New global climate mitigation efforts to sharply reduce greenhouse gas (GHG) emissions will probably not save these islands, given the considerable time lag between past emissions and future effects. As the Kiribati President also pointed out in his speech at the UN, '[m]itigation efforts, therefore, will not be able to reverse our situation – a situation that we have recently acknowledged and to which my Government has provided a long term strategy for our people's future survival'.[11]

Several examples illustrate that this is not an exaggerated warning. In the case of the people of Lateu Village on the island of Tegua, they were relocated to higher ground after their coastal homes were devastated by repeated large surging waves. New land had to be identified by the Vanuatu government to relocate the some 100 displaced people (Boehm, 2006). Likewise, the people of PNG's Carteret Islands, numbering fewer than 2,000, are on the verge of becoming dislodged from their traditional lands and way of life. Three of the islands in the Carteret archipelago are almost under water, and are no longer fit for human habitation. In Labutali, 1,500 people had to be relocated inland from their coastal village because of surging waves. Even communities at higher elevations face unwelcome environmental changes. For instance, in the highlands of PNG, malaria, which was unknown to the people in these areas, has become a new threat to their health (Mueller, et al., 2005). Sorcery is being blamed as the cause of death for people who are dying of malaria because highlanders are not familiar with this illness.

The fragile ecosystems and biodiversity of the South Pacific will almost certainly be further adversely affected by the effects of climate change. PNG, which contains about 6–8 per cent of the world's biological diversity (PNG Department of Environment and Conservation, 2007, p. 4), stands to lose many of its endemic species due to changes in climatic patterns. The smaller PICT may also lose much of their biodiversity, especially if sea levels rise as predicted (Nurse, 1998, p. 343). Marine ecosystems are also threatened. Scientists forecast that warmer seas will bleach coral reefs – the most biologically rich marine areas – in the South Pacific and other oceans of the planet, rendering such areas nearly devoid of life (Buddemeier, Kleypas and Aronson, 2004). Some bleaching of Australia's Great Barrier Reef has already been documented (Berkelmans and Oliver, 1998).

These and other climate change effects are thus beginning, and will probably happen regardless of the actions of the peoples of the South Pacific. Although they are long familiar with foreign economic and cultural intrusion into their lives (Crocombe, 2001),[12] climate change provides an unprecedented and potentially much more ominous challenge. At a community level, most people have little knowledge about climate change. This point was recently noted by a commentator in PNG (Kolma, 2008):

> [a]s the age of global warming and carbon trading finally reaches our shores, our people are aware of the wider significance of climate change by instinct. The majority of our population lives off nature as they have done for millenniums. They read nature to tell the seasons and the weather. They need no scientists to tell them about the hot days, shifting water levels, tell-tale changes in plant and animal lives and movement. They know the climate is changing. They just do not appreciate why it is happening. That is what scientists can tell them.

If given an opportunity to contribute to climate policy and its implementation, however, indigenous peoples in the South Pacific may play a useful role. Edwards (2000, p. 260) argues that the knowledge and understanding of indigenous peoples of their local environments should be harnessed, with their consent, 'to reduce their vulnerability to climate change impacts'. They have 'learnt through direct experience to cope with extremes of climate' (Edwards, 2000, p. 260). Indigenous environmental knowledge could provide researchers with valuable information to track how the climate is affecting plants, animals and entire ecosystems. In PNG, several indigenous groups have even organised themselves as 'climate change societies', providing a forum for public discussions and information sharing.

Of course, greater local knowledge of climate change alone will be unlikely greatly to assist Pacific islanders wishing to mitigate a threat caused by economic activities in distant parts of the world. Rather, at best, they can hope to adapt better to the impacts of a changing climate. The following section examines how some communities are already starting to adapt in the face of rising sea levels.

Indigenous Peoples Adaptations to Climate Change

Indigenous customary laws and traditional knowledge and practices have assisted in the relocation of people from inundated coastal areas. In the cases of the Carterets, Lateu and Labutali, the relocations were made possible because of several local factors. These included family lineages (kinship), customary norms of social responsibility and reciprocity, and local land tenure arrangements.

The people of Carterets started to relocate to Tinputz, in the northern part of Bougainville island, with the assistance of the Catholic Church and a local non-governmental organisation, Tulele Peisa. This process began with very little or no formal governmental support.[13] The Carterets islanders were resettled after gaining the approval of the local landowners and the Church. In November 2008, the government of the Autonomous

Region of Bougainville began to play a more active role in the relocation process by assisting 40 families of the Carterets atolls in moving to the main island of Bougainville.

The landowners of the main island, Bougainville, gave their approval to the Carterets islanders' relocation because they are related by blood through family kinship. Most of the people living on Carterets are related to the denizens of Bougainville island. The mainlanders also felt a social responsibility based on customary law to look after their island relatives. Culturally, the mainlanders know that if they are ever in trouble, the island settlers will reciprocate the good will (Lokobau, 2008). The Carterets people were given title and access to the land because of their consanguinity with their new community. As one of the officers of the Autonomous Region of Bougainville plainly put it, 'the administration would resettle the families in north Bougainville because of similarities in cultures' (Lokobau, 2008).

In the Lateu case, similar arrangements occurred to enable the relocation of the coastal denizens on Tegua Island. The government was involved in the relocation process. Another case similar to the Lateu case is that of the Labutali people of Huon Peninsula, in PNG. Originally coastal dwellers, they had to relocate two kilometres inland due to surging storms and rapid soil erosion caused by rising sea levels (World Vision, 2008b, p. 5). These people were, like the Carterets islanders, relocated without any assistance from the government.

A similar set of circumstances facilitated the relocation of the Labutali people. Their resettlement was made easier because of kinship ties to the new host community, which was expected to show hospitality and compassion. Several specific factors were at work in this case. Firstly, the land that they were relocated to was owned by a clan or a number of clans from their village or neighbouring villages. With governmental assistance, they could have either bought the land or leased it from the land owners. However, given the gravity of the situation, the land owners felt obliged to allow the resettlement on their land at no cost. Secondly, given that these people have the same customs and practices, the potential for conflicts relating to access and use of the customary land, and nearby marine resources, was lessened. Thirdly, as a close knit cultural group, the strong social capital that they have helped them to cope with any detrimental effects that climate change might have on the more vulnerable members (such as children, orphans, widows and old people) of the community.

These cases highlight the role that indigenous people can play in addressing one of the adverse effects of climate change. Apart from the Lateu case, the two examples in PNG show that even without any formal governmental assistance, indigenous people can utilise their customary laws and traditional practices to adapt to a changing environment. It must be acknowledged, however, that these examples involve relatively close neighbours; where relocation involves moving to another country, the issues will invariably be more complex and difficult. For instance, the emergence of 'climate refugees' creates a problem with international legal consequences (Williams, 2008). Which countries should accept these people? What rights would they have under international law? These are critical issues that will need to be addressed by PICT and their neighbours before too long.

4. CLIMATE LAW AND THE SOUTH PACIFIC

Introduction

At the UN's 63rd General Assembly Third Committee Meeting, in New York in November 2008, representatives of the South Pacific called on the world's major emitters of GHGs quickly to reduce their pollution. They stated that climate change has 'serious implications for sustainable development, energy, human rights, security, gender and many other questions of concern'.[14] But what should the countries of the South Pacific do about climate change when it is a problem of global magnitude and created largely by economic activities in other regions? Primarily, they must address the challenges of climate *adaptation.* Currently, few states in the South Pacific have enacted laws or policies that address climate change. Their existing initiatives are mostly inchoate policy proposals or rudimentary legislative frameworks.

As of early 2009, only three countries in this region – Fiji, Western Samoa and PNG – had made some initial progress. The Fijian government adopted a National Climate Change Policy Framework in late 2007, after some five years of preparation and consultation.[15] The policy, which focuses on adaptation issues in vulnerable areas, aims to build collaboration among governmental and nongovernmental organisations so that climate concerns are integrated into all relevant areas of government policy. It will be implemented through a range of existing environmental legislation covering land use planning, forests management and sustainable development.

The PNG government established in late 2008 an Office of Climate Change and Carbon Trading.[16] Its primary responsibility is to provide and coordinate policy with regard to protection and management of the country's extensive forestry resources as a valuable international carbon sink which the government believes can reap economic benefits for the country through carbon trading and compensation. In particular, the Office is intended to help secure royalty payments to the forest-owing communities that wish to participate in carbon sink trading. Also, the PNG Department of Planning and National Monitoring completed a draft Carbon Trade Policy in 2005, but in 2008 the government shelved the policy in favour of developing a new climate change policy and enabling legislation.[17] The shift was influenced largely by PNG authorities' new focus on participating in the international initiative known as Reduce Emissions from Deforestation and Degradation (REDD), which is discussed in detail in another chapter by Claire Stockwell and others in this book.

In the case of Western Samoa, the government has tasked the Department of Lands, Survey and Environment with securing funding from international donors to develop a climate change policy. The government wishes to develop a policy framework that targets the reduction of carbon emissions from various sources and to market the credits internationally. The policy formulation process will begin in 2009.[18] Among other PICT, however, there is little evidence of policy, institutional or legal action on climate change. Yet, these countries' existing environmental governance systems may be sufficiently adaptable to address some climate change issues, such as through land use planning regulations and forestry management laws (Boer, 1996).

At a regional level, several environmental conventions have been negotiated in the South Pacific. None, however, specifically address climate change or have much indirect relevance. The 1986 Convention for the Protection of the Natural Resources and Environment of the South Pacific Region, which entered into force in 1990, focuses on traditional pollution threats such as the dumping of waste at sea.[19] The 1976 Convention on the Conservation of Nature in the South Pacific, known as the 'Apia Convention', also came into effect in 1990.[20] It is also not particularly relevant to climate change; apart from the fact that it has been ratified by only a minority of PICT, its provisions relating to the preservation of natural areas based on the Western protected areas model make it inappropriate for many customary land ownership arrangements, rendering its implementation difficult (Boer, Ramsay and Rothwell, 1998, p. 115). It is also largely superseded by the international Convention on Biological Diversity, as noted above.

The Bali Roadmap

Any regional response to climate change must surely dovetail with international action on the subject. The resolution of the UN Framework Convention on Climate Change (UNFCCC)'s Conference of Parties (COP)

13th Session at Bali in 2007, commonly known as the 'Bali Roadmap', provided a general nonbinding set of agreements for the next round of international negotiations on climate change (Ott, Sterk and Watanabe, 2008). The Bali Action Plan calls for parties to the UNFCCC to continue dialogue on a number of important matters including, mitigation actions, adaptation measures, reduced emissions from deforestation and forest degradation, policy formulation, and technology transfer. The Action Plan also established an Ad Hoc Working Group on Long-term Cooperative Action to coordinate the implementation of the Bali Roadmap.

While the international climate regime requires developed countries to take the leading role in reducing their GHG emissions, all states are expected to develop national climate change policies, programmes and relevant regulatory frameworks to address climate change. Furthermore, through the Clean Development Mechanism (CDM), developing countries, including those in the South Pacific, may participate in projects to assist those nations to meet their emission reduction obligations. The PICT have all ratified the Kyoto Protocol, but few have any experience with the CDM (Haites and Aslam, 2000). To do so, each must establish a Designated National Authority to initiate and monitor CDM projects. While the current CDM project eligibility rules do not readily accommodate land use management projects, such as forestry projects to enhance carbon sinks, there are calls for the proposed Copenhagen Protocol to provide a framework for projects in this sector. As well, some developing countries, including PNG, want a framework for financial compensation in exchange for the developing countries' agreement to offset emissions through, for example, conserving or planting large areas of forests (Saulei and Genolagani, 2007).

Few countries in the South Pacific are ready to participate in CDM projects, let alone to develop domestic legal regimes to address climate change. Even the larger countries such as PNG, which should have taken the lead in this matter, have been struggling to adopt a climate change policy and law for several years (Saulei and Genolagani, 2007). The lack of an institutional, policy and regulatory framework places the region's indigenous peoples in a more precarious position.

Any CDM project intended for the South Pacific, or other scheme under the proposed Copenhagen Protocol, will inevitably involve indigenous peoples. Such projects, particularly if they relate to the management of carbon sinks, would affect their land rights and the environment on which their livelihoods depend. Without a climate change policy and legal framework, indigenous people will likely encounter difficulties actively participating in or accepting CDM or other climate-related projects. Some of the pertinent issues that would arise with any such projects of concern to indigenous peoples include: project ownership (project, land, security, etc.); technology transfer; capacity-building; revenue generation and sharing; resettlement of displaced populations; agriculture and food security; and environmental protection. Invariably, any CDM project would be located on or affect traditional land, and therefore issues such as community consent, involvement and benefit would need to be resolved satisfactorily. In general, most of these issues have not been adequately dealt with at the national level in the South Pacific. Only PNG and Samoa are beginning to tackle these issues seriously.

At a regional level, SPREP has initiated some activities to improve the knowledge and technical capacities of PICT to deal with climate change.[21] These include the Pacific Islands Climate Change Assistance Program, introduced in 1997; the Pacific Islands Framework on Climate Change, Climate Variability and Sea Level Rise, introduced in 2000; and the Capacity Building for the Development of Adaptation Measures (2002–2005). The first programme was designed to strengthen the capacities of several countries in terms of training, institutional capacity and planning for meeting the national reporting commitments under the UNFCCC. The second programme aimed to promote action and strengthen partnerships at all levels so as to help islanders understand and respond to climate change and its impacts. The third programme was designed to integrate climate change adaptation into national and sectoral planning and budgeting and, further, to increase communities' capacity to adapt to climate-related risks and vulnerabilities.

Two other regional initiatives being managed by SPREP that specifically focus on mitigation and adaptation are the Pacific Islands Greenhouse Gas Abatement through Renewable Energy Project, and the Pacific Adaptation to Climate Change Project. Both initiatives are funded by the Global Environment Facility. The first project, begun in 2007 as a five year programme, aims to remove technical, institutional, financial, market, policy and awareness barriers to the widespread and productive utilisation of feasible renewable energy technologies in the PICT. The second aims to enhance the resilience of a number of key development sectors, such as food production, water resources management and the coastal zone.

In 2005, regional leaders also adopted the Pacific Islands Framework for Action on Climate Change 2006–2015 (PIFACC), and directed SPREP to develop an action plan to implement it. The PIFACC 'is intended to promote links with, but in no way supersedes, more specific regional and international instruments and plans across specific sectors that link to weather and climate including water; agriculture; energy; forestry and land use; health; coastal zone management; marine ecosystems; ocean management; tourism and transport'.[22] Essentially, the PIFACC aims to ensure that 'Pacific island people build their capacity to be resilient to the risks and impacts of climate change with the key objective to deliver on the expected outcomes'[23] by observing six key principles, namely:

1. Implementing adaptation measures;
2. Governance and decision-making;
3. Improving understanding of climate change;
4. Education, training and awareness;
5. Contributing to global GHG reduction; and
6. Partnerships and cooperation.

By adopting these six principles, the PICT leaders envisaged that a series of outcomes will be achieved by 2015, in line with the Millennium Development Goals and the Johannesburg Plan of Implementation. Thus for instance, by 2015, in relation to:

- Principle 1: adaptation measures in vulnerable priority areas supported by existing data sets and traditional knowledge, or new data developed in some instances as necessary, will be achieved.
- Principle 2: climate change considerations will be mainstreamed into national policies, planning processes, plans and decision-making at all levels and across all sectors.
- Principle 3: technical data sets will be integrated with relevant climatic, environmental, social and economic information and data sets, and traditional knowledge for risk management.
- Principle 4: human capacity to identify and integrate economic, scientific and traditional knowledge into adaptation and GHG reduction practices will be strengthened.
- Principle 5: CDM initiatives will be developed and implemented, where appropriate.
- Principle 6: climate change related assistance from development partners will be coordinated and harmonised to maximise benefits to Pacific Island Countries and Territories.

The success of the PIFACC ultimately depends on the climate change strategies that can be implemented at the country level, with the support of international development partners. In a meeting in 2008, the leaders of the Pacific Island Forum adopted the Niue Declaration on Climate Change.[24] The Niue Declaration is meant to be the bridging platform for the various regional initiatives and the region's response to the Bali Roadmap. The leaders made two major commitments under the Declaration, namely:

1. to develop Pacific-tailored approaches to combating climate change, consistent with their ability actively to defend and protect their own regional environment; and
2. to advocate and support the recognition, in all international fora, of the urgent social, economic and security threats caused by the adverse impacts of climate change and sea level rise to our territorial integrity and continued existence as viable dynamic communities, and the potential for climate change to impact on national and international security.

The Niue Declaration also calls on the region's international development partners to assist PICT to deal with climate change adaptation and mitigation challenges through greater technical and financial support. The Declaration also implores SPREP to meet the individual needs of its member countries through its mandated roles, which include consolidating and distributing information on climate change, and increasing Pacific Island countries' capacity to manage their engagement in the UNFCCC. The President of Kiribati declared that the Niue Declaration:

> is a reflection of our concerted regional efforts to respond to the adverse impacts of climate change which had been severely felt across our region. The Declaration marks a new chapter in initiating the first ever high level climate change declaration for our part of the world and calls for collective efforts to fight the growing threat of climate change to many aspects of our lives and future as nations.[25]

Yet, most PICT have yet to demonstrate concrete work to deal with climate change. The current commitments adopted by PICT leaders are very broad and open-ended, and there is no clear link between the Niue Declaration, the PIFACC and the Pacific Plan,[26] which is the roadmap for development in the South Pacific. Instead, the text of the Niue Declaration reveals the high dependence of the region on its international development partners (mainly the OECD countries). Although the Niue Declaration does not refer to the Bali Roadmap explicitly, the Declaration's references to technology transfer, investment in adaptation measures and other climate policy issues that dovetail with some aspects of the Bali plan.

The leaders of the South Pacific envisage that with the support of their international development partners they can obtain relevant knowledge and technology to engage actively in climate adaptation and mitigation programmes. They also recognise that their regional initiatives are not having the desired impacts. Implicit in the Niue Declaration is awareness that existing national policy and legal frameworks concerning climate change are inadequate or nonexistent. New initiatives are needed. One issue that is starting to generate particular interest in the South Pacific, particularly in the heavily forested countries of Melanesia, is known as REDD.

Reduce Emissions from Deforestation and Degradation (REDD)

Although the regional response to climate change embedded in the PIFACC and the Niue Declaration does not focus on REDD, some countries in Melanesia with extensive intact forests believe that REDD provides them with an opportunity to be major players in international action on climate change while providing economic benefits for their peoples.

For PNG, REDD is about forest conservation, protection and rehabilitation in order to enhance their carbon sink capacity. The principal argument by the PNG government and its partners in the Coalition of Rainforest Nations[27] is that they should be 'compensated'[28] for protecting and preserving their forests that provide global climate benefits. Central to this argument is the fact that deforestation is the 'second single greenhouse gas source behind energy production, being responsible for about 20% of human GHG emissions' (Dutschke and Wolf, 2007, p. 4). By conserving and protecting its forests, PNG and the other rainforest-rich countries can contribute meaningfully to the reduction in GHG emissions.

REDD was introduced into the international climate change discussions at COP 11 (UNFCCC, 2005). At COP 13 in Bali, REDD was elevated to a key part of the Bali Roadmap, and it is likely that this will be reflected in the eventual Copenhagen Protocol. In practical terms, REDD requires the designation of forests for conservation and the creation of mechanisms to compensate host countries, particularly developing countries. This process is not simple, as it involves the resolution of a host of other issues such as: assessment of forest types; ownership and management of forests; access (prior informed consent) of forest owners and users; distribution of compensation; security of tenure; marketing of carbon credits; financing; and effective control of the causes of forest degradation. Further, as the UNFCCC is designed to promote sustainable development, the question of how REDD will contribute to that goal in host countries and local communities must also be addressed at both the regional and national levels. The role of indigenous and other local peoples in the REDD mechanism remains uncertain. In the case of PNG and the other countries of Melanesia, the involvement of indigenous peoples will be critical to the success of any REDD project. As indigenous peoples own about 90 per cent of the land in PNG, Vanuatu, Solomon Islands and Fiji, their active and full participation in any CDM or REDD project is crucial. The manner in which they are to be engaged and on what terms and conditions are issues that still need to be clarified by South Pacific governments.

While the South Pacific nations will continue to assert a common regional position in future climate law negotiations, REDD will become of greater concern to some PICT. Likely, PNG and other countries in Melanesia will push strongly for the inclusion of REDD in any post-2012 agreement. However, a strong South Pacific regional approach to the climate policy debate should not be compromised. The PICT can cooperate through many regional forums, thereby creating a strong regional identity. Their interests on climate change are also being addressed through the Small Islands Developing States (SIDS) Network,[29] the Alliance of Small Island States (AOSIS)[30] and the long-standing Group of 77.[31]

5. CONCLUSION: KEY CHALLENGES

Indigenous peoples of the South Pacific are threatened by climate change, requiring innovative policy and legal responses at the domestic, regional and international levels. Already under siege from a range of environmental and social stresses, indigenous communities in the South Pacific are now confronted with the threat of climate change. It could greatly exacerbate existing problems of poverty, poor health, land loss and scarcity, and food insecurity. For the South Pacific, as tiny contributors to GHG emissions, the primary policy focus should be improving resilience to *adapt* to global warming (Barnett, 2001). To address this situation, piecemeal approaches by individual governments will not work. A regional approach to climate change is essential to coordinate action, share best practices, and reach economies of scale in designing solutions.

So far, the regional response to climate change has been sporadic and noncommittal. Two immediate actions are imperative. First, and foremost, is the practical implementation of the regional commitments made under the Pacific Islands Framework for Action on Climate Change 2006–2015 and the Niue Declaration of 2008 through an integrated climate change policy and regulatory framework in each nation. Second, the PICT must recognise, promote and apply customary law and practices to help indigenous people cope with climate change. Such local empowerment will be particularly important in forest management for REDD projects and in relocations of communities displaced by rising seas or other climatic impacts.

The 10-year PIFACC will be reviewed in 2010. Only three countries in the South Pacific, Fiji, PNG and Western Samoa, have so far made some tangible progress on formulating a national climate change regulatory framework. The general lack of national action in implementing the PIFACC may be attributed to two main barriers – insufficient institutional and technical capacities, and lack of financial support.[32] The strong desire of Pacific island leaders for international development partners, as evident in the Niue Declaration, suggests that overcoming these barriers will be a seminal part of the PIFACC review in 2010. Most likely, they will look to the closest developed countries, namely Australia and New Zealand, as well as multilateral organisations. Climate policy in Australia and New Zealand has fluctuated greatly in the past decade, reflective of changes in the governing political parties (Gillespie and Burns, 2000; Bonyhady and Christoff, 2007).

While a regional approach will likely be insufficient without international assistance, such aid can be a mixed blessing. For developing countries, such as those in the South Pacific, it can create further dependence on foreign powers, which may be abused to advance their neocolonial interests over the welfare of vulnerable peoples (Edwards, 2000, p. 259). Therefore, policy-makers in the South Pacific must also address climate change at the local level, such as by providing for a greater role for indigenous communities in CDM and REDD projects, to ensure that international initiatives dovetail legitimately with local needs.

Community-based approaches that safeguard the interests of indigenous peoples must be incorporated into multi-faceted strategies that encompass poverty reduction, public health, agriculture and food security, and forest management. Governments which act unilaterally may face community antagonism, as the PNG government has recently experienced from community landowners incensed by its plans to auction the country's forests to an international carbon market (Radio Australia, 2008). At the same time, governments can play a crucial role in helping local communities in their negotiations with unscrupulous foreign investors. Also, where communities are faced with irreversible threats, such as rising sea levels, government-assisted resettlement programmes are usually necessary. Even here, however, local communities must be closely involved to ensure that displaced people are relocated into a hospitable setting. The governments of the PICT must formulate policy and legal frameworks that give protection not only to the new settlers, but also to the original inhabitants of land proposed for occupation by the resettled.

NOTES

* School of Law, University of Papua New Guinea.
1. Although Australia and New Zealand are part of the South Pacific, the focus of

this chapter is the developing Pacific Island countries and not these two developed countries.

2. E.g., PNG Constitution (s. 9 and sch. 2.1), Vanuatu Constitution (ss 49 and 74), Marshall Islands Constitution (Art. X), Samoa Constitution (ss 100 and 111), Solomon Islands Constitution (s. 76 and sch. 3); Cook Islands Act 1915 (s. 422), Kiribati Constitution (preamble), Constitution of the Federated States of Micronesia (Art. V(2)).
3. 1760 UNTS 79; (1992) ILM 31, 818.
4. I and three colleagues (Professor Donna Craig – Macquarie University, Centre for Environmental Law, Mr Yoli Tom'taval – South Pacific University Law Faculty, and Dr Justin Rose – New England University Law School) conducted a series of case studies in Melanesia and Pohnpei examining the integration of customary law and state law in the management of the environment and natural resources from 2006–2007. The project, entitled 'Indigenous Governance of Natural Resources in Melanesia: A Project to Develop Legal Capacity-Building Strengthening Community-Based Institutions, Customary Laws and Environmental Management Approaches', showed that customary law still plays a vital role in the management of the environment and natural resources.
5. International Labour Organisation (ILO), Convention No. 169 (1989) ILM 28, 1382.
6. UN Doc A/61/L.67, 7 September 2007.
7. SPREP, 'Climate Change, Variability and Sea Level Change' www.sidsnet.org/pacific/sprep/topic/climate.htm (visited 1 April 2009).
8. See the various case studies on how traditional agricultural practices have contributed to modern scientific knowledge (Morrison, Geraghty and Crowl, 1994).
9. For the plight of the Tuvaluans see www.wwf.org.au/articles/climate-refugees-in-a-drowning-pacific and www.tuvaluislands.com/warming.htm (visited 15 March 2009).
10. Quoted at www.undemocracy.com/generalassembly_63/meeting_9 (visited 15 March 2009).
11. Ibid.
12. A synopsis of the dilemma faced by indigenous people in Melanesia in holding onto their traditional cultures and embracing modernity is vividly presented by Narokobi (1983).
13. Although the PNG national government allocated US$1.3 million for the relocation of the Carterets and other island people affected by climate change in 2007, none of this money has trickled down to the communities.
14. Quoted in *Pacific Magazine* (5 November 2008).
15. See www.fiji.gov.fj/publish/page_10746.shtml (visited 15 March 2009).
16. See http://pidp.eastwestcenter.org/pireport/2008/October/10-16-07.htm (visited 15 March 2009).
17. I have been actively involved in this process and I am very familiar with the issues in PNG, my home country.
18. Personal discussion with the officers from the Department in Suva, Fiji on 27 November 2008.
19. (1986) ILM 26, 25.
20. (1990) Australian Treaty Series No. 4.
21. See generally www.sprep.org.
22. The Pacific Islands Framework for Action on Climate Change 2006–2015, p. 2. The document can be downloaded from the SPREP website, at www.sprep.org (visited 15 March 2009).
23. Ibid., p. 3.
24. The Pacific Islands Forum is an annual summit of leaders and observers of the 16 South Pacific Island countries. The 2008 meeting was held in Alofi, Niue, from 19–20 August.
25. The President made this statement to the 63rd UN General Assembly. Reported in *Ami Tautaeka* 8(40) (3 October 2008).
26. The Pacific Plan is a strategic development programme that seeks to strengthen regional

cooperation and integration in the South Pacific. The Plan was adopted by the South Pacific Forum leaders in 2006. The three pillars of the Pacific Plan are: (i) sustainable development; (ii) good governance; and (iii) regional security.
27. See www.rainforestcoalition.org/eng (visited 15 March 2009).
28. The term 'compensation' is probably not the right term to use in relation to the REDD mechanism. However, until an appropriate term is accepted at the international level, its use will suffice for the time being.
29. See www.sidsnet.org (visited 15 March 2009).
30. See www.sidsnet.org/aosis (visited 15 March 2009).
31. See www.g77.org (visited 15 March 2009).
32. These two barriers are not limited only to climate change but also extend to other cross-cutting issues such as environmental law-making and policy development. See Clarke, Millar and Sollberger (2008), p. 70.

REFERENCES

Alley, R. (1999), 'The South Pacific's Environmental Policy Tensions', *Public Administration and Development* 19(2), 137.

AusAID (ed.) (2008a), *Making Land Work: (Vol. 1) Reconciling Customary Land and Development in the Pacific* (AusAID).

AusAID (ed.) (2008b), *Making Land Work: (Vol. 2) Case Studies on Customary Land and Development in the Pacific* (AusAID).

Barnett, J. (2001), 'Adapting to Climate Change in Pacific Island Countries: The Problem of Uncertainty', *World Development* 29(6), 977.

Barnett, J. (2007), 'Food Security and Climate Change in the South Pacific', *Pacific Ecologist*, Winter, 32.

Berkelmans, R. and J.K. Oliver (1999), 'Large-scale Bleaching of Corals on the Great Barrier Reef', *Coral Reefs* 18(1), 55.

Boehm, P. (2006), 'Global Warning: Devastation of an Atoll', *The Independent*, 30 August.

Boer, B (ed.) (1996), *Environmental Law in the South Pacific* (IUCN).

Boer, B., R. Ramsay and D.R. Rothwell (1998), *International Environmental Law in the Asia Pacific* (Kluwer Law International).

Bonyhady, T. and P. Christoff (2007), *Climate Law in Australia* (Federation Press).

Brookfield, H. (1989), 'Global Change and the Pacific: Problems for the Coming Half-Century', *The Contemporary Pacific*, 1(1–2), 1.

Buddemeier, R.W., J.A. Kleypas and R.B. Aronson (2004), *Coral Reefs and Global Climate Change* (Pew Center on Global Climate Change).

Burns, W. (2000), 'The Impact of Climate Change on Pacific Island Developing Countries in the 21st Century', in A. Gillespie and W.C.G. Burns (eds), *Climate Change in the South Pacific: Impacts and Responses in Australia, New Zealand, and Small Island States* (Kluwer Academic Publishers).

Charters, C. (2009), 'Indigenous Peoples and International Law and Policy', in B.J. Richardson, S. Imai and K. McNeil (eds), *Indigenous Peoples and the Law: Comparative and Critical Perspectives* (Hart Publishing, 2009).

Charters, C., L. Malezer and V. Tauli-Corpuz (2009), *Indigenous Voices: The UN Declaration on the Rights of Indigenous Peoples* (Hart Publishing).

Clarke, P., I. Millar and K. Sollberger (eds) (2008), *Capacity Building for Environmental Law in the South Pacific* (IUCN).

Connell, J. (1993), 'Climate Change: A New Security Challenge for the Atoll Islands of the Southwest Pacific', *Journal of Commonwealth and Comparative Politics* 31(2), 195.

Crocombe, R. (2001), *The South Pacific* (University of the South Pacific).

Dang, L. (1999), 'Country Report: Federated States of Micronesia' in Secretariat of the South Pacific Commission (ed.), *Symposium on the Protection of Traditional Knowledge and Expressions of Indigenous Cultures in the Pacific Islands* (South Pacific Commission).

Dutschke, M. and R. Wolf (2007), *Reducing Emissions from Deforestation in Developing Countries* (Deutsche Gesellschaft für Technische Zusammenarbeit).

Edwards, M. (2000), 'Parochialism and Empowerment: Responding to Ecocolonialism and Gloablisation in the Southwest Pacific', in A. Gillespie and W.C.G. Burns (eds), *Climate Change in the South Pacific: Impacts and Responses in Australia, New Zealand, and Small Island States* (Kluwer Academic Publishers).

Fingleton, J. (1982), 'Land Policy in Papua New Guinea', in D. Weisbrot, A. Paliwala and A. Sawyer (eds), *Law and Social Change in Papua New Guinea* (Butterworths).

Flannery, T. (1994), *The Future Eaters: An Ecological History of the Australasian Lands and People* (Reed Books).

Garnaut, R. (2008), *Garnaut Climate Change Review: Interim Report to the Commonwealth, State and Territory Governments of Australia* (Government of Australia).

Ghai, Y. (ed.) (1988), *Law, Politics and Government in the Pacific Island States* (University of the South Pacific).

Gillespie, A. and W.C.G. Burns (eds), *Climate Change in the South Pacific: Impacts and Responses in Australia, New Zealand, and Small Island States* (Kluwer Academic Publishers).

Giraud-Kinley, C. (1999), 'The Effectiveness of International Law: Sustainable Development in the South Pacific Region', *Georgetown International Environmental Law Review* 12, 125.

Haites, E. and Aslam, M. (2000), *The Kyoto Mechanism and Global Climate Change: Coordination Issues and Domestic Policies* (PEW Centre on Global Climate Change).

Hay, J.E. (1996), *Regional Assessment of the Vulnerability of Pacific Islands to the Impacts of Global Climate Change and Accelerated Sealevel Rise* (South Pacific Regional Environment Programme).

Haynes, C.P. (1981), 'Succession to Land in Papua New Guinea' (paper presented at the Australian Universities Law Schools Association Conference at Macquarie University).

Hoegh-Guldberg, O., et al. (2000), *Pacific in Peril: Biological, Economic and Social Impacts of Climate Change on Pacific Coral Reefs* (Greenpeace).

Johannes, R.E. (ed.) (1989), *Traditional Ecological Knowledge: A Collection of Essays* (IUCN).

Ketan J. and J. Muke (2001), *National Heritage Management: A Site Management Plan for the Kuk World Heritage Project in Papua New Guinea* (University of Papua New Guinea Press).

Kolma, F. (2008), 'Cashing in on Carbon Trading', *The National*, 3 November.

Kwa, E. (2005), *Traditionalizing Sustainable Development: The Law, Policy and Practice in Papua New Guinea,* PhD thesis (University of Auckland).

Lokobau, J. (2008), '40 Atolls Families First to Relocate', *The National*, 6 November.
Marks, K. (2008), 'Paradise Lost: Climate Change Forces South Sea Islanders to Seek Security Abroad', *The Independent*, 6 June.
Morrision, J., P. Geraghty and L. Crowl (eds) (1994), *Land Use and Agriculture: Science of Pacific Island Peoples (Vol. II)* (Institute of Pacific Studies).
Mueller, I., et al. (2005), 'Epidemic Malaria in the Highlands of Papua New Guinea', *American Journal of Tropical Medicine* 72(5), 554.
Narokobi, B. (1983), *Life and Leadership in Melanesia* (University of the South Pacific).
Ntumy, M. (eds) (1993), *South Pacific Islands Legal Systems* (Hawaii University Press).
Nunn, P. (2001), 'Sea-level Changes and their Effects on Tropical Pacific Island Environments: The Next Hundred Years', in M. Grzechnik and J. Chittleborough (eds), *Pacific Islands Conference on Climate Change, Climate Variability and Sea Level Rise: Linking Science and Policy* (National Tidal Facility).
Nurse, L., et al (1998), 'Small Island States', in R. Watson, M. Zinyowera and R. Moss (eds), *The Regional Impacts of Climate Change: An Assessment of Vulnerability* (Intergovernmental Panel on Climate Change).
Ott, H.E., W. Sterk and R. Watanabe (2008), 'The Bali Roadmap: New Horizons for Global Climate Policy', *Climate Policy* 8, 91.
Papua New Guinea Department of Environment and Conservation (2007), *National Biodiversity Strategy and Action Plan 2007* (Department of Environment and Conservation).
Radio Australia (2008), 'PNG Landowners, Government at Loggerheads Over Carbon Trading', News, 17 November.
Richardson, B.J. (2009), 'The Ties That Bind: Indigenous Peoples and Environmental Governance', in B.J. Richardson, S. Imai and K. McNeil (eds), *Indigenous Peoples and the Law: Comparative and Critical Perspectives* (Hart Publishing).
Sack, P. (eds) (1982), *Pacific Constitutions* (Australian National University).
Saulei, S. and J. Genolagani (2007), 'Carbon Trading Opportunity' (paper presented at the Divine Word University Conference – Papua New Guinea's Prospect for Sustainable Development, Divine Word University).
Thomas, L. (2007), *Money Grows on Trees* (The Nature Conservancy).
United Nations Development Programme (UNDP) (2008), *Human Development Report 2007/2008. Fighting Climate Change: Human Solidarity in a Divided World* (UNDP).
United Nations Economic and Social Commission for Asia and the Pacific (ESCAP) (1999), *Impact Assessment and Regional Response Strategies for Climate Change* (ESCAP).
United Nations Framework Convention on Climate Change (2005), *Reducing Emissions from Deforestation in Developing Countries: Approaches to Stimulate Action*, FCCC/CP/2005/MISC.1.
Veitayaki, J. (1998), 'Traditional Marine Resource Management Practices Used in the Pacific Islands: An Agenda for Change', *Ocean and Coastal Management* 37(1), 123.
Vidal, J. (2005), 'Pacific Atlantis: First Climate Change Refugees', *Guardian*, 25 November.
Williams, A. (2008), 'Turning the Tide: Recognizing Climate Change Refugees in International Law', *Law and Policy* 30(4), 503.

World Vision (2006), *How are the Neighbours? The Millennium Development Goals & Our Region 2006* (World Vision).
World Vision (2008a), *Preparing Coastal Communities in Asia for Future Catastrophes: Asia Pacific Disaster Report* (World Vision).
World Vision (2008b), *Disaster Monitor: Papua New Guinea* (World Vision).

PART 2

Expanding the scope of the climate change legal regimes

6. Supporting adaptation in developing countries at the national and global levels

Jolene Lin*

1. INTRODUCTION

International climate change negotiations and treaty-making have historically focused on the mitigation of greenhouse gas (GHG) emissions. There was concern during the Kyoto Protocol negotiations that too much discussion about adaptation to climate change would send the wrong signal of fatalistic acceptance of the impacts of climate change and detract attention from efforts to create legally binding emission reduction obligations amongst developed countries (Parry, et al., 1998, p. 741).[1] Further, countries proposing adaptation at the negotiating table ran the risk of indicating a lack of commitment towards limiting their GHG emissions and being branded 'closet polluters' (Burton, 2003, p. 3). The adaptation approach was therefore associated with a 'fatalistic and optimistic view', akin to a 'do nothing' strategy, but also with 'faith in scientific progress' as technical and technological developments were considered to provide adaptation solutions (Tarlock, 1992, p. 172).

However, in the face of irrefutable scientific evidence that climatic changes are already underway, as well as forecasts of the potentially dire physical impacts of global warming, adaptation has become an equally important issue within climate change discourse and policy.[2] The United Nations Framework Convention on Climate Change (UNFCCC) Secretariat has devoted significant resources towards promoting and supporting adaptation activities in developing and least developed countries; development agencies are beginning to 'mainstream' adaptation within their existing activities; non-governmental organisations (NGOs) are spearheading local capacity-building initiatives and developing projects to help vulnerable communities adapt to climate change (Klein, 2008; Gigli and Agrawala, 2007).

This chapter examines the institutional and normative frameworks at the international and state levels that support developing countries' adaptation efforts.

At the international level, international financial institutions such as the World Bank, regional institutions and the United Nations (UN) agencies address adaptation issues in the course of carrying out their official functions. Every international body appears to have a finger in the adaptation pie, so to speak. As a result, the international response may appear to be highly disparate, chaotic, potentially duplicative, and lacking proper coordination and focus on helping communities and peoples adapt to the impacts of climate change. This chapter argues that, on the contrary, there is significant coherence in the international framework. There is a division of labour amongst various bodies, with the UNFCCC Secretariat serving as a focal point of all international negotiations pertaining to climate change, international adaptation policies, funding and information exchange amongst Member States.

At the state level, based on a broad survey that the author has conducted on East Asian countries, most developing countries have inadequate institutional mechanisms and legal frameworks to address climate change adaptation largely because of resource constraints and the lack of expertise. Those countries that have more advanced or developed adaptation strategies generally have these strategies because they are significant recipients of international donor assistance (and therefore support from the international community to carry out studies and projects) or are more economically well-off. After all, Asia is comprised of many nations at different stages of socio-economic development. Further, various countries within the region have different priorities, some being more vulnerable than others to climate change impacts such as rising sea levels and changes in rainfall patterns (UNFCCC Secretariat, 2007b).[3] However, across the board, the countries surveyed lack proper institutional and normative frameworks that are important for coordinating and developing adaptation responses at the state level. This chapter will examine three case studies (Vietnam, China and Laos) for their varying responses towards climate change, and it argues for the strengthening of institutional and normative frameworks within these countries to carry out adaptation work.

Section 2 of this chapter begins with a brief discussion of what we mean by the term 'adaptation' and then sets out the normative framework governing adaptation to climate change. The UNFCCC and Kyoto Protocol contain a number of provisions on adaptation which create the legal basis on which much of the international community's work, especially that of the UNFCCC Secretariat, rests. Section 3 then broadly surveys some of the international organisations involved in adaptation work. It is argued that, far from a chaotic and disorganised array of agencies involved in adaptation work, the picture that emerges is one of division of labour amongst different bodies, with the UNFCCC Secretariat serving as a key focal point. Section 4 argues that many developing countries have not put in place proper institutional arrangements to coordinate adaptation policies and projects, which may lead to 'maladaptation'. Countries are urged to develop these local institutional structures to coordinate with and leverage on the support of the international institutional framework. In this way, developing countries will be better placed to adapt to the impacts of climate change effectively. Section 5 of this chapter discusses the direction in which adaptation policy is headed, particularly in the context of the Bali Roadmap and the post-2012 climate change negotiations. Section 6 concludes.

2. DEFINING ADAPTATION AND ITS NORMATIVE FRAMEWORK

Terminology

In general, the term 'adaptation' is applied in the climate change literature to describe any action taken to adjust to changing climatic conditions, whether in natural or socio-economic systems (Klein and Tol, 1997). Some definitions of climate-related adaptation found in the literature include the following: '[a]daptation to climate is the process through which people reduce the adverse effects of climate on their health and well-being, and take advantage of the opportunities that their climatic environment provides' (Burton, 1992);[4] '[a]daptation to climate change includes all adjustments in behaviour or economic structure that reduce the vulnerability of society to changes in the climate system' (Smith, et al., 1996);[5] and the Intergovernmental Panel on Climate Change (IPCC) defines adaptation as 'the adjustment in natural or human systems in response to actual or expected climatic stimuli or their effects, which moderates harm or exploits beneficial opportunities' (Parry, et al., 2007).

For policymakers, the concept of 'planned adaptation' may provide a more useful reference point. Planned adaptation 'requires a strategic policy decision based on an awareness that climatic conditions have changed or are about to change and that action is needed to return to or maintain a desired state' and may be reactive (taking place after the impacts of climate change have become manifest) or anticipatory (taking place before the impacts are apparent) (UNFCCC Secretariat, 2007b, p. 5). Examples of planned adaptation include building flood protection dykes in response to anticipated rises in sea levels, introducing short-period rice varieties in agricultural areas affected by seawater intrusion and drought, reducing the risks of flood damage by regulating development on flood plains, and promoting flood-proof building design.[6]

'Adaptation' is not defined in the UNFCCC and therefore must be understood in relation to defined terms such as 'climate change' and the 'adverse effects of climate change'.[7] Further, the UNFCCC does not contain one provision addressing adaptation in a comprehensive manner. Instead, many articles bear upon the issue, creating a web of inter-related responsibilities, approaches and mechanisms for addressing adaptation needs (Mace, 2006).

The principle of 'common but differentiated responsibility' (CDR principle) guides the allocation of responsibilities and obligations amongst signatory Parties to the UNFCCC, in the contexts of both GHG mitigation and adaptation.[8] Broadly speaking, the UNFCCC seeks to ensure that the needs of 'vulnerable' countries (which often also have less adaptive capacity because of financial constraints and heavy reliance on climate-sensitive economic activities such as agriculture) are adequately identified and addressed, and the costs of adaptation are borne by those who are most financially able and most responsible for causing climate change. As a result, the convention draws the distinction amongst 'developed', 'developing' and 'least developed countries', and between countries that are 'vulnerable' and 'particularly vulnerable' to the impacts of climate change. Article 3 of the UNFCCC sets out the guiding principles that flow from applying the CDR principle, chiefly: '[d]eveloped country Parties' are expected to 'take the lead in combating climate change' and '[t]he specific needs and special circumstances of developing country Parties, especially those that are particularly vulnerable to the adverse effects of climate change, and of those Parties, especially developing country Parties, that would have to bear a disproportionate or abnormal burden under the Convention, should be given full consideration'. The UNFCCC further distinguishes countries with different physical characteristics; Article 4(8) highlights the needs of small island countries, countries with low-lying coastal areas, countries prone to floods, droughts, and desertification and those with fragile eco-systems.

Provisions of the UNFCCC and the Kyoto Protocol on Adaptation

Article 4(1) of the UNFCCC requires signatory Parties to formulate national and, where appropriate, regional programmes to address adaptation to climate change, as well as to mainstream adaptation concerns into domestic policy-making.[9] Article 4(4) embodies the international community's recognition of the disparity in the adaptive needs and capacities between developed and developing countries and the urgent need to provide assistance to the latter. It states, '[t]he developed country Parties and other developed Parties included in Annex II shall also assist the developing country Parties that are particularly vulnerable to the adverse effects of climate change in meeting costs of adaptation to those adverse effects' and serves as an important legal basis for securing funding from developed country Parties for adaptation work in developing countries. Article 4(8) reinforces this recognition of the unique position of developing countries by requiring Parties to the UNFCCC to 'give full consideration to what actions are necessary under the Convention, including actions related to funding, insurance and the transfer of technology, to meet the specific needs and concerns of developing country Parties arising from the adverse effects of climate change'. Finally, Article 4(9) responds to the difficulties that least developed countries (LDCs) face in adapting to climate change by providing that '[t]he Parties shall take full account of the specific needs and special situations of the least developed countries in their actions with regard to funding and transfer of technology'.

While GHG mitigation is the focus of the Kyoto Protocol, this international agreement contains two important provisions relating to adaptation. Article 10(b) requires all Parties to '[f]ormulate, implement, publish and regularly update national and, where appropriate, regional programmes containing . . . measures to facilitate adequate adaptation to climate change'. Importantly, Article 12(8) provides the funding mechanism of the Adaptation Fund. The Fund is financed by a levy (or 'share of proceeds') of 2 per cent of the Certified Emission Reductions (CERs) issued for a Clean Development Mechanism (CDM) project activity.[10] CDM activities in LDCs are exempt from the share of proceeds, so that these projects can be more competitive relative to projects in other non-Annex I Parties.[11] The Adaptation Fund is further discussed below at section 5.

An Adaptation Protocol?

Some scholars have argued that the slow development of adaptation policy is at least partly due to the lack of explicit provisions in the UNFCCC, which makes it difficult for policymakers and scholars to identify where best to address adaptation to climate change under the UNFCCC.[12] Negotiators have also argued that the absence of a legal provision in the UNFCCC solely on adaptation, coupled with the fact that adaptation is mentioned only five times in the text of the Convention, provides an insufficient basis to address adaptation under the UNFCCC (Schipper, 2006,

p. 90). In order to overcome this gap, a strategy that has been adopted is to discuss adaptation in the context of other Articles in the UNFCCC that address related issues, particularly 'developing-country issues' such as capacity building and technology transfer. This approach has been criticised for its piece-meal nature and, furthermore, has broadened the scope of adaptation to encompass development issues such that adaptation has been re-branded as a developing-country rallying cry (Schipper and Boyd, 2006, p. 75; Schipper, 2006).[13] In short, the lack of a clear definition of adaptation in the international climate change regime, further confused by its association with other aspects of the UNFCCC, has created a barrier to policy development on adaptation.

One might then suggest that a remedy would be adopting an 'Adaptation Protocol' or a set of new treaty provisions that clarify the rights and obligations of various Member States in relation to climate change adaptation, just as there are specific treaty provisions on mitigation in the Kyoto Protocol.[14] It may be argued that an Adaptation Protocol might have been desirable even as recently as the early 2000s, but such a protocol is unnecessary at this stage, when adaptation is clearly set to stay on the UNFCCC agenda. Today, there is a far stronger international consensus that adaptation and mitigation are not mutually exclusive options, but rather different responses to climate change that are necessary because mitigation alone will not reduce the need for societies to adapt to climatic changes which are already underway (Lin, 2008).[15] This policy consensus is likely only to grow stronger, given the momentum that adaptation has gained in policy formulation, research and the international climate change negotiations for a successor instrument to the Kyoto Protocol (Lin, 2008).

Therefore, while the absence of a strong normative framework has caused adaptation policy on the international level to develop in a less than satisfactory manner, better scientific understanding, the different political consensus on adaptation today, and the momentum that adaptation has gained as a development issue are factors that have combined to lift adaptation from its inferior status vis-à-vis mitigation. Today, what hinders advancement on adaptation policy has more to do with the lack of expert knowledge on adaptation and the paltry financial commitment by the international community which clearer provisions on adaptation in the legal texts will not resolve. This is because the main obstacle is the lack of political will and not the absence of a legal basis for action. The existing web of responsibilities and mechanisms created by the UNFCCC and the Kyoto Protocol provides a sufficient legal framework for Member States to reach a burden-sharing agreement on adaptation financing. An Adaptation Protocol is therefore not a panacea.

3. INTERNATIONAL INSTITUTIONAL FRAMEWORK SUPPORTING ADAPTATION

There are a number of reasons why a strong international framework is important for supporting countries' efforts to adapt to the impacts of climate change. First, international organisations facilitate cooperation and information exchange amongst policymakers, local stakeholders and experts by hosting workshops and meetings, and establishing information databases. These initiatives increase our shared knowledge about adaptation, which is still at the embryonic stage, and help build capacity to manage adaptation challenges. Secondly, vulnerable communities require significant financial and technical support to carry out adaptation activities. International organisations play a crucial role in providing financial support for adaptation activities. Thirdly, international legal obligations reinforce political commitment and executive action amongst Member States to address adaptation concerns. Fourthly, organisations at the international and regional levels can and do facilitate more work on adaptation at the local level by providing platforms for the development of local adaptation policies and strategies.

Because climate change affects almost every aspect of social and economic life, and adaptation in particular cuts across various policy issues, such as sanitation and health care, disaster relief, food security, poverty eradication and sustainable development, many international organisations that specifically address these issues have also had to address climate change adaptation or, in the jargon, to 'mainstream adaptation' in their working agendas and policymaking. These organisations include the Food and Agriculture Organization of the United Nations (FAO), the World Health Organization (WHO), the Asian Development Bank and the World Bank. Let us take a closer look at what some of these organisations do in relation to climate change.

The FAO is dedicated to eradicating hunger. Apart from providing a forum for countries to discuss food security issues, the FAO conducts research and assists countries to improve agriculture, forestry and fisheries practices. Agriculture, forestry and fisheries are some of the most climate-sensitive economic sectors and climate change is likely to have serious impact on food production. While there is considerable uncertainty about how projected climatic changes will play out locally, adaptation strategies that increase preparedness and resilience are necessary. The FAO carries out considerable research on climate change and food security, and contributes to adaptation by providing data and analyses on soil and water conditions, biomass and crop systems, monitoring climate variability, providing technical advice to member countries through consultations, participating in technical meetings and technical services to donor programmes, and cooperating with the UNFCCC Secretariat and other agencies.[16]

The WHO is the directing and coordinating body on health issues within the UN system. As part of its responsibility for providing leadership on global health issues, the WHO has had to address climate change as it has the potential to affect human health in a number of ways, for example, by altering the geographic range and seasonality of certain infectious diseases (World Health Organization, 2000). Apart from serving as a source of information and expertise, the WHO has also been instrumental in harnessing regional cooperation towards addressing climate change and health issues.[17] For example, the WHO and the governments of countries including Bangladesh, Cambodia, Fiji, Indonesia, Laos, New Zealand, Korea, Sri Lanka and the Philippines have developed a 'Regional Framework for action to protect human health from effects of climate change in South East Asia and Pacific Region', which could serve as a template for action in other parts of the world (World Health Organization, 2007). The significance of the impact of climate change on human health and international public health policy was underscored by the passing of the 'Climate Change and Health' resolution by the World Health Assembly in May 2008.[18] Amongst other things, the resolution calls on the WHO to strengthen its work in raising awareness of the health implications of climate change, and to support capacity-building and research in health protection from climate change. The resolution also calls upon Member States to develop health measures to be integrated into adaptation plans.

As an international development bank, the World Bank is actively involved in climate change mitigation and adaptation because of the impact of climate change on its mission of reducing poverty in developing countries. Climate change also has the potential to hamper the achievement of many of the Millennium Development Goals as well as the Bank's advisory work in climate-sensitive sectors such as agriculture, energy and water (The World Bank, 2008). The World Bank has therefore recognised climate change to be both an environmental and development issue (Evans, 2007). Some of the ways in which the Bank seeks to contribute to adaptation efforts include the mainstreaming of adaptation issues in the Country Assistance Strategies, developing climate risk insurance models and products, working with the Global Facility for Disaster Risk Reduction and Recovery, and providing technical assistance to adaptation projects.[19]

The Asian Development Bank (ADB) is the only regional, intergovernmental development bank based in Asia. Apart from pursuing its mandate of fighting poverty in Asia and the Pacific, the ADB also actively supports environmental management efforts in these countries. Climate change mitigation and adaptation have ostensibly become key areas of ADB's work programme because climate change risks 'present a real threat to ongoing poverty reduction and economic development efforts' (Asian Development Bank, 2007, p. 22). The ADB has therefore put an adaptation programme in place. This programme operates on three fronts: (1) promoting national adaptation planning through better analysis of the impacts of climate change at the local and national levels as well as the identification of cost-effective 'climate proofing' options;[20] (2) providing technical assistance for project level 'climate proofing' of existing infrastructure and future project designs; and (3) supporting specific adaptation investments as defensive measures against anticipated climate change impacts.[21] Both the ADB and the World Bank are executing and implementing agencies of the Global Environmental Facility (GEF) respectively and therefore have direct access to GEF climate change funding.[22] The adaptation activities undertaken by the development banks so far have been on a project basis and focus on practical measures rather than institutional capacity-building.[23]

The discussion above provides a sample of the numerous international organisations involved in climate change adaptation. At first glance, there may appear to be significant overlap without a coordinating 'framework', so to speak, as these activities appear to be highly disparate, chaotic, potentially duplicative and lacking proper coordination and focus. However, this is not really the case. Each agency, such as the FAO and the WHO, carries out some form of adaptation programme in relation to its core area of work and expertise. Each agency thereby brings to the table highly specialised knowledge about climate change in a specific area (for example, cropping practices and public health), which together create a multi-faceted response to climate change.

The role of the UNFCCC Secretariat in all this has been to provide a focal point for all climate change-related issues, by serving as the forum for international climate change negotiations, developing international adaptation policies, funding and information exchange amongst Member States.[24] The Secretariat also actively establishes dialogue with other UN agencies and invites these agencies to share their research findings and experience.[25] We may summarise the contribution of the UNFCCC Secretariat as: (1) serving as a knowledge hub that assists communities and countries that may not have the resources to attain information which is crucial for carrying out adaptation (by collecting, generating and disseminating information and know-how); and (2) a key driving force behind adaptation initiatives at the state level through its programmes such as the National Adaptation Programmes of Action (NAPAs) for LDCs and the Nairobi Work Programme (as well as through its capacity-building programmes and promotion of regional and international cooperation).[26] Briefly, the purpose of NAPAs is to provide a process for LDCs to identify current climatic risks and priority activities to address urgent and immediate adaptation needs. The Nairobi Work Programme on Impacts, Vulnerability and Adaptation to climate change is a five-year programme (2005–2010) that aims to help countries improve their understanding and assessment of the impacts of climate change and to make informed decisions on practical adaptation actions and measures.[27]

4. INSTITUTIONAL AND LEGAL FRAMEWORKS AT THE STATE LEVEL

International institutions necessarily have to work with governments at the state and local levels in order to develop country-specific adaptation policies and to carry out adaptation projects 'on the ground'. The 'country-driven' approach advocated by the UNFCCC reinforces the role that countries are expected to play, especially in relation to integrating adaptation into national sustainable development and poverty reduction strategies.[28] Legal and institutional frameworks to facilitate the integration of adaptation strategies across all sectors of national policymaking are therefore necessary. However, as the following case studies of Vietnam, China and Laos show, most developing countries do not have adequate institutional arrangements to facilitate bottom-up stakeholder participation and inter-agency cooperation. This may lead to maladaptation, whereby the adaptation strategy is poorly selected or is unsuitable because of unforeseen consequences, poor use of available resources, or even governance gaps because various institutions at the state level are not communicating/coordinating their actions which will have a negative impact on policy implementation. Countries ought to develop these local structures to coordinate with and leverage on the support of the international institutional framework. In this regard, it can be argued that imposing a legal obligation (under the UNFCCC) on Member States to establish a focal point for adaptation issues within their governments may be necessary.

Vietnam

Located in Southeast Asia, Vietnam is one of the most disaster-prone countries in the world, especially concerning water-related disasters. Vietnam is primarily a rice-based agricultural economy, with agriculture contributing about 24 per cent of the country's Gross Domestic Product

(GDP).[29] Currently, over 70 per cent of the population lives in rural and low-lying coastal areas that are susceptible to water-related natural disasters.[30] Vietnam is highly vulnerable to the impacts of a sea level rise. A recent World Bank study on the potential impacts of sea level rise on 84 coastal developing countries demonstrates that a one metre increase in sea level would affect about 5 per cent of Vietnam's land area, affect 11 per cent of the population, have a negative impact on 7 per cent of its agriculture sector, and reduce GDP by 10 per cent (Dasgupta, et al., 2007).

The Ministry of Natural Resources and Environment (MoNRE) is the national focal agency for climate change-related activities in Vietnam.[31] The National Office for Climate Change and Ozone Protection was established within the MoNRE to serve as the 'National Focal Point' for the implementation of the UNFCCC and the Vienna Convention on the Protection of the Ozone Layer (and the Montreal Protocol). Technical expert teams, including one for vulnerability and adaptation to climate change, have been established to assist the implementation of climate change projects.[32]

Vietnam has also established a National Team on Climate Change which includes representatives of the various ministries, including from the key portfolios of finance, industry and justice.[33] In terms of policy development, the National Programme to implement the UNFCCC addresses adaptation issues through broad national strategy and policy recommendations. In addition, the 'National Environment Protection Strategy for the Period 2001–2010' addresses many issues that are linked to climate change, for example, rational resources utilisation, waste management, controlling air, land and water pollution, afforestation in watersheds, and mitigating GHG emissions and ozone-depleting substances. Many climate change-related policies and programmes have therefore been introduced or launched as part of this Strategy.

While there are inter-agency mechanisms in place to promote policy coordination, the Vietnamese institutional framework does not engage with local communities and NGOs to promote bottom-up approaches to address adaptation. Also, the framework is highly geared towards policy coordination at the level of central government, and does not address the involvement of local and regional governments.

The Lao People's Democratic Republic (Lao PDR)

Lao PDR is classified as a Land-Locked Least Developed Country (LLDC) within the UNFCCC.[34] In Laos' First National Communication submitted under the UNFCCC in accordance with its treaty obligations under Articles 4.1 and 12, no assessment of the country's vulnerability to the impacts of climate change was conducted. General observations about Lao PDR's high economic dependency on its natural resources and agriculture and the country's consequent vulnerability to the impacts of climate change were made.[35] Further, Laos has not submitted its National Adaptation Programme of Action (NAPA) to the UNFCCC, and there is no information in the public domain indicating the development of a national programme of action to the best of the author's knowledge.[36] As recognised in Laos' First National Communication, conducting a vulnerability and impact assessment is vital in order to take appropriate adaptation measures. As an LLDC, Laos is likely to be at high risk to certain climatic impacts, particularly in relation to water and forest resources.[37]

Usually, the process of developing a NAPA helps to raise awareness about the importance of climate change adaptation, and brings together community, national and international stakeholders to develop adaptation programmes and policies. This will also be accompanied by the establishment of certain legal and institutional arrangements to support adaptation strategies. It is unsurprising that these arrangements are not well established in Laos, given that the country has not embarked on developing a NAPA. Broadly, the Science Technology and Environment Agency (STEA) is responsible for managing and protecting the environment.[38] Climate change-related activities also fall within its responsibilities. The National Greenhouse Gas Inventory Committee (NGIC) is chaired by the STEA chairman and provides overall guidance on climate change policy. The Technical Working Group (TWG), which consists of representatives from each government ministry, is responsible for preparing the country's GHG inventory and exploring actions to address climate change.[39] The lack of information in the public domain made it difficult for the author to ascertain the exact functions and nature of the work of these bodies. However, from the National Communication, it is clear that adaptation is not directly addressed by the STEA and its associated bodies.

China

Home to approximately one-fifth of the world's population and a rapidly industrialising, huge economy, China has become one of the most pivotal actors in international climate policy debates, including on adaptation issues. Observations of climate trends and variability show that China is already experiencing the impacts of climate change. Briefly, annual rainfall has declined since the late 1990s in the northern regions, and average temperatures have risen over the past 50 years (more pronounced in winter than in summer). Short-duration heatwaves have become more frequent since the late 1990s, as well as increasingly warmer days and nights during this period (Cruz, et al., 2007).[40] China is particularly vulnerable to the impacts of climate change on water supply, agriculture and its coastal and marine systems.

Established in 1998, the National Coordination Committee on Climate Change (NCCCC) is comprised of 17 ministries and agencies.[41] It formulates and coordinates China's climate change-related policies and measures, and provides guidance for central and local governments' response to climate change. From 2001, the NCCCC organised the compilation of the Initial National Communication on Climate Change of the People's Republic of China, and presented the report to UNFCCC at the tenth session of the Conference of the Parties (COP10) in December 2004. In June 2007, an inter-ministerial leading group chaired by the Premier was established.[42] This group has external functions related to the UNFCCC and domestic functions related to the implementation of two mandatory domestic targets on energy intensity (20 per cent reduction) and emissions reduction of major pollutants (10 per cent) for 2006–2010.[43] The National Development and Reform Commission (NDRC) is the lead agency for external (UNFCCC) affairs, together with the foreign affairs, science and technology, environment and metrological agencies. The Ministry of Environmental Protection and NDRC are the leading agencies coordinating the implementation of the domestic energy intensity and emissions reduction targets.

Amongst the countries surveyed in this chapter, China is the only one which has clear arrangements for local government involvement in climate change programmes. Local governments have been tasked with enhancing the organisation and leadership on local responses to climate change, and to formulate and implement local climate change programmes as a matter of priority.[44] There have been proposals to set up a regional administration system to coordinate climate-related work, build up local expertise, initiate proper climate change policy and measures according to local conditions such as geographical environment, climatic conditions and levels of economic development, and strengthen coordination between national and local governments to ensure the smooth implementation of policies and measures.[45] Nonetheless, challenges remain as to how such a regional system will work in tandem with the existing environmental protection bureaucracy which itself has difficulties achieving policy coordination between bureaux at different levels of government (Lieberthal, 1997; Lin, 2004, p. 617).[46]

Observations on Institutional Frameworks for Adaptation

At the very least, all three Asian countries surveyed above have the institutional infrastructure capable of translating their international commitments under the UNFCCC and the Kyoto Protocol into domestic action. This probably remains the primary function of the climate change institutional framework in most countries, though with the increasing need for concerted domestic action to tackle climate change, we may witness future adjustments in the institutional arrangements from the focus on external communications (with the UNFCCC) to domestic coordination (amongst agencies and between levels of government).

Of the case studies, due to different domestic circumstances and international obligations, China possibly has more advanced institutional capacity than Vietnam and Laos. However, all three countries can benefit from the creation of more channels to involve various interest groups such as NGOs, academia and the private sector.[47] NGOs can, for example, initiate campaigns to raise public consciousness on climate change, assist in monitoring policy implementation and provide assistance to vulnerable communities, while academia can provide the scientific information and other know-how that policymakers require for domestic policy formation.

As the level of government that is closest to human activity, adaptation institutional frameworks should engage local governments more closely than is evident at the moment. This is important for successful adaptation as local governments are more likely to have the information and understanding of local conditions and needs. Engaging municipalities is not peculiar to adaptation policy. It is an approach towards environmental governance and sustainable development in general. As stated in Agenda 21:

> [l]ocal authorities construct, operate and maintain economic, social and environmental infrastructure, oversee planning processes, establish local environmental policies and regulations, and assist in implementing national and subnational environmental policies. As the level of governance closest to the people, they play a vital role in educating, mobilizing and responding to the public to promote sustainable development.[48]

Therefore, while international institutions are necessary to support adaptation policy through research, financial assistance and technical advice, attention must also be paid to the role of municipal governments in formulating and implementing climate change policy.

One approach is for more partnerships between international organisations and local governments, and for international organisations such as the UNFCCC Secretariat, the UN Environment Programme (UNEP) and FAO to give greater support to local government adaptation initiatives.[49] Such organisations have so far focused on cities and climate change mitigation, and global cities have come together to form networks such as the C40 initiative to share best practices on issues such as energy efficiency, waste management and transport policy.[50] These developments have occurred because of the realisation that local-level initiatives can sometimes engender more powerful change in a society's environmental consciousness and behaviour. Cities are also major energy consumers, as well as centres of innovation that can advance mitigation solutions (OECD, 2008).

It is arguable that the post-2012 climate change treaty should include provisions that will render it mandatory for states to empower (politically and financially) their local governments to undertake adaptation action. This will provide the UNFCCC Secretariat and other UN agencies with a clear legal basis for requiring Member States to put in place the domestic institutional frameworks necessary for bottom-up stakeholder participation and inter-agency cooperation. On the other hand, local governments are limited in certain respects, such as their restricted jurisdiction over trans-boundary natural resources and fiscal resources. Further, corruption and bureaucratic power struggles may render local governments the cause of, not the solution to, poor adaptation. In such cases, increasing local government involvement necessitates broader governance reforms such as tackling corruption and strengthening government institutions, which are challenges that lie beyond merely addressing adaptation. Recognising that local governments may not be relied upon in all cases to implement adequate or better-informed adaptation policy at the local level is important for designing an appropriate institutional framework to coordinate the functions of different levels of government. The incentives and interests of the various actors have to be identified, and the necessary checks and balances between the different levels and branches of government built into this framework. At the international level, returning to the idea of including provisions on local government involvement in the post-2012 treaty, legal recognition of the role of local governments provides normative force (and a legal basis for practical action) for involving a key domestic actor in adaptation policy.

Generally, climate change appears to be treated as a stand-alone development and environmental issue rather than being integrated into existing national sustainable development measures and policies. It is not difficult to see why this might be the case. Many developing countries do not have institutional structures and policy plans to address sustainable development per se, although they may seek to pursue sustainable development through existing bureaucratic arrangements for environmental management/protection (i.e., environmental agencies and related sectoral agencies such as those responsible for forestry, fisheries and natural resources). As such, there are usually no existing sustainable development structures or plans into which adaptation might be integrated. On the other hand, countries have had to set up certain institutional mechanisms to fulfil their UNFCCC legal obligations and it would have been more expedient for policymakers to place adaptation under the overall climate change framework.

There are also other difficulties with mainstreaming climate change, particularly adaptation, within sustainable development. Climate change and sustainability have developed in parallel discourse for a long time, despite the intellectual case for integration having been made, largely because climate change discourse was originally framed as a natural science problem and then subsequently divorced from larger sustainability issues as it began to receive arguably excessive economic and political attention (Swart, et al., 2003). Sustainable development, on the other hand, has been framed much more through problem-driven social science addressing current economic, social and environmental problems at the local level (Swart, et al., 2003). The linkages between the two discourses have therefore been weak, hindering the exploration of policy synergies. Some scholars suggest that another barrier has been 'mainstreaming overload', as competing agendas such as gender, governance and environment vie for integration within development activities (Agrawala, 2005). Many development projects also tend to be funded on relatively short time horizons such as three to five years, which may not provide the most suitable vehicle for implementing long-term adaptation strategies (Agrawala, 2005). A detailed analysis of the challenges of integrating climate change policies and sustainable development policies is beyond the scope of this chapter.[51] For present purposes, adaptation should not be pursued separately from sustainable development, as climate policies can affect wider sustainable development objectives and vice versa. Further, developing countries have limited financial resources, and leveraging on the linkages between adaptation and sustainable development policy can be a cost-effective and efficient way to proceed.

5. CURRENT DEVELOPMENTS

As mentioned earlier, there is an international consensus on the importance of adaptation, the vulnerability of developing countries to the impacts of human-induced climate change, and the need for international cooperation to address the challenges posed by adaptation. This consensus is reflected in Decision 1/CP. 13, otherwise known as the 'Bali Action Plan'.[52] The Bali Action Plan sets out the broad elements that will guide negotiations for the post-2012 climate change regime or, in the official jargon, to 'launch a comprehensive process to enable the full, effective and sustained implementation of the Convention through long-term cooperative action, now, up to and beyond 2012'.[53] Paragraph 1(c) of the Bali Action Plan identifies adaptation as one of the five key building blocks of the international climate change regime. However, it should be noted that this consensus is not free of controversy. As the saying goes, the devil is in the detail. There remains significant contention over key issues such as financing of adaptation projects, the inadequacy of the existing funding mechanisms, technology transfer, and so on. Section 5 briefly examines how adaptation financing is being addressed within the context of the Bali Roadmap and the post-2012 climate change negotiations.

Developing countries have long been frustrated over the lack of an effective burden-sharing mechanism in the UNFCCC and the Kyoto Protocol that would provide adequate financial resources for adaptation projects. The financing mechanisms that have been developed under the auspices of the UNFCCC and the GEF (i.e., the Special Climate Change Fund (SCCF), the Least Developed Countries Fund (LDCF) and the GEF Trust Fund) have proven to be bureaucratically cumbersome and prone to 'capture' by donor-country interests, and the concept of 'incremental costs' that the GEF uses to allocate funding has been problematic in the context of adaptation (Mace, 2005).

The Adaptation Fund, projected to be operational only from 2012, has been the source of some optimism. Some scholars believe that the Adaptation Fund could significantly exceed bilateral donations and become the main conduit for adaptation financing (Sopoaga, et al., 2007). The key difference between the Adaptation Fund and the other two funds (the SCCF and the LDCF) is that the Adaptation Fund is not dependant on financial contributions by developed countries. The Adaptation Fund is principally replenished through a levy on the CERs generated by the CDM. The 2 per cent share of proceeds will be collected directly by the CDM Executive Board and transferred to the Adaptation Fund for monetisation (Adaptation Fund, 2008). The revenues that will be generated from the 2 per cent levy until 2012 are projected to be between US$160 and US$950 million, while funding presently given to or pledged by donor countries to the LDCF and the SCCF is about US$170 million (Müller, 2007).

A positive outcome at the Conference of the Parties/Meeting of the Parties (CMP) conference in Bali, 2007, was the relative ease with which the governance structure of the Adaptation Fund was agreed upon. A major point of contention had been whether the Adaptation Fund should be managed by the GEF. While the European Union, Japan and other industrialised countries had viewed the choice of the GEF as a self-evident one, the developing countries rejected the choice of the GEF because of their unhappiness with the way the GEF has managed the SCCF and the LDCF. However, the CMP eventually decided at the Bali conference that the Adaptation Fund would be governed by an independent board directly accountable to the CMP and assisted by a secretariat and a trustee.[54] Furthermore, parties eligible for funding will not be required to carry out adaptation projects via Implementing or Executing Agencies of the GEF. Instead, they may submit their project proposals directly to the Adaptation Fund Board.[55] For many, it was seen as a 'major victory for the developing world in setting a new governance system for funding of adaptation activities' (Schalkwyk, 2007).

The Adaptation Fund, SCCF, LCDF and GEF Trust Fund are important sources of funding, but, given that the estimated costs of adaptation far exceed currently available funding, it still remains urgently necessary for climate change policymakers to devise ways to increase financial support for adaptation in vulnerable countries (UNFCCC, 2007a).[56] Switzerland has proposed a global carbon tax with an exemption for countries whose annual per capita emissions are less than 1.5 tonnes of carbon dioxide. The resources generated would flow into a multilateral fund for adaptation and insurance along with a national climate change fund (UNFCCC, 2008a, para. 14(f)). Another proposal being considered under the processes launched by the Bali Action Plan consists of extending the share of proceeds to assist in meeting the costs of adaptation to joint implementation and emissions trading under the Kyoto Protocol (UNFCCC, 2008a). Norway has proposed that adaptation should be financed through auctioning a share of the Assigned Amounts of all Annex-I Member States (UNFCCC, 2008a, para. 14(d)). These options will have implications for the development of the international carbon market, as it is assumed that introducing a levy will always have the effect of shifting market activity from the levied mechanism to the non-levied one, amongst other things, and will therefore require careful consideration. Apart from taxing the Kyoto Protocol mechanisms, another proposal is to impose levies on highly-polluting activities such as aviation and marine transportation.[57] Such levies would be in accordance with the polluter pays principle. Further, it can be argued that if levies are to be imposed at all, they should be placed on highly polluting activities to reflect their real social costs, and not on the Kyoto mechanisms which deliver environmental benefits. It remains to be seen whether any of these proposals will receive serious consideration in Copenhagen in 2009.

6. CONCLUSION

The need for communities to increase their adaptive resilience will become greater as we begin to experience more impacts of climate change.

Successful adaptation, however, is a challenging process which takes long-term planning and cannot be achieved overnight. In this regard, proper institutional and normative frameworks have to be in place at the international and state levels to support the adaptation efforts of developing countries. While it appears that many international organisations are involved in adaptation work and, as the old saying goes, 'too many cooks spoil the broth', this chapter argues that there is actually a fairly coherent division of labour amongst the international agencies which all lend their specific expertise on climate change as experienced within their official mandates. The UNFCCC is a crucial focal point with a coordinating role within the UN system. At the state level, the survey of three countries, Vietnam, Laos and China, shows that they lack the institutional mechanisms and legal frameworks that are important for coordinating and developing adaptation responses at the state level. They and other countries are urged to develop these local institutional structures to coordinate with and leverage on the support of the international institutional framework. In this way, developing countries will be better placed to adapt to the impacts of climate change effectively.

NOTES

* Jolene Lin is Assistant Professor at the Faculty of Law, University of Hong Kong.

1. Policy discussion on climate change in the 1980s did include adaptation as an important aspect. Over the years, however, mitigation became favoured as the global response and adaptation appeared to be relegated to local responses to specific changes brought upon by climate change. This focus on mitigation efforts was concretised by the Kyoto Protocol which created legally binding mitigation targets for developed countries. Another reason why adaptation received less attention than mitigation of greenhouse gases in earlier international negotiations is that there was insufficient knowledge on differentiating between the impact of human-induced climate change and natural climate change variability. See Parry, et al. (1998), p. 741.
2. The IPCC Fourth Assessment Report has highlighted the urgency of adaptation; see www.ipcc.ch (visited 15 March 2009). Under a business-as-usual scenario, the Earth's temperature could rise by 3 degrees Celsius this century. Even in the case of a 1 to 2.5 degree Celsius rise in global temperature, the IPCC predicts serious effects, including reduced crop yields in tropical areas, spread of climate sensitive diseases like malaria, and sea level rises that could inundate entire island states.
3. For an overview of the vulnerabilities and adaptation options in Asia see UNFCCC Secretariat (2007b).
4. See Burton (1992), definition reproduced in Klein and Tol (1997).
5. See Smith, et al. (1996), definition reproduced in Klein and Tol (1997).
6. These examples are drawn from information available on the Australian Government, Department of Climate Change website: www.greenhouse.gov.au/impacts/coasts.html#impacts (visited 15 March 2009).
7. Ibid.
8. For discussion about the principle of Common but Differentiated Responsibility see, e.g., Cullet, 2003; Stone, 2004, p. 276.

9. For discussion about Art. 4(1) see Hall, 2005, p. 59.
10. Decision 17/CP. 7, FCCC/CP/2001/13/Add.1, Modalities and procedures for a clean development mechanism as defined in Art. 12 of the Kyoto Protocol. However, financing for the Adaptation Fund is not limited to the revenues from the 2 per cent CDM levy. Paragraph 2 of Decision 10/CP. 7 states that 'the adaptation fund shall be financed from the share of proceeds on the clean development mechanism project activities and other sources of funding'.
11. See Decision 17/CP. 7, FCCC/CP/2001/13/Add.1, Modalities and procedures for a clean development mechanism as defined in Art. 12 of the Kyoto Protocol.
12. See, for example, Yamin and Depledge (2005).
13. For discussion about the piece-meal nature of the approach of linking adaptation to other provisions in the UNFCCC see Schipper and Boyd (2006). For discussion about the re-branding of adaptation as a developing-country issue see Schipper (2006).
14. See, e.g., Müller (2006).
15. The author examined international legal documents available in the public domain which have direct bearing on the legal, regulatory and institutional frameworks necessary for the implementation of adaptation measures, including resolutions of the UN General Assembly, reports by the UN Secretary-General, decisions of the UN Environment Programme (UNEP) Governing Council, decisions of the UNFCCC Conference of the Parties (COP) and the Kyoto Protocol Conference of the Parties/Meeting of the Parties (CMP), decisions of relevant intergovernmental organisations, and declarations or official statements by political organisations, and concluded that there is a significantly stronger political consensus today on the urgency of adaptation.
16. See the FAO homepage 'Policy Framework', www.fao.org/climatechange/49372/en/. Also see the 'Food Security' homepage www.fao.org/climatechange/49357/en/ for a list of publications on climate change and food security.
17. The WHO has published numerous technical reports on climate change-related impacts on public health policy. See, e.g., Kovats, et al. (2003); Ebi, et al. (2005).
18. Sixty-first World Health Assembly, WHA61.19, Agenda item 11.11, 24 May 2008. The resolution can be found online at www.who.int/gb/ebwha/pdf_files/A61/A61_R19-en.pdf (visited 15 March 2009).
19. The Global Facility for Disaster Reduction and Recovery (GFDRR) is a partnership of the International Strategy for Disaster Reduction (ISDR) system to support the implementation of the Hyogo Framework for Action (HFA). The HFA, endorsed by the United Nations General Assembly in Resolution 60/195, is the primary international agreement for disaster reduction: see http://gfdrr.org/index.cfm?Page=About%20the%20GFDRR&ItemID=2 (visited 1 April 2009). For information about the World Bank Country Assistance Strategies see Projects and Operations page on the World Bank, at http://web.worldbank.org (visited 1 April 2009).
20. There are 13 countries in the Asia and Pacific region eligible for financial support for adaptation activities through the Least Developed Countries Fund, which is administered by the GEF. As a GEF agency, ADB can assist with the implementation of National Adaptation Programs of Action (NAPAs). See Asian Development Bank, 2007, p. 22.
21. See, e.g., D. Ponzi, 'Technical Assistance (Financed by the Government of Canada) For the Climate Change Adaptation Program for the Pacific', Asian Development Bank, 2002.
22. See the list of UNEP executing and implementing agencies, online: www.gefweb.org/Partners/Exe_Agencies/exe_agencies.htm and www.gefweb.org/participants/Implementing_Agencies/implementing_agencies.html (visited 15 March 2009).
23. See case studies in Asian Development Bank, 'Climate Proofing: A Risk-Based Approach to Adaptation' (2005).
24. For a succinct summary of the official role of the Secretariat see CDM Rulebook, Bodies, UNFCCC Secretariat, at http://cdmrulebook.org/PageId/76 (visited 15 March 2009).

25. See, e.g., 'Information on Methods and Tools for Impact, Vulnerability and Adaptation Assessments: Submissions from Relevant Organisations', Subsidiary Body for Scientific and Technological Advice, 27th Session, Bali, 3–11 December 2007, FCCC/SBSTA/2007/MISC.13.
26. NAPAs were initiated with Decision 28/CP. 7 as part of the Marrakesh Accords of 2001.
27. See UNFCCC, http://unfccc.int/adaptation/sbsta_agenda_item_adaptation/items/3633.php (visited 15 March 2009).
28. See, for example, Decision 5/CP.7, Decision 6/CP. 9 and Decision 3/CP.11 which refer to the importance of a 'country-driven approach'.
29. See United Nations Asian and Pacific Centre for Agricultural Engineering and Machinery, at www.unapcaem.org/ppt/vn-01.htm (visited 15 March 2009).
30. See UNDP Viet Nam, 'Viet Nam at a Glance: Human Development Overview', www.undp.org.vn/undpLive/Content/UNDP/About-Viet-Nam/Viet-Nam-at-a-Glance?languageId=1 (visited 15 March 2009).
31. Vietnam's Initial National Communication at pp. 84–87; Socialist Republic of Viet Nam, Ministry of Natural Resources and Environment, 'Viet Nam Initial National Communication under the United Nations Framework Convention on Climate Change' (2003).
32. See the National Office for Climate Change and Ozone Protection website: www.noccop.org.vn/index.html (visited 15 March 2009).
33. See Vietnam's Initial National Communication, pp. 84–87.
34. See 'List of Least Developed Countries', UNFCCC website, http://unfccc.int/cooperation_and_support/ldc/items/2666.php (visited 15 March 2009). The country's 2004 GDP per capita stood at US$378 (IPCC Fourth Assessment Report: Asia, Table 10.1).
35. See Lao PDR's First National Communication, p. 95.
36. See NAPAs official homepage: http://unfccc.int/adaptation/napas/items/2679.php (visited 15 March 2009).
37. Droughts normally associated with ENSO years have occurred in Laos, and the droughts in 1997 to 1998 caused massive crop failures and water shortages and forest fires in various parts of Laos, the Philippines and Indonesia: see Table 10.3, *Summary of observed changes in extreme events and severe climate anomalies*, IPCC Fourth Assessment Report 2007: Asia, p. 476.
38. See Lao PDR's First National Communication, p. 39.
39. Figure 1.10 in Lao PDR's First National Communication.
40. See Cruz, et al. (2007), pp. 469–506, Tables 10.2 and 10.3.
41. See National Development and Reform Commission of the People's Republic of China, *China's National Climate Change Programme* (2007), p. 12.
42. See State Council Document (2007) No. 18, 'Circular on the Establishment of National Climate Change Leading Group & National Energy Conservation and Emission Reduction Leading Group of the State Council'.
43. See State Council Document (2006) No. 28, 'Decision on Strengthening Energy Conservation Work'.
44. Ibid., at p. 56.
45. Ibid.
46. See Lieberthal (1997), *China Environment* Series 1, at p. 3. Also see the discussion of the institutional and legal framework of China's environmental management regime in Lin, 2004, p. 617 (Part II).
47. See discussion of the value of public–private partnerships in adaptation in chapter 3 of Agrawala and Fankhauser (2008).
48. See 'Local Authorities' Initiatives in Support of Agenda 21' (chapter 28 of Agenda 21) at www.un.org/esa/sustdev/documents/agenda21/english/agenda21chapter28.htm (visited 15 March 2009).
49. This would be in accordance with recommendations in Agenda 21; see para. 28(4) of ibid.

50. See, e.g., ICLEI-Local Governments for Sustainability's 'Cities for Climate Protection' Campaign, 2008, www.iclei.org; the C40 initiative www.c40cities.org (visited 15 March 2009).; and UNEP's Cities and Climate Change programme, 2008, www.unep.org/urban_environment/Issues/climate_change.asp (visited 15 March 2009).
51. See, e.g., Cohen, et al. (1998), pp. 341–71; Markandya and Halsnaes (2002); Robinson and Herbert, 2001, pp. 130–48; Smit, et al. (2001).
52. See Decision 1/CP. 13, Document Ref.: FCCC/CP/2007/6/Add.1 (14 March 2008).
53. See para. 1, Decision 1/CP. 13, Document Ref.: FCCC/CP/2007/6/Add.1 (14 March 2008).
54. See Draft decision -/CMP3, para. 6.
55. See Draft decision -/CMP3, paras 28, 29.
56. The UNFCCC (2007a) estimates that the overall additional investment and financial flows needed for adaptation in 2030 will amount to several tens of billions of US dollars. In particular, about US$14 billion in investment and financial flows are estimated to be needed for agriculture, forestry and fisheries, about US$11 billion required for water supply infrastructure (85 per cent of which will be needed in non-Annex I Parties), and the costs of treating diseases due to climate change are estimated at US$5 billion in 2030.
57. See Submission from Tuvalu, 'An International blueprint on Adaptation', Doc. Ref. FCCC/CP/2007/MISC.2, 8 December 2007.

REFERENCES

Adaptation Fund (2008), 'Monetization of Certified Emission Reductions for the Adaptation Fund', AFB/B.2/9, www.adaptationfund.org/images/AFBB2-9-Monetization.pdf (visited 14 November 2008).

Agrawala, S. (ed.) (2005), *Bridge Over Troubled Waters: Linking Climate Change and Development* (OECD).

Agrawala, S. and S. Fankhauser (eds) (2008), *Economic Aspects of Adaptation to Climate Change: Costs, Benefits and Policy Instruments* (OECD).

Asian Development Bank (ADB) (2007), *Climate Change: ADB Programs – Strengthening Mitigation and Adaptation in Asia and the Pacific* (ADB).

Burton, I. (1992), *Adapt and Thrive* (Canadian Climate Centre).

Burton, I. (2003), 'Mitigation and Adaptation: The Case for Separation', *Paper Presented at Mitigation and Adaptation: Towards a Mutual Agenda Workshop* (Essen, 15–16 May 2003).

Cohen, S., et al. (1998), 'Climate Change and Sustainable Development: Towards Dialogue', *Global Environmental Change* 8(4), 341.

Cruz, R.V., et al. (2007), 'Asia', in M.L. Parry, et al. (eds), *Climate Change 2007: Impacts, Adaptation and Vulnerability. Contribution of Working Group II to the Fourth Assessment Report of the Intergovernmental Panel on Climate Change* (Cambridge University Press).

Cullet, P. (2003), *Differential Treatment in International Environmental Law* (Ashgate Publishing).

Dasgupta, S., et al. (2007), *The Impact of Sea Level Rise on Developing Countries: A Comparative Analysis*, Policy Research Working Paper No. 4136 (World Bank).

Ebi, K.L., N.D. Lewis and C.F. Corvalán (2005), *Climate Variability and Change and Their Health Effects in Small Island States: Information for Adaptation Planning in the Health Sector* (World Health Organization).

Evans, J.W. (2007), 'Climate Change is a Development Issue: Towards Climate-resilient Development', *Annual Review* (World Bank).

Gigli, S. and S. Agrawala (2007), *Stocktaking of Progress on Integrating Adaptation to Climate Change into Development Co-operation Activities* (OECD).

Hall, L. (2005–06), 'Technology Transfers under the United Nations Framework Convention on Climate Change', *Colorado Journal of International Environmental Law and Policy* 17(1), 59.

Klein, R. and R. Tol (1997), *Adaptation to Climate Change: Options and Technologies* (UNFCCC Secretariat).

Klein, R. (2008), 'Adaptation: Needs, Financing and Institutions', *Breaking the Climate Deadlock Briefing Paper* (The Climate Group).

Kovats, S., K.L. Ebi and B. Menne (2003), 'Methods of Assessing Human Health Vulnerability and Public Health Adaptation to Climate Change', *Health and Global Environmental Change Series No. 1* (World Health Organization).

Lieberthal, K. (1997), 'China's Governing System and Its Impact on Environmental Policy', *China Environment Series* 1, 3.

Lin, J. (2004), 'Assessing the Dragon's Choice: The Use of Market-based Instruments in Chinese Environmental Policy', *Georgetown International Environmental Law Review* 16(4), 617.

Lin, J. (2008), 'The Legal and Institutional Measures for Effective Implementation of Measures for Adaptation to the Impacts of Climate Change in the Asia Pacific Region: Part I' (contracted research for the Centre of Environmental Research, Training and Information, Sri Lanka).

Mace, M.J. (2005), 'Funding for Adaptation to Climate Change: UNFCCC and GEF Developments Since COP-7', *Review of European Community and International Environmental Law* 14(3), 225.

Mace, M.J. (2006), 'Adaptation under the UNFCCC', in W.N. Adger, et al. (eds), *Fairness in Adaptation to Climate Change* (MIT Press).

Markandya, A. and K. Halsnaes (eds) (2002), *Climate Change and Sustainable Development* (Earthscan).

Müller, B. (2006), 'Montreal 2005: What Happened, and What it Means', Paper No. EV35 (Oxford Institute for Energy Studies).

Müller, B. (2007), 'The Nairobi Climate Change Conference: A Breakthrough for Adaptation Funding', *Oxford Energy & Environment Comment Working Paper* (Oxford Institute for Energy Studies).

Organisation for Economic Co-operation and Development (OECD) (2008), 'Cities and Climate Change', www.oecd.org/document/27/0,3343,en_2649_34361_39760027_1_1_1_1,00.html (visited 6 February 2009).

Parry, M., et al. (1998), 'Adapting to the Inevitable', *Nature* 395(6704), 741.

Parry, M.L., et al. (eds) (2007), *Climate Change 2007: Impacts, Adaptation and Vulnerability. Contribution of Working Group II to the Fourth Assessment Report of the Intergovernmental Panel on Climate Change* (Cambridge University Press).

Robinson, J.B. and D. Herbert (2001), 'Integrating Climate Change and Sustainable Development', *International Journal of Global Environmental Issues* 1(2), 130.

Schalkwyk, M. (2007), 'Press Statement on 11 December 2007', quoted in B. Muller (2008), 'Bali 2007: On the Road Again!' (Oxford Climate Policy).

Schipper, E. (2006), 'Conceptual History of Adaptation in the UNFCCC Process', *Review of European Community and International Environmental Law* 15(1), 82.

Schipper, E. and E. Boyd (2006), 'UNFCCC COP 11 and COP/MOP 1: At Last, Some Hope?', *Journal of Environment and Development* 15(1), 75.

Smit, B. and O. Pilifosova (2001), 'Adaptation to Climate Change in the Context of Sustainable Development and Equity', in J.J. McCarthy, et al. (eds), *Climate Change 2001: Impacts, Adaptation and Vulnerability* (Cambridge University Press).
Smith, J.B., et al. (eds) (1996), *Adapting to Climate Change – Assessments and Issues* (Springer-Verlag).
Sopoaga, E., et al (2007), *On the Road to Bali: Operationalising the Kyoto Protocol Adaptation Fund* (International Institute for Environment and Development).
Srinivasan, S. (2005), 'Adaptation to Climate Change: A Critical Challenge for Asian Development', *What's New from IGES (IGES newsletter)*, (Institute for Global Environmental Strategies).
Stone, C.D. (2004), 'Common but Differentiated Responsibilities in International Law', *American Journal of International Law* 98(2), 276.
Swart, R., J. Robinson and S. Cohen (2003), 'Climate Change and Sustainable Development: Expanding the Options', *Climate Policy* 3: *Supplement 1 S19-S40* (special supplement on climate change and sustainable development) at S20 and S21.
Tarlock, A.D. (1992), 'Now, Think Again About Adaptation', *Arizona Journal of International and Comparative Law* 9(2), 169.
United Nations Framework Convention on Climate Change (UNFCCC) Secretariat (2007a), 'Investment and Financial Flows Relevant to the Development of an Effective and Appropriate International Response to Climate Change', http://unfccc.int/cooperation_and_support/financial_mechanism/items/4053.php (visited 6 February 2009).
UNFCCC Secretariat (2007b), 'Background Paper: Impacts, Vulnerability and Adaptation to Climate Change in Asia', *Asian Regional Workshop on Adaptation* (Beijing, China, 11–13 April).
UNFCCC Secretariat (2008a), 'Funding Adaptation In Developing Countries: Extending the Share of Proceeds Used to Assist in Meeting the Costs of Adaptation; and Options Related to Assigned Amount Units of Parties Included in Annex I to the Convention', Technical Paper FCCC/TP/2008/6.
UNFCCC Secretariat (2008b), 'Report on the Workshop on Investment and Financial Flows to Address Climate Change', FCCC/AWGLCA/2008/CRP.3.
World Bank (2008), 'Climate Change is a Development Issue', http://go.worldbank.org/VRETHAGHE0 (visited 6 February 2009).
World Health Organization (WHO) (2000), *Climate Change and Human Health: Impact and Adaptation* (WHO).
World Health Organization (2007), *Regional Framework for Action to Protect Human Health from Effects of Climate Change in South East Asia and Pacific Region* (WHO).
Yamin, F. and J. Depledge (2005), *The International Climate Change Regime: A Guide to Rules, Institutions and Procedures* (Cambridge University Press).

7. Designing a REDD mechanism: the TDERM triptych

Claire Stockwell, William Hare and Kirsten Macey*

1. INTRODUCTION

At present, the international climate regime does not address deforestation in developing countries. Deforestation was discussed during the negotiation of the United Nations Framework Convention on Climate Change (UNFCCC)[1] and its Kyoto Protocol,[2] but was included only in developed countries' commitments.[3] In 2005, the Conference of the Parties (COP) to the UNFCCC began to examine anew how incentives for reducing emissions from deforestation and forest degradation (REDD) could be included in the regime (UNFCCC, 2005, 2006b, p. 18). REDD was also included as part of the issues to discuss for a post-2012 climate agreement when full negotiations were launched at the end of 2007 in Bali, Indonesia.[4] These negotiations are set to conclude at COP 15 in Copenhagen, Denmark, in December 2009. As deforestation accounts for around 20 per cent of global greenhouse gas (GHG) emissions, it is imperative that this source of emissions be addressed (Rogner, et al., 2007).

Many countries and NGOs have developed proposals on how a REDD mechanism could work. Discussions to date (September 2008) have focused primarily on methodologies for measuring emission reductions and whether financing for the mechanism should be derived from market or non-market means. Unfortunately this 'market vs. fund' debate dominates the entire discussion of the broader policy issues around the design of a REDD mechanism. While many proposals address other issues, they do so often only in a cursory fashion (see Hare and Macey, 2007, for the main proposals). At the third negotiating session of 2008 in Accra, Ghana, countries began discussing 'policy approaches and positive incentives' for REDD (UNFCCC, 2008b). While some new proposals were tabled, full consideration of the policy question is still lacking. The purpose of this chapter is to contribute to a broadening of that discussion to encompass other important policy issues.

Section 2 reviews the causes and environmental impacts of deforestation. Section 3 outlines the main issues that will have to be resolved for an effective REDD mechanism. Section 4 examines lessons learned from the Clean Development Mechanism (CDM) and how these may influence policy choices in a REDD mechanism. Section 5 outlines our proposal for a Tropical Deforestation Emission Reduction Mechanism Triptych and explains how it addresses the policy considerations raised in the earlier sections.

2. DEFORESTATION: FRAMING THE ISSUE

Almost 4 billion hectares (ha) or 30 per cent of the world's land area is covered in forest. From 2000 to 2005, the world lost some 13 million ha of forest per year due to deforestation (UN FAO, 2006). The overall loss in total forest cover was only about half that, due to tree planting, landscape restoration and natural regeneration of forests on abandoned land. Nonetheless, the UN Food and Agriculture Organization (UN FAO) concluded that deforestation is continuing at 'an alarmingly high rate'. Only two countries, Brazil and Indonesia, are responsible for two-thirds of this net forest cover loss, at 3.1 and 1.8 million ha respectively (ibid.).

Deforestation contributes to both great loss of biodiversity and climate change. Tropical forests are some of the most biologically diverse regions of the world, while deforestation is responsible for around 20 per cent of GHG emissions globally (Millennium Ecosystem Assessment, 2005; Rogner, et al., 2007). Climate change itself will also affect forest regions negatively, exacerbating this downward spiral (Millennium Ecosystem Assessment, 2005).

The causes of deforestation are complex and vary from country to country as well as within countries over time (Geist and Lambin, 2001). They can be divided into two categories: direct or proximate, and indirect or underlying. Direct causes are physical in nature and refer to the actual process of deforesting. These include agricultural expansion, wood extraction and infrastructure development (for example dams, roads and human settlements). Indirect or underlying causes refer to circumstances or policies that promote or drive the direct causes of deforestation. These include demographic, economic, technological, policy or institutional, and cultural forces. Much research focuses on just one or a handful of causes. Few comprehensive reviews exist. It is therefore difficult to get a clear picture of the drivers (Contreras-Hermosilla, 2000; Geist and Lambin, 2001; UNFCCC, 2006a; Kanninen, et al., 2007, pp. 15–27).

The prominence of the different causes varies regionally (Geist and

Lambin, 2001; Lambin and Geist, 2003). In Latin America, pasture creation for cattle ranching as well as small-scale farming aided by road construction are the dominant proximate causes of deforestation (Nepstad, et al., 2001, 2002; Fearnside, 2005; Barreto, et al., 2006). While government policies have facilitated migration and subsidised agricultural production, insecure property rights compound the situation (de Oliveira, 2008). Commercial logging is a more prevalent cause in Asia, followed by cultivation of cash crops, particularly oil palm and small-scale farming (Sunderlin and Resosudarmo, 1996; Matthews, 2002; Leimgruber, et al., 2005; Uryu, et al., 2008). In recent years, the skyrocketing demand for pulp in China and crude palm oil in Europe has only made the situation worse (Kanninen, et al., 2007, p. 18). Corruption, insecure land tenure, weak enforcement of forestry laws, transmigration and government support for estate agriculture (cash crops) and logging are the main underlying causes (Matthews, 2002; McMorrow and Talip, 2001). In West and Central Africa, logging, fuelwood collection for domestic use and subsistence agriculture are the most frequent proximate causes (Geist and Lambin, 2001; Dessie and Kleman, 2007). Weak institutions (lack of enforcement and mismanagement), migration and population are the dominant underlying causes in this region.

A major review of case studies of deforestation concluded that: 'no universal policy for controlling tropical deforestation can be conceived. Rather, a detailed understanding of the complex set of proximate causes and underlying driving forces affecting forest cover changes in a given location is required prior to any policy intervention.' (Geist and Lambin, 2002, p. 150).

To reduce emissions from deforestation will be no different. The Center for International Forestry Research delivered a similar message at COP 13 when it warned that a REDD mechanism would fail unless policy makers were able to grasp the root causes of deforestation and the need for different solutions to address these.[5]

The challenge is to develop a mechanism that takes into account these regional differences and interacting causes and allows for the implementation of a variety of actions involving a number of actors at different levels. While a formidable challenge in its own right, and one that must be overcome in order to reduce emissions, it is only one of many challenges to designing an effective international response.

3. DESIGNING A REDD MECHANISM

The design of any REDD mechanism will be complex, given the complex nature of the problem. Both methodological and broader policy issues will have to be addressed. The way these broader questions are addressed will have implications for methodological arrangements, and vice versa. Methodological issues have been addressed elsewhere and will not be reviewed here save for a few comments about definitions and baselines (see Hare and Macey, 2007, for an in-depth discussion). This section outlines six key issues:

- The nature of REDD commitments;
- Financing (how the resources are raised, how much is required, how resources are disbursed, when and to whom);
- Activities undertaken (including the types of activities, whether they occur at the national or sub-national level and who controls their implementation);
- Liability issues;
- Capacity building; and
- Institutional arrangements.

It is necessary first to say a few words about the scope and objectives of the mechanism.

Scope and Objectives

REDD stands for reducing emissions from deforestation and forest degradation, yet defining precisely what is meant by deforestation and forest degradation is quite challenging. The Kyoto Protocol defines deforestation as the direct human-induced conversion of forested land to non-forested land.[6] Forest degradation is not defined in the Protocol. The Intergovernmental Panel on Climate Change was tasked with developing a definition but could not reach consensus on a single one (Penman, et al., 2003). Forest degradation relates to a reduction in the quality, not the size, of a forest area, including the amount of carbon it can store. Inclusion of degradation in a REDD mechanism poses several challenges beyond this definitional issue. It could create loopholes through which countries with significant quantities of degraded lands could continue to deforest by offsetting the associated emissions with afforestation and reforestation projects on their degraded forest lands. A broad definition of deforestation would address most, if not all, of the impacts from forest degradation and fragmentation without burdening the current negotiations with debates over new concepts that could reopen discussions on afforestation and reforestation. In this chapter and in our proposal for a REDD mechanism we focus on deforestation, but deforestation broadly defined. That said, further work is needed on the implications of including forest degradation separately in a REDD mechanism.

Some countries would also like to expand the mechanism to include forest conservation or land-use, land-use change and forestry (LULUCF) in general (UNFCCC, 2008d). REDD should in our view be dealt with separately from these issues. Specific and targeted action will be required to reduce deforestation rates and associated GHG emissions. This action will be significantly different from measures designed to support conservation, afforestation, reforestation or other land use changes. It can best be supported by its own mechanism, though this mechanism could form part of a larger approach to LULUCF.

The goals of a REDD mechanism should include at a minimum:

- Reducing emissions from deforestation and forest degradation;
- Protecting biodiversity;
- Respecting the rights of indigenous peoples and vulnerable communities; and
- Ensuring equitable distribution of benefits both within and between countries.

The goals of the mechanism should be articulated in advance, since a mechanism the sole purpose of which is to reduce emissions could be structured quite differently from a mechanism the purpose of which is also to protect biodiversity and the rights of indigenous peoples.

Nature of Commitments

Commitments to address emissions from deforestation may be incentive-based or obligation-driven, and may target both developing and developed countries.[7] A REDD mechanism may give developing countries incentives to act in the form of payment for the successful reductions of emissions, or oblige them to act by imposing some kind of REDD target. It may give developed countries incentives to contribute to a fund or purchase REDD credits that would count towards their general emission reduction obligations, or it could require them to finance REDD in developing countries through a variety of means (direct payment, market-linked methods like auctioning, etc.) or meet a certain portion of their emission reduction targets by supporting REDD activities.

The transnational or global character of some of the underlying causes of deforestation make it difficult for national governments to address REDD on their own. Commitments to reduce emissions need not therefore be limited to the developing countries in which deforestation occurs. Agreement to work collectively on some of these underlying causes could form part of any country's REDD commitments. Countries could for example agree to adopt national legislation preventing the sale of illegally or unsustainably harvested timber products. Such action could be coordinated through existing initiatives such as the Forest Law Enforcement and Governance (and Trade) (FLEG/T) initiatives of the World Bank and the European Commission.

Financing

A central focus of the REDD debate has been on whether financing for the mechanism should be derived from market or non-market means. Proponents of the market approach argue that it is the only way to raise the necessary level of financing. While there are many estimates of the likely costs of REDD, it is clear that the scale of resources is in the order of billions of dollars per year (Stern, 2007; UNFCCC, 2007a). Supporters of a non-market approach, namely the establishment of a fund to finance REDD activities, point to the potential negative effect REDD credits may have on the carbon market. Because REDD credits are expected to be inexpensive and plentiful, they could flood the market and lower the price of carbon unduly if not properly controlled.

The market/non-market debate is only the first of several orders of issues to consider with respect to financing. Others include to whom the money will be paid, when, how it will be transferred and how much is required. Distribution of benefits to relevant actors is a key equity consideration and will likely determine the success or failure of REDD activities on the ground. Another important issue is how to ensure that those undertaking activities have access to stable sources of funding to both initiate and sustain them. Finally, the choice of the source of funds will have implications for a number of other elements of the mechanism relating to accounting methodologies, reporting and verification of emissions reductions, and liability (Hare and Macey, 2007).

Activities

Because of the complexity and interconnectedness of the proximate and underlying causes of deforestation, reducing emissions from deforestation and degradation will require action at a number of levels (national, regional, local) and involve numerous actors (for example indigenous peoples, landholders, local communities, private investors, and municipal governments). On this there is broad agreement. The contentious issue is how sub-national actors figure into the equation. One possibility is that national governments will coordinate and delegate all REDD activities within their territories, allowing sub-national actors to implement activities, but making the national government the only entity to interact with the international mechanism. Alternatively, national governments may allow sub-national actors to interact directly with an international mechanism (that is, sub-national actors would both implement and control activities). Under this option, it would be necessary to consider the baselines that would be used for sub-national activities, leakage concerns and the liability these actors may face. Overall reductions must be measured at the national level to minimise leakage. This does not however dictate the level at which REDD activities may occur.[8] Multiple baselines may be required: national to determine the actual reductions achieved, and sub-national to indicate the success of these activities within the national bubble.

Different sources of financing may be better suited to supporting different types of REDD activities. This may depend upon who controls the implementation of the activity. Activities the success of which is dependant upon government performance tend to be less attractive to outside investment and may not be supported by all types of market-based financial mechanisms. Some activities, especially those aimed at addressing the underlying causes of deforestation for instance by reforming tax or regulatory regimes to remove perverse incentives, can be undertaken only by government. Other activities, such as payment for environmental services, could be implemented by governments or private actors, but outside investors may only be interested in supporting the latter. Of course, if payment is sought after the reductions are achieved, performance risk is no longer a concern and financing may therefore be obtained by market or non-market means. This assumes however that the entities involved can secure up-front financing to undertake REDD activities, an assumption that may not hold in all circumstances.

Liability

The question of liability arises in relation to the consequences that ensue if deforestation rates increase after financial support has been provided. Assigning liability is most complex for a compliance-driven system, that is a system in which developed countries may use emissions reductions achieved through REDD activities to count towards their emission reduction obligations. In a compliance-driven system, one must determine whether the buyer or seller of compliance units is liable for addressing the increasing rates of deforestation or forest degradation. This liability can take many forms. If the buyer is liable, he/she will need to secure alternative REDD units in order to remain in compliance with his or her obligations. The seller could be required to repay any funds received to date. If the seller is liable, he/she will need to make up for the shortfall in compliance units. This could be through other units produced by the same REDD activity or those obtained from other sellers. There could also be a penalty involved in failing to deliver on the compliance units promised.

Assigning liability is also complicated when sub-national actors are allowed to implement activities. Sub-national actors are responsible for and assume the risks associated with their own REDD activities. If these activities are unsuccessful, they will not receive any credits or payment. One must determine however who assumes the risk or is liable for national level emissions in cases where sub-national activities have been successful in reducing emissions, but overall national level emissions continue to rise. Should sub-national actors receive the credits or payment for their reductions while the national government is held liable for the discrepancy with national emission trends? That is, the national government would be required to pay or provide sub-national actors with credits from an international market. Or as part of the cost of undertaking REDD activities, should sub-national actors assume the risk that national level emissions may not decrease, and thus that they may not receive credit or payment even if their actions are successful?

A number of mechanisms could be employed to counteract increasing deforestation rates, including a reserve of credits or some other insurance scheme. Situations in which the rate of deforestation and the associated GHG emissions increase because of non-human agents such as fire, disease or insect infestation will require special consideration. Financial support should be provided only when the atmosphere experiences a benefit, but by the same token governments or other actors should not be held accountable for activities beyond their control.

Other Key Issues

The remaining key issues for the design of a REDD mechanism are institutional arrangements and capacity considerations. Institutional arrangements could range from simple to complex depending on governance and methodological choices. The UNFCCC Secretariat could potentially oversee a system of bilateral transfers of financial support for reported reductions with little modification from the current structure. An amalgamated fund would require an entity to manage and administer its resources, while a market structure would require at a minimum a supervisory entity analogous to the CDM Executive Board. Institutional arrangements will also be necessary to involve and protect the rights of indigenous peoples and vulnerable communities.

Countries in which deforestation occurs are at various stages of development and have varying abilities to measure and achieve emission reductions. These varying capacities must be acknowledged and taken into account to design an effective and equitable REDD mechanism. This may warrant different modalities for different sets of countries.

Overall, many permutations are possible for the design and governance of a REDD mechanism. All we have done here is to outline some of the key issues. With the various options on the table, we next examine past experience with forest-related initiatives and other international mechanisms to elucidate what combination of options might work best.

4. LESSONS LEARNED FROM THE CLEAN DEVELOPMENT MECHANISM AND BEYOND

There are 40 legally binding instruments and 19 non-legally binding agreements and processes relating to forests at the international or regional level, managed by 33 different organisations (UNFF, 2004). While not all of these initiatives address deforestation directly, lessons learned from their implementation may be useful in designing the most effective REDD mechanism. A mid-term review of the World Bank's Forest Law Enforcement and Governance programme (initiated in 2001 to combat illegal logging) found that many countries had experienced difficulty implementing the necessary reforms due to the complexity of the issue and the politics involved (Contreras-Hermosilla, 2007). Governments will likely encounter similar difficulties removing perverse incentives or developing positive ones to reduce emissions from deforestation. Research related to timber certification has shown that small-scale producers and community groups have not participated extensively in certification processes due to the complexity of the rules (Bass, et al., 2001). These actors may face similar barriers to participation in a REDD mechanism if REDD rules are overly complex.

Forest-related programmes such as timber certification often have an 'on the ground' component and an institutional support structure. How they have fared at the local level may be informative for the design of particular REDD activities. Experience with institutional structures is instructive for both domestic and international governance of the system. While some lessons learned may be more relevant for the development of domestic programmes, discussion at the international level can facilitate knowledge transfer and ensure that REDD rules are most conducive to the successful implementation of activities on the ground.

Forest-related initiatives are not the only areas that may provide insight for REDD. Other multilateral mechanisms may also be worth examining, given the potential scale of financial flows and use of market mechanisms. Within the climate regime, the Clean Development Mechanism (CDM) is a prime example. The CDM is one of the Kyoto Protocol's project-based flexibility mechanisms (see generally Lee, et al., 2004). The CDM can provide insight into how the use of markets interacts with a mechanism's other policy aspects such as achieving multiple (non-market) objectives and equitable regional distribution of projects. The governance of the CDM may also provide insight into which institutional arrangements work best in the context of the climate regime. The remainder of this section focuses on lessons learned from the CDM. A fuller discussion of other forest-related initiatives would be prudent, but we do not pursue it here.[9]

The CDM's Sustainable Development Benefits?

The CDM has the twin objectives of contributing to sustainable development in the host country and reducing GHG emissions. To be just and to avoid undermining the environmental integrity of the post-2012 framework a REDD mechanism must also have multiple objectives, namely to reduce emissions from deforestation, protect biodiversity and avoid impinging upon the livelihoods and rights of indigenous peoples and vulnerable communities. As the CDM is market-based, it is important to determine whether or not it has delivered on both its market and non-market objectives.

Unfortunately, many studies have shown that a trade-off between the two objectives exists whereby the desire to produce low-cost emission reductions trumps a project's potential contribution to sustainable development (Cosbey, et al., 2006; Boyd, et al., 2007; Olsen, 2007; Sutter and Parreño, 2007; Olsen and Fenhann, 2008). This conclusion should come as no surprise. Sustainable development benefits are not a monetised component of the project and thus of little value to investors except as a possible niche market. Though many attempts were made to develop international criteria for assessing sustainable development during the negotiation of the Protocol, individual host countries were ultimately given the prerogative to define and operationalise this component (Kelly and Helme, 2000). As countries seek to attract investment there is pressure to minimise or simplify the sustainable development component of their project approval process, leading to a potential 'race to the bottom' (Sutter, 2003). While information on rejected CDM projects is often unavailable, there appear to be very few instances of projects being rejected on the grounds of lack of sustainable development benefits (Burian, 2006; UNDP, 2006). There is furthermore little incentive to ensure that even those sustainable development benefits outlined in the Project Design Document materialise, as there is no requirement to monitor or verify their attainment in contrast to the achievement of GHG reductions.[10] As one leading author concludes, 'the real problem is that the CDM works perfectly' (Olsen, 2007, p. 67). A market mechanism will produce only that which has economic value (in this case GHG reductions), unless the design of the mechanism clearly requires other deliverables in a measurable, reportable and verifiable manner.

Regional Distribution

The Marrakesh Accords, the rulebook of the Kyoto Protocol, emphasise the need for 'equitable geographic distribution of clean development mechanism project activities at regional and sub-regional levels'.[11] Much has been written about the unequal distribution of CDM projects. A brief review of the CDM project pipeline demonstrates why (Silayan, 2005).[12] China and India dominate the market, accounting for around two-thirds of the pipeline in terms of both number of projects and projected certified emission reductions (CERs) produced by 2012 (see Tables 7.1 and 7.2). The top ten countries account for about 90 per cent of the pipeline in terms of number of projects and projected CER production, while sub-Saharan Africa has only 1.4 per cent of the projects and is projected to generate around 2.6 per cent of the credits.[13] Of the 143 developing countries that had ratified the Protocol as of May 2008, only 68 have at least one project in the pipeline. Of the 46 Least Developed Countries that have ratified the Protocol, only 11 have at least one project in the pipeline. The CDM pipeline represents only a snapshot of the how the CDM market could develop; whether projects will be successfully registered and ultimately implemented remains uncertain. Yet a similar picture emerges if we assess the number of credits issued to date: four countries (India, China, South Korea, and Brazil) account for 89.4 per cent.[14]

Most deforestation takes place in just two countries, Brazil and Indonesia (UN FAO, 2006). There is also a great range in capacity to reduce deforestation (and the associated emissions) and monitor these reductions. Furthermore, historical data are lacking for many countries (Kanninen, et al., 2007). It is therefore reasonable to suppose that a REDD mechanism would face similar distributional challenges as has the CDM.

Public Participation

The CDM is managed by an Executive Board (EB) which approves projects and issues credits. The CDM contains limited modalities for

Table 7.1 Top ten countries by number of projects

Host Country	Total Number of Projects	Percentage	Cumulative Percentage
1. China	1173	35.3%	35.3%
2. India	914	27.5%	62.8%
3. Brazil	277	8.3%	71.1%
4. Mexico	180	5.4%	76.5%
5. Malaysia	115	3.5%	80.0%
6. Philippines	71	2.1%	82.1%
7. Indonesia	65	2.0%	84.1%
8. Chile	52	1.6%	85.7%
9. Thailand	45	1.4%	87.1%
10. South Korea	44	1.3%	88.4%

Source: J. Fenhann (2008), 'UNEP Risø Centre's CDM/JI Pipeline Analysis and Database', May 2008 version, http://cdmpipeline.org.

Table 7.2 Top ten countries by projected CER credits in 2012

Host Country	Total Number of Projects	Percentage of CERs by 2012	Cumulative Percentage
1. China	1173	53.6%	53.6%
2. India	914	14.8%	68.4%
3. Brazil	277	6.8%	75.2%
4. South Korea	44	4.0%	79.2%
5. Mexico	180	2.6%	81.8%
6. Malaysia	115	2.4%	84.2%
7. Chile	52	1.5%	85.7%
8. Indonesia	65	1.5%	87.2%
9. Argentina	24	1.2%	88.4%
10. Nigeria	2	1.0%	89.4%

Source: J. Fenhann (2008), 'UNEP Risø Centre's CDM/JI Pipeline Analysis and Database', May 2008 version, http://cdmpipeline.org.

public participation, namely access to information from various sources (the EB, the UNFCCC Secretariat and Designated Operational Entities) and an invitation to submit comments.[15] These provisions alone are inadequate to ensure effective public participation in CDM decision making. More troubling however is the fact that not even these limited provisions are being met. An assessment of registered projects has found that the manner in which stakeholders are invited to submit comments and how these comments are considered are insufficient (Schneider, 2007).

During the negotiation of the Marrakesh Accords, environmental non-governmental organisations argued that the CDM should conform to the principles of public participation agreed to in the Convention on Access to Information, Public Participation in Decision-Making and Access to Justice in Environmental Matters (Aarhus Convention),[16] namely that the public should have access to the relevant information, be able to participate in the decision-making and be able to appeal to an independent and impartial body for review.[17] These arguments are even more valid in a REDD context. Buy-in from the local communities is often critical to the success of forest-related projects (WWF, 2004; Molnar, et al., 2004). At a minimum, public participation mechanisms for REDD should be in line with emerging norms in international law.

Concluding Remarks

Experience with the CDM demonstrates that non-market objectives, like sustainable development benefits and equitable regional distribution, are unlikely to be met in a market setting without intervention. Furthermore the limited public participation provisions of the CDM have met with even more limited success. These are valuable lessons for the design of a REDD mechanism.

5. A TROPICAL DEFORESTATION EMISSION REDUCTION MECHANISM

At COP 13 in 2007, Hare and Macey tabled a proposal on behalf of Greenpeace for a Tropical Deforestation Emission Reduction Mechanism (TDERM) (Hare and Macey, 2007). The Mechanism is characterised by a hybrid market-linked fund, the functioning of which would meet both climate and biodiversity objectives and allow for the participation of all developing countries in which deforestation occurs, regardless of their level of capacity. This section elaborates on the modalities of the Mechanism in light of the policy discussion in section 3 and the lessons learned from the CDM in section 4.

The scope of the mechanism is limited to deforestation, though with a definition of deforestation that is broad enough to address most of the impacts of degradation.[18] The TDERM would require developed countries to purchase and hold a minimum number of Tropical Deforestation Emission Reduction Units (TDERUs) towards compliance with their commitments under an amended Kyoto Protocol (or a new Protocol under the UNFCCC depending on the outcome of the current round of negotiations). The number would be equivalent to a certain percentage of their second commitment period targets (that is a percentage of their base year times the length of the commitment period). Linking the TDERM to compliance with developed countries' emission reduction obligations is meant to signal that they are committed to reducing deforestation globally.

Institutional Arrangements

At the international level the mechanism would be overseen by an Executive Committee (ExComm), under the authority and guidance of the COP or Conference of the Parties serving as the meeting of the Parties to the Kyoto Protocol (COP/MOP). At the domestic level each participating developing country would establish a REDD Designated National Authority (REDD DNA). The ExComm would be a professional body with permanent staff, to avoid the capacity issues that have plagued the CDM EB. It would be supervised by a Board comprised of government, civil society and indigenous representatives. Civil society and indigenous representatives should be included, given the link REDD activities may have with peoples' livelihoods and to accord with the principles of public participation and free, prior and informed consent, as agreed in the Aarhus Convention and the UN Declaration on the Rights of Indigenous Peoples.[19] The ExComm would be responsible for selling TDERUs to developed countries, paying developing countries for the reductions achieved and reviewing annual reports from all Parties. The REDD DNA would be responsible for developing the REDD Strategy Paper (see below), coordinating the implementation of government REDD activities at the national level, approving sub-national activity if any, and reporting to the ExComm. Country interaction with the ExComm would be along one of three tracks: the TDERM Triptych.

The three track approach reflects the varying levels of capacity in countries in which deforestation occurs and allows for universal participation. Countries adopting a Track 1 approach would be those that have sufficient technical capacity to undertake adequate monitoring, reporting and verification of emission reductions from REDD activities. National baselines would be used, but countries would be free to allow sub-national investments. Approval of sub-national activities would be the prerogative of the participating country's REDD DNA. Track 2 countries would be those that are unable to fulfill all of the monitoring, reporting and verification requirements of Track 1, but still possess a relatively high level of technical capacity. These countries would be able to undertake REDD initiatives themselves as well as allow sub-national investment. While countries would approve sub-national activities, this approval will be supervised by the ExComm in a manner similar to Joint Implementation (JI) Track 2 oversight.[20] Track 3 countries would be those that currently have little technical capacity or do not possess the data required for accurate monitoring, reporting and verification. As with Track 2, any sub-national activities would have to be approved by the ExComm in addition to the participating country, but Track 3 countries would be able to use simplified modalities. Track 3 may also be appropriate for countries with low rates of deforestation which anticipate minimal REDD investment and thus do not feel that a large expenditure of human and financial resources is justified.

International standards would need to be developed to ensure the protection of biodiversity and the rights of indigenous communities in the implementation of REDD activities. These standards should draw from principles already agreed in other fora, such as the Convention on Biological Diversity[21] and the UN Declaration on the Rights of Indigenous Peoples. Without international standards a REDD mechanism is likely to repeat a CDM-like experience in which only the emission reduction objective is met. There can be no justification moreover for mandating international standards for measuring GHG reductions but not for the mechanism's other goals. The standards would apply to all REDD activities, both national and sub-national. Conformity with the standards would be assessed initially during the approval process and monitored for the duration of the activity.

REDD Strategy Paper

An assessment of over 150 case studies of deforestation concluded that 'no universal policy for controlling tropical deforestation can be conceived. Rather, a detailed understanding of the complex set of proximate causes and underlying driving forces affecting forest cover changes in a given location is required prior to any policy intervention' (Geist and Lambin, 2002, p. 150). Such an understanding is the first and most formidable challenge faced in the implementation of an effective REDD mechanism. Under the TDERM, all countries would be required to develop and submit a REDD Strategy Paper to the ExComm. The Paper would review the direct and underlying causes of deforestation in the country and outline the various policies and measures the country plans to undertake to address these causes. The Paper could also outline areas in which the country would welcome investment and/or involvement by sub-national actors, if any. Countries would also have to outline how these activities will conform with provisions to ensure the protection of biodiversity and the rights of indigenous and forest peoples. It is on the basis of the activities outlined in the Paper that the ExComm would disburse any up-front financing to countries. If the ExComm were to find areas of concern in the Strategy Paper, either it or a UN implementing agency would work with the country to resolve the issue.[22]

Financing

Financing of the TDERM would be driven by developed country compliance, as developed countries would be required to meet a fraction of their targets with TDERUs. The price of these units would either be established by auction or be linked to the market price for other Kyoto units. Assuming a carbon price of €20/t CO_2e for example, a 2 per cent REDD target from a 1990 base year would yield €9 billion/year (see Table 7.3). A minimum level would be established to ensure that the Mechanism is adequately financed. TDERUs would be capped at a certain level to ensure that these units do not undermine the market price of carbon.

Many uncertainties, including establishing accurate baselines, supporting additional (beyond business-as-usual) activities and avoiding international leakage, are difficult to resolve simply by system design. To address these concerns, the ExComm would apply a discount rate to the units it issues. Different discount rates would be applied to units issued under the three tracks. Developing countries able to meet a higher level of reporting,

Table 7.3 Illustrative examples of values of different TDER limits

% of 1990 base year developed country industrial gas emissions (22.8 $GtCO_2$e/yr)	Value of TDERUs € Bn/yr at 20€/tCO_2e	TDERUs allowed $MtCO_2$e/yr (Discounted by a factor of 3)[23]	Actual deforestation emission reductions $MtCO_2$e/yr	Deforestation reduction (in million hectares) (150 tCO_2e /ha)	% of deforestation reduction in comparison to average[24]
1%	4.6	228	685	1.24	10%
2%	9.1	456	1369	2.49	19%
3%	13.7	684	2054	3.73	29%
4%	18.2	912	2738	4.97	38%
5%	22.8	1140	3423	6.22	48%

Source: Hare and Macey 2007 (footnotes omitted).

monitoring and verification (thus increasing the certainty of their measured reductions) would be subject to a lower discount rate. This would create an incentive to improve reporting, monitoring and verification over time to secure greater financial benefits.

While the ExComm would sell units to developed countries near the average market rate, it would pay developing countries a price closer to the actual cost of reducing emissions. This price differential would be used to cover the ExComm's expenses, support capacity building efforts, such as the development of the REDD Strategy Papers, and finance an insurance scheme. The design of the system assumes that REDD units will cost less to produce than other market credits. Studies have shown this assumption to be reasonable (for example Sohngen, et al., 2008). The portfolio of REDD activities would be reviewed regularly by the COP or COP/MOP to ensure that the reductions for which the ExComm is issuing units are actually being achieved.

Sub-national actors would participate in the mechanism if allowed by the participating government. The ExComm would pay national governments or sub-national actors directly for reductions achieved after discounting. Governments would report regularly on the state of their emissions at the national level and any activities undertaken. Sub-national actors would report on reductions achieved through their activities. Sub-national actors would be paid for their reductions first, with the national government receiving any leftover. This would create an incentive to apply strict guidelines in the approval of sub-national activities, as sub-national activities that suffer from leakage or other problems but are still successful in obtaining reductions within the boundaries of their activities will reduce the amount of the overall payment made to the participating country for its own efforts.

The different discount rates of the various Tracks may influence the viability of sub-national level activities, especially for Track 3. This is fair. If a full market approach were adopted, the countries in Track 3 would not have the requisite capacity to participate. Furthermore, these countries are likely to be less attractive to investors generally due to their investment climates. The TDERM would strive to achieve equitable regional distribution of REDD activities and funds. A disproportionate share of the ExComm's capacity building funds would likely be invested in Track 2 and 3 countries to prepare them for greater participation in the Mechanism. While Track 3 countries may have less private investment, this does not bar the government from launching its own initiatives such as enhanced enforcement in parks or removal of perverse tax incentives. These types of activities are likely to be unattractive to private investment anyway as their outcomes rely on government performance. By building these design elements into the system, the TDERM strives to avoid repeating the experience under the CDM.

Financing for REDD activities may be capital-intensive in the initial stages of implementation. Project financing, especially for small-scale and more sustainable development-oriented projects, has proven challenging in the CDM (Ellis and Kamel, 2007). Payment for reductions achieved by participating developing countries (and sub-national actors) would only be made after verification that the reductions had been achieved, but loans or other means of financial support could be provided *ex ante*. To ensure accountability and transparency, funds would be transferred to a REDD trust fund in the participating country. The trust fund idea is not new: it has been used in debt-for-nature swap programmes and to support domestic programmes involving payment for environmental services. Access to the capacity building resources of the ExComm could be made directly by the national government or through a UN implementing agency on its behalf. The Kyoto Protocol's Adaptation Fund is a precedent for this type of direct access by national governments (UNFCCC, 2008c, p. 7; Adaptation Fund Board, 2008). Modalities for financial support of sub-national activities should also be considered.

Liability Issues

Under TDERM, national governments would bear the ultimate liability for national-level emissions, but the extent of liability would vary depending on the track. The ExComm's insurance scheme and capacity building resources, resources made available through the price differential on compliance units, would cover instances where emission reductions at the national and sub-national level are out of sync. If for example national baselines remain unchanged or increase despite REDD activities, participating countries would not receive payment from the Mechanism as they would have achieved no reductions. In such cases successful sub-national activities would receive payment, but from the ExComm's insurance scheme not the general budget. No units would be issued or recorded in relation to this payment. There would be no justification for stopping payment as these activities were approved by the national government (and the ExComm in the case of Tracks 2 and 3) and are subject to discount rates. Such cases would serve as an incentive to both entities to develop more robust REDD Strategies and adopt more rigorous reporting to reduce uncertainties in estimating emissions from deforestation.

The extent of liability of national governments varies depending on the chosen track. The higher the track, the greater the liability. Governments in Track 1 would be required to pay back the financial support received if national emissions increased during the relevant accounting period. Subsequent reductions achieved would be eligible for support from the Mechanism. Governments in Tracks 2 and 3 would not be required to pay back the support received, but they would be barred from participating in the Mechanism, including approval of new sub-national activities, until emissions from deforestation returned to previous levels. These countries, particularly Track 3 countries, would have access to some of the ExComm's capacity building resources to assist in this endeavour. Some means must be developed to ensure that governments are not held responsible for events that are not caused by human activity and are beyond their control (for example some forest fires).

Ombudsman

An Ombudsman position would be created to monitor compliance with international standards and serve as an appeal mechanism for sub-national actors. Experience with the sustainable development aspect of the CDM has shown that the market is not an effective mechanism to achieve non-market objectives. Furthermore, stakeholder involvement is lacking. REDD activities, specifically those designed to address the direct causes of deforestation, are intrinsically linked to land. Land tenure conflicts have led to violent disputes in the past (Alston, et al., 2000; de Oliveira, 2008). A REDD mechanism has the potential to affect indigenous peoples and vulnerable communities even more than the CDM, and therefore warrants higher standards and greater participation by all stakeholders.

Standards are worthless without the means to enforce them. Stronger stakeholder consultation provisions should be included in the approval process of sub-national activities than those that exist for the CDM. Affected communities and stakeholders should be able to petition the Ombudsman in cases in which the standards are not being met. The Ombudsman should have the ability to halt payment to activities that do not conform to the standards. The Ombudsman should also issue an annual report detailing how standards are being met and identify those entities – both national and sub-national – that are in non-compliance. This report should be submitted to the COP or COP/MOP for consideration.

States and sub-national actors should have the right to appeal ExComm decisions to the Ombudsman. A similar appeals procedure has been suggested as a means to improve the CDM (UNFCCC, 2008a, Annex III). Sub-national actors would not be able to appeal to the Ombudsman in relation to Track 1 activities, as this approval process would be the sole prerogative of the national government. The Ombudsman could however play a mediation role in such instances.

Capacity Building

All developing countries will require some level of capacity building (CB), both institutional and technical, in order to participate in a REDD mechanism. Many commentators have discussed the need for capacity building or readiness activities to take place before 2013 (UNFCCC, 2007b). Given the time it took countries to establish DNAs under the CDM, the increased technical complexity and demands of measuring and monitoring REDD activities, and the lack of available data and research in many areas, this is an insufficient amount of time to develop the requisite capacity to participate fully in REDD or develop an effective national REDD strategy. Building such capacity requires concerted effort on a dedicated track rather than piecemeal capacity building projects. The capacity building element of the Mechanism would ensure that countries are able to participate and assist them in graduating to a higher track. The goal would be for Track 3 countries to be able to participate in Track 2 activities by the third commitment period.

Developed countries should also be required to contribute financially to capacity building activities in addition to the funds raised by the ExComm through its price differential on units. These activities would contribute to the measuring, reporting and verifying of capacity building required under the Bali Action Plan.[25] Developed countries could choose to contribute their CB-designated resources to the ExComm to disburse or undertake their own bilateral CB-related activities. If the latter approach were adopted, countries would need to report on these activities annually to the ExComm. A similar approach has been used in the Multilateral Fund for the Implementation of the Montreal Protocol (Multilateral Fund Secretariat, 2008, p. 97).

Graduation

A REDD mechanism will have to evolve with the climate regime, including the increasing capacity of countries to act and the changing nature of their commitments. The TDERM's three tracks would respond to increased capacity (internal graduation). The TDERM is also flexible enough to respond to the changing nature of countries' commitments (external graduation). During the second commitment period, we assume that all developing countries for which deforestation is a concern would participate in the TDERM. In the third commitment period, some of the more advanced developing countries may take on greater commitments, similar to those undertaken currently by developed countries. The TDERM is designed to respond to the uncertainties associated with measuring and monitoring and the challenges associated with a market approach. There is however no reason why a country willing and able to adhere to Article 3, paragraphs 3 and 4 of the Kyoto Protocol should not be allowed direct access to the carbon market. Such a country would not participate in the TDERM. Adoption of the same LULUCF rules that govern developed countries should not prejudge other types of commitments this developing country might undertake.

The minimum number of units required for developed country compliance would ensure adequate capitalisation to support the activities of the countries in the TDERM. The maximum number would ensure that countries outside the Mechanism do not flood the market. To set the minimum sufficiently high, one would need to know which countries are not participating in the Mechanism ahead of time. As the overall size of the TDERM would be smaller, there would be proportionally fewer resources for capacity building, insurance and the maintenance of the ExComm. This might affect some of the activities the TDERM is able to undertake.

6. CONCLUDING REMARKS

It is imperative that the international climate regime develop some method of reducing emissions from deforestation in developing countries, which represent 20 per cent of global GHG emissions. To date, discussions have focused on the methodological aspects of REDD. The little discussion there has been of 'policy approaches and positive incentives' has been narrow and incomplete. Deforestation has long been on the agenda of the international community, yet rates of deforestation remain 'alarmingly high'. If the climate regime is serious about addressing REDD, the discussion needs to be broadened to consider the full range of policy issues. Leaving these to the end of the negotiations may result in repeating some of the problems encountered with the CDM. Numerous lessons should be learned from past experience with other forestry initiatives and from the structure and governance of other international mechanisms.

This chapter has outlined the key policy issues that need to be considered in the development of an effective REDD mechanism. It has also examined one area that may provide insight for the design: the CDM. Through a hybrid market-linked fund, the Tropical Deforestation Emissions Reduction Mechanism strives to overcome two of the CDM's major failures: inability to meet non-market goals and highly uneven distribution of projects. The TDERM's three track approach takes into consideration the varying levels of capacity of the countries in which deforestation occurs and ensures that all countries will be able to participate in and benefit from the Mechanism from the beginning. It is each country's sovereign prerogative to decide how to address REDD, but if it wishes to secure international financial support for these activities it should be required to meet international standards to protect biodiversity and the rights of indigenous peoples. There can be no justification for mandating international standards for measuring GHG reductions, but not these other parameters. The CDM's failure to generate large sustainable development benefits for developing countries due to the lack of standards is a key lesson here. The standards and the Ombudsman who oversees compliance with them would be the first steps to ensuring the distribution of benefits to relevant actors. Should a state allow sub-national actors to participate in the Mechanism, the ExComm's authority to pay actors directly would also assist in this regard. The minimum developed country targets would ensure that the fund is adequately capitalised, while the maximum cap on units would ensure that the market price of carbon is not adversely affected. The discount rate applied to TDERUs would minimise the concern about additionality and leakage, while the insurance scheme would guard against volatility in deforestations rates.

Developing comprehensive strategies to address both the proximate and underlying causes of deforestation is the first challenge of REDD and a primary focus for national governments. The key challenge for the international community is to design a mechanism that will support these activities with sufficient and stable financing in a manner that allows developing countries of varying capacities to share in these resources, and protects biodiversity and the rights of indigenous peoples. Neither is a simple or straightforward task. Billions of dollars have already been spent trying to halt deforestation. The climate cannot afford for billions more to be squandered while millions of hectares are converted from sinks to sources. There is only one shot to get REDD right.

NOTES

* Claire Stockwell is a consultant on climate change law and policy. William Hare is a Visiting Scientist at the Potsdam Institute for Climate Impact Research and a Director of Climate Analytics. Kirsten Macey is a policy officer with Climate Analytics.

1. (1992) ILM 31, 848 ('UNFCCC').
2. Kyoto Protocol to the United Nations Framework Convention on Climate Change, 11 December 1997, (1998) ILM 37, 22 ('Kyoto Protocol').
3. Developed countries must include deforestation activities when accounting for emissions and removals from land use, land use change and forestry (LULUCF) (ibid., Art. 3.3). By contrast, due to concerns over leakage (deforestation activities moving outside the project boundaries), baseline uncertainties (deforestation that would have occurred in the absence of the project) and the potential to undermine incentives to

reduce fossil fuel emissions by generating a large volume of low cost credits, the types of forest-related activities that could be undertaken in developing countries via the Clean Development Mechanism (CDM), one of the Protocol's flexibility mechanisms, were limited to afforestation and reforestation (Hare and Macey, 2007). The Convention contains a general obligation to 'promote . . . the conservation and enhancement, as appropriate, of sinks and reservoirs . . . including forests' (UNFCCC, supra note 1, Art. 4.1(d)), but this provision has proven to be insufficient to reduce emissions from deforestation.

4. Conference of the Parties to the UNFCCC, 'Bali Action Plan', Decision 1/CP.13, UN Doc. FCCC/CP/2007/6/Add.1, p. 3 (Bali, 3–15 December 2007) ('Bali Action Plan'). A negotiating track for industrialised countries to deepen their commitments under Kyoto has been in place since 2005. The first commitment period of the Kyoto Protocol ends in 2012. The session in Bali launched a second track under the Convention to discuss mitigation action by developing countries and re-engagement of the US in the international regime.
5. Center for International Forestry Research (CIFOR) (2007), 'New Report Warns Failure to Understand Root Causes of Deforestation Imperils New Efforts to Curb Forest-Based Carbon Emissions', Media Release, www.cifor.cgiar.org/PressRoom/MediaRelease/2007/2007_12_07_redd.htm, 7 December (visited 10 February 2009).
6. Conference of the Parties serving as the meeting of the Parties to the Kyoto Protocol, 'Land use, land-use change and forestry', Decision 16/CMP.1, UN Doc. FCCC/KP/CMP/2005/8/Add.3, p. 5 (Montreal, 28 November–10 December 2005), Annex, para. 1(d).
7. Developed countries are obliged to address deforestation that occurs within their territory under the Kyoto Protocol: see supra note 3, thus this section only considers developed countries' incentive-based or obligation-driven commitments to assist developing countries to reduce emissions from deforestation.
8. National baselines do not however capture international leakage, which occurs when deforestation moves from one country to another. Other means will be needed to address this type of leakage.
9. Due to space constraints, this chapter considers only some of the lessons from the CDM that could be relevant to the design and implementation of a REDD mechanism. Others include defining additionality, access to project finance, transaction costs, modalities for small-scale projects, the ability for communities to engage in the CDM, the efficiency and effectiveness of third party verification, and privileges and immunities of CDM Executive Board members.
10. Conference of the Parties serving as the meeting of the Parties to the Kyoto Protocol, 'Modalities and procedures for a clean development mechanism as defined in Article 12 of the Kyoto Protocol', Decision 3/CMP.1, UN Doc. FCCC/KP/CMP/2005/8/Add.1, p. 17 (Montreal, 28 November–10 December 2005), para. 53.
11. Conference of the Parties to the UNFCCC, 'Modalities and procedures for a clean development mechanism, as defined in Article 12 of the Kyoto Protocol', Decision 17/CP.7, UN Doc. FCCC/CP/2001/13/Add.2, p. 20 (Marrakesh, 29 October–10 November 2001).
12. The pipeline consists of all CDM projects from the validation stage (2,122 projects) through those seeking registration (169) to those already registered (1,033). Numbers in parentheses represent the number of projects in each stage as of 1 May 2008. All projects must be registered with the CDM Executive Board to receive credit for reducing GHG emissions. Data are from J. Fenhann (2008), 'UNEP Risø Centre's CDM/JI Pipeline Analysis and Database', http://cdmpipeline.org (visited 1 April 2009). The database is updated monthly. The 1 May 2008 version is on file with the authors.
13. Ibid.
14. Ibid.
15. See Eddy and Wiser 2002 for a detailed review of public participation in relation to the CDM project cycle and the relevant provisions.
16. 25 June 1998, 2161 UNTS 447 ('Aarhus Convention').

17. Climate Action Network, 'Public Participation in the CDM and JI: Climate Action Network (CAN) Recommendations' (21 July 2000), www.climatenetwork.org/climate-change-basics/cop-9docs/CAN-pubpartrec.pdf (visited 18 February 2009).
18. See the discussion about the challenges of including forest degradation in a mechanism at the beginning of section 3, above.
19. UN General Assembly Resolution 61/295, Declaration on the Rights of Indigenous Peoples (13 September 2007), UN Doc. A/61/L.67/Annex.
20. Conference of the Parties serving as the meeting of the Parties to the Kyoto Protocol, 'Guidelines for the implementation of Article 6 of the Kyoto Protocol', Decision 9/CMP.1, UN Doc. FCCC/KP/CMP/2005/8/Add.2, p. 7 (Montreal, 28 November–10 December 2005), para. 24. JI is the Kyoto Protocol's other project-based flexibility mechanism.
21. 5 June 1992, (1992) ILM 31, 818.
22. The Multilateral Fund for the Implementation of the Montreal Protocol follows a similar approach with respect to the development of its country programmes (ICF Consulting, 2004).
23. Discounting is a pragmatic way to address concerns over uncertainties in establishing accurate baselines, measuring reductions, supporting additional activities and accounting for international leakage. In column 3, a discount factor of 3 has been applied to the emission reductions reported (column 4). These discounted units represent the total number of units that should be allowed into the system to be used for compliance purposes. We have used a discount factor of 3 as an example; no specific discount factor is proposed or advocated here and should be the subject of further work.
24. This column shows the percentage by which the rate of deforestation would be reduced under the TDERM compared to the gross global average rate of deforestation (13 million hectares per year) as estimated by the UN FAO (2006).
25. Bali Action Plan, supra note 4.

REFERENCES

Adaptation Fund Board (2008), *Draft Provisional Operational Policies and Guidelines for Parties to Access Resources from the Adaptation Fund*, Doc. AFB/B.3/8 (Adaptation Fund Board).

Alston, L., et al. (2000), 'Land Reform Policies, the Sources of Violent Conflict, and Implications for Deforestation in the Brazilian Amazon', *Journal of Environmental Economics and Management*, 39, 162.

Barreto, P., et al. (2006), *Human Pressure on the Brazilian Amazon Forests* (World Resources Institute).

Bass, S., et al. (2001), *Certification's Impacts on Forests, Stakeholders and Supply Chains* (International Institute for Environment and Development).

Boyd, E., et al. (2007), *The Clean Development Mechanism: An Assessment of Current Practice and Future Approaches for Policy* (Tyndall Centre for Climate Change Research).

Burian, M. (2006), *The Clean Development Mechanism, Sustainable Development and its Assessment*, Report No. 264 (Hamburg Institute of International Economics).

Contreras-Hermosilla, A. (2000), *The Underlying Causes of Forest Decline* (Center for International Forestry Research).

Contreras-Hermosilla, A. (2007), *Forest Law Enforcement and Governance Program: Review of Implementation* (World Bank).

Cosbey, A., et al. (2006), *Making Development Work in the CDM: Phase II*

of the Development Dividend Project (International Institute for Sustainable Development).
de Oliveira, J. (2008), 'Property Rights, Land Conflicts and Deforestation in the Eastern Amazon', *Forest Policy and Economics*, 10, 303.
Dessie, G. and J. Kleman (2007), 'Pattern and Magnitude of Deforestation in the South Central Rift Valley Region of Ethiopia', *Mountain Research and Development*, 27, 162.
Eddy, N. and G. Wiser (2002), 'Public Participation in the Clean Development Mechanism of the Kyoto Protocol', in C. Bruch (ed.), *The New 'Public': The Globalization of Public Participation* (Environmental Law Institute).
Ellis, J. and S. Kamel (2007), *Overcoming Barriers to Clean Development Mechanism Projects* (Organisation for Economic Co-operation and Development and International Energy Agency).
Fearnside, P. (2005), 'Deforestation in Brazilian Amazonia: History, Rates, and Consequences', *Conservation Biology*, 19, 680.
Geist, H. and E. Lambin (2001), *What Drives Tropical Deforestation? A Meta-Analysis of Proximate and Underlying Causes of Deforestation based on Subnational Case Study Evidence* (LUCC International Project Office).
Geist, H. and E. Lambin (2002), 'Proximate Causes and Underlying Driving Forces of Tropical Deforestation', *BioScience*, 52, 143.
Hare, B. and K. Macey (2007), *Tropical Deforestation Emission Reduction Mechanism: A Discussion Paper* (Greenpeace).
ICF Consulting (2004), *External Evaluation of the Financial Mechanism of the Montreal Protocol* (ICF Consulting).
Kanninen, M., et al. (2007), *Do Trees Grow on Money? The Implications of Deforestation Research for Policies to Promote REDD* (Center for International Forestry Research).
Kelly, C. and N. Helme (2000), *Ensuring CDM Project Compatibility with Sustainable Development Goals* (Center for Clean Air Policy).
Lambin, E. and H. Geist (2003), 'Regional Differences in Tropical Deforestation', *Environment*, 45, 22.
Lee, M.-K., et al. (2004), *CDM Information and Guidebook* (2d edn, UNEP Risø Centre).
Leimgruber, P., et al. (2005), 'Forest Cover Change Patterns in Myanmar (Burma): 1990–2000', *Environmental Conservation*, 32, 356.
Matthews, E. (ed.) (2002), *The State of the Forest: Indonesia* (Forest Watch Indonesia and Global Forest Watch).
McMorrow, J. and M.A. Talip (2001), 'Decline of Forest Area in Sabah, Malaysia: Relationship to State Policies, Land Code and Land Capability', *Global Environmental Change*, 11, 217.
Millennium Ecosystem Assessment (2005), *Ecosystems and Human Well-Being: Biodiversity Synthesis* (World Resources Institute).
Molnar, A., et al. (2004), *Who Conserves the World's Forests? Community-Driven Strategies to Protect Forests and Respect Rights* (Forest Trends and Ecoagriculture Partners).
Multilateral Fund Secretariat (2008), *Multilateral Fund for the Implementation of the Montreal Protocol: Policies, Procedures, Guidelines and Criteria*, April 2008 version (United Nations Environment Programme).
Nepstad, D., et al. (2001), 'Road Paving, Fire Regime Feedbacks, and the Future of Amazon Forests', *Forest Ecology and Management*, 154, 395.

Nepstad, D., et al. (2002), 'Frontier Governance in Amazonia', *Science*, 295, 629.

Olsen, K. (2007), 'The Clean Development Mechanism's Contribution to Sustainable Development: A Review of the Literature', *Climatic Change*, 84, 59.

Olsen, K. and J. Fenhann (2008), 'Sustainable Development Benefits of Clean Development Mechanism Projects: A New Methodology for Sustainability Assessment based on Text Analysis of the Project Design Documents submitted for Validation', *Energy Policy*, 36, 2819.

Penman, J., et al. (eds), *Definitions and Methodological Options to Inventory Emissions from Direct Human-induced Degradation of Forests and Devegetation of Other Vegetation Types* (Intergovernmental Panel on Climate Change).

Rogner, H.-H., et al. (2007), 'Introduction', in B. Metz, et al. (eds), *Climate Change 2007: Mitigation. Contribution of Working Group III to the Fourth Assessment Report of the Intergovernmental Panel on Climate Change* (Cambridge University Press).

Schneider, L. (2007), *Is the CDM Fulfilling its Environmental and Sustainable Development Objectives? An Evaluation of the CDM and Options for Improvement* (WWF).

Silayan, A. (2005), *Equitable Distribution of CDM Projects Among Developing Countries*, Report No. 255 (Hamburg Institute of International Economics).

Sohngen, B., et al. (2008), 'Avoided Deforestation as a Greenhouse Gas Mitigation Tool: Economic Issues', *Journal of Environmental Quality*, 37, 1368.

Stern, N. (2007), *The Economics of Climate Change: The Stern Review* (Cambridge University Press).

Sunderlin, W. and I. Resosudarmo (1996), *Rates and Causes of Deforestation in Indonesia: Towards a Resolution of the Ambiguities* (Center for International Forestry Research).

Sutter, C. (2003), *Sustainability Check-Up for CDM Projects: How to Assess the Sustainability of International Projects under the Kyoto Protocol* (Wissenschaftlicher Verlag).

Sutter, C. and J. Parreño (2007), 'Does the Current Clean Development Mechanism (CDM) Deliver its Sustainable Development Claim? An Analysis of Officially Registered CDM Projects', *Climatic Change*, 84, 75.

United Nations Development Programme (UNDP) (2006), *The Clean Development Mechanism: An Assessment of Progress* (UNDP).

United Nations Food and Agriculture Organization (UN FAO) (2006), *Global Forest Resources Assessment 2005: Progress Towards Sustainable Forest Management* (UN FAO).

United Nations Framework Convention on Climate Change (UNFCCC) (2005), *Reducing Emissions from Deforestation in Developing Countries: Approaches to Stimulate Action*, UN Doc. FCCC/CP/2005/MISC.1 (UNFCCC).

United Nations Framework Convention on Climate Change (UNFCCC) (2006a), *Background Paper for the Workshop on Reducing Emissions from Deforestation in Developing Countries – Part II: Policy Approaches and Positive Incentives*, Working paper No. 1(b) (UNFCCC).

United Nations Framework Convention on Climate Change (UNFCCC) (2006b), *Report of the Conference of the Parties on its Eleventh Session*, UN Doc. FCCC/CP/2005/5 (UNFCCC).

United Nations Framework Convention on Climate Change (UNFCCC) (2007a), *Report on the Analysis of Existing and Potential Investment and Financial*

Flows relevant to the Development of an Effective and Appropriate International Response to Climate Change, Dialogue Working Paper No. 8 (UNFCCC).
United Nations Framework Convention on Climate Change (UNFCCC) (2007b), *Views on Issues Related to Further Steps under the Convention related to Reducing Emissions from Deforestation in Developing Countries: Approaches to Stimulate Action*, UN Doc. FCCC/SBSTA/2007/MISC.14 (UNFCCC).
United Nations Framework Convention on Climate Change (UNFCCC) (2008a), *Report of the Ad Hoc Working Group on Further Commitments for Annex I Parties under the Kyoto Protocol on its resumed Fifth Session*, UN Doc. FCCC/KP/AWG/2008/3 (UNFCCC).
United Nations Framework Convention on Climate Change (UNFCCC) (2008b), *Report of the Ad Hoc Working Group on Long-term Cooperative Action under the Convention on its Third Session*, UN Doc. FCCC/AWGLCA/2008/12 (UNFCCC).
United Nations Framework Convention on Climate Change (UNFCCC) (2008c), *Report of the Conference of the Parties serving as the Meeting of the Parties to the Kyoto Protocol on its Third Session*, UN Doc. FCCC/KP/CMP/2007/9/Add.1 (UNFCCC).
United Nations Framework Convention on Climate Change (UNFCCC) (2008d), *Views on Outstanding Methodological Issues related to Policy Approaches and Positive Incentives to Reduce Emissions from Deforestation and Forest Degradation in Developing Countries*, UN Doc. FCCC/SBSTA/2008/MISC.4 (UNFCCC).
United Nations Forum on Forests (UNFF) (2004), *Recent Developments in Existing Forest-related Instruments, Agreements, and Processes*, Background Document No. 2 (UNFF).
Uryu, Y., et al. (2008), *Deforestation, Forest Degradation, Biodiversity Loss and CO_2 Emissions in Riau, Sumatra, Indonesia* (WWF Indonesia).
WWF (2004), *Are Protected Areas Working? An Analysis of Forest Protected Areas* (WWF).

8. The role of marine 'forests' and soils as carbon sinks: enhanced bio-sequestration as a mitigation strategy to help avoid dangerous climate change

Robert Fowler*

1. INTRODUCTION

One of the most critical questions facing humanity today is: 'What will it take to avoid dangerous climate change'?[1] This question is at the forefront of negotiations currently being undertaken by the international community concerning a new intentional regime with respect to climate change. The adoption of the United Nations Framework Convention on Climate Change[2] (UNFCCC 1992) in 1992 was a tentative first step towards the development of comprehensive and effective strategies to combat climate change. As a 'framework' Convention, it inevitably avoided specific targets. The Kyoto Protocol represented a significant step forward, by establishing initial targets for the reduction of CO_2 emissions by some 39 countries listed in Annex 1 to the UNFCCC.[3] But these targets were never envisaged to address climate change fully; rather, they were developed as an interim measure that it was hoped would result in the reduction of global CO_2 emissions by 5 per cent from their 1990 levels by 2012.

At the Thirteenth Conference of the Parties (COP) to the UNFCC in Bali, Indonesia in December 2007, an Action Plan was agreed with the aim of developing a 'post-Kyoto' agreement at the 15th COP in Copenhagen in December 2009. The principal objective of this agreement, according to the Bali Action Plan, would be to establish a 'shared vision for long-term cooperative action, including a long-term global goal for emissions reductions, to achieve the ultimate objective of the Convention' (UNFCCCa, 2008). Thus, the negotiation process leading to Copenhagen has two broad objectives:

- first, to establish the targets and timetables relating to reductions in the emission of CO_2 and other greenhouse gases (GHGs) that can achieve a 'safe' long-term level of concentration of these gases in the Earth's atmosphere; and
- second, to identify and implement the specific strategies by which the agreed targets and timetables can be achieved.

This chapter considers how each of these objectives might be achieved, and includes an assessment of what contribution might reasonably be expected from developing countries in this regard. It is not focused therefore on the existing international legal regime for addressing climate change, except insofar as elements such as the Clean Development Mechanism (CDM) under the Kyoto Protocol might provide a suitable vehicle for the delivery of new or additional strategies to deliver the fundamental goal of stabilisation of GHG concentrations at a safe level.

With respect to the first objective concerning targets and timetables, section 2 of this chapter addresses in some detail the science of climate change in order to identify what target for the stabilisation of GHG concentrations should be adopted by the international community. Reaching an agreement on this target is a necessary precursor to the task of identifying specific targets and timetables for GHG emissions reductions. More specifically, the chapter will describe the scientific assessments contained in the IPCC's Fourth Assessment Report in late 2007 (IPCCa, 2007) and some scientific reports that have emerged since then which suggest that much more stringent reductions targets are necessary than those canvassed by the IPCC. The need to take into account ongoing growth in emissions in developing countries when setting reductions targets for developed countries is also discussed.

In relation to the second objective of identifying strategies that will deliver the agreed reduction targets, it will be suggested in section 2 of this chapter that international negotiators should give greater attention to the role of enhanced bio-sequestration as a mitigation strategy, in light of recent scientific evidence that indicates a potentially significant capacity for bio-systems to absorb CO_2 at greater levels than at present. In particular, the possible contribution of sea-grasses and soils in this regard will be discussed. Once again, the possible involvement of developing countries in such strategies will also be considered.

The underlying theme of this chapter is that there is a need to find the means of bridging the substantial gap between what the scientific evidence indicates is necessary to avoid dangerous climate change and the level of commitment to emission reductions that is evident currently on the part of both developed and developing countries. This gap is being exacerbated by the rapid development of new and disturbing scientific observations, in particular that a 'tipping point' may already have been reached with respect to global warming (Hansen, et al., 2008). It is in this context that enhanced bio-sequestration may offer an additional pathway to the ultimate goal of avoiding dangerous climate change, thereby also offering opportunities to developing countries in particular to limit their future emissions whilst also continuing to improve living conditions for their citizens.

2. ESTABLISHING TARGETS AND TIMETABLES TO AVOID 'DANGEROUS' CLIMATE CHANGE

The IPCC's Fourth Assessment Report

Causes and effects of climate change

The most obvious scientific reference for the negotiators of the post-Kyoto agreement is the Fourth Assessment Report of the IPCC, released in November 2007 just prior to the UN Climate Change Conference in Bali (IPCC, 2007a). Its principal conclusion concerning the existence of human-induced global warming is not qualified in the same manner as in earlier reports:

> Warming of the climate system is unequivocal, as is now evident from observations of increases in global average air and ocean temperatures, widespread melting of snow and ice and rising global average sea level. (IPCCb, 2007, p. 12).

The IPCC also concluded that: '[m]ost of the observed increase in globally-averaged temperatures since the mid-20th century is *very likely* due to the observed increase in anthropogenic GHG concentrations' (IPCC, 2007b, p. 5).

In assessing the possible extent of human-induced global warming, the IPCC canvassed six scenarios for GHG emissions from 2000 to 2100 and projected the likely global average temperature change associated with each scenario. These indicate potential increases of between 1.8 and 4.0 degrees Celsius (° C) from 2000 levels by 2100 as a 'best estimate', but within a 'likely' range of 1.1–6.4° C (IPCCb, 2007, p. 5 – Figure SPM5). As will be seen below, there is a widely-held view that it will be necessary to restrict warming to around 2° C to have a reasonable chance of avoiding dangerous climate change.

An important aspect of the IPCC report is its predictions with respect to sea-level rise associated with global warming, and in this regard the IPCC actually retreated from some of its earlier, tentative estimates. The

2007 report predicts that sea-levels may rise by somewhere between 18 and 59 centimeters by 2100 (IPCC, 2007b, p. 8 – Figure SPM1) and virtually excludes the possibility of significant melting of the Antarctic ice sheet. It also does not envisage an immediate, major threat from the melting of the Greenland ice sheet: '[c]urrent models suggest virtually complete elimination of the Greenland ice sheet and a resulting contribution to sea level rise of about 7m if global average warming were sustained for millennia in excess of 1.9 to 4.6° C relative to pre-industrial values' (IPCC, 2007b, p. 12).

The IPPC (ibid., p. 13) does, however, acknowledge the possibility that sea-levels could rise more rapidly over the coming centuries:

> Partial loss of ice sheets on polar land could imply metres of sea level rise, major changes in coastlines and inundation of low-lying areas, with greatest effects in river deltas and low-lying islands. Such changes are projected to occur over millennial time scales, but more rapid sea-level rise on century time scales cannot be excluded.

It will be suggested below that this aspect of the IPCC assessment is particularly open to challenge in terms of its failure to take into account more recent evidence of melting of ice sheets in the polar regions, with the consequence that its predictions of likely sea-level rise during the current century may be seriously inadequate.

Setting a global GHG stabilisation target

For the purpose of considering targets and timetables for the stabilisation of GHG concentrations, the most relevant information is provided in the final section of the IPCC's Summary for Policymakers, which addresses the 'long-term perspective'. A specific link to Article 2 of the UNFCCC is provided at the commencement of this section through the observation that determining what constitutes 'dangerous anthropogenic interference with the climate system' involves value judgments, whilst also noting that 'science can support informed decisions on this issue' (IPCC, 2007b, p. 18). This is an important caveat inasmuch as it suggests that the ultimate choice of mitigation targets and strategies will involve a subjective, political decision rather than one based solely on the scientific evidence.

This section of the Summary for Policymakers employs six 'stabilisation scenarios', each of which constitutes a range for the concentration of CO_2 at long-term, stable levels (see Table SPM6). Category I comprises the lowest mitigation scenario (350–400 parts per million (ppm)), and essentially represents the status quo, given that CO_2 levels were measured at 384 ppm in 2007.[4] The highest mitigation scenario (Category VI) envisages a future concentration of CO_2 of between 660–790 ppm.

In order to account for other GHGs (methane, nitrous oxide and others), each of the stabilisation scenarios is converted into a CO_2-equivalent (CO_2-eq) concentration. For Category I, a range of 445–490 ppm CO_2-eq is indicated, with the report noting in a footnote to Table SPM.6 that 'the best estimate of total CO_2-eq concentration in 2005 for all long-lived GHGs is about 455 ppm' (IPCC SPM, 2007, p. 20). Thus, a stabilisation target of approximately 450 ppm CO_2-eq – which has been widely promoted by environmental non-government organisations since the release of the IPCC report – would require a lowering in concentration of GHGs from current levels, following a temporary higher peak due to the momentum in climate change. The concept of 'peak and decline' as a basis of achieving a global stabilisation target is explained in the following statement in the IPCC Summary (IPCC, 2007b, p. 19): '[i]n order to stabilise the concentration of GHGs in the atmosphere, emissions would need to peak and decline thereafter. The lower the stabilisation level, the more quickly this peak and decline would need to occur.' Table SPM6 identifies 'peaking years' for each of the six stabilisation scenarios, with 2015 being identified as the latest year by which emissions can peak if the Category I stabilisation scenario is to be achieved.

In choosing a particular stabilisation level, the most significant consideration is widely regarded to be the likely average temperature increase that will be associated with such a level. Figure SPM.11 of the IPCC's Summary for Policymakers provides an illustration of the relationship between the six stabilisation targets and the likely equilibrium global average temperature increase above pre-industrial levels. It suggests that the 'best estimate' for Category I is an average increase of just over 2° C; for Category 2, which assumes a concentration of about 550 ppm CO_2-eq, the best estimate is in the vicinity of 3° C. Considerable reliance has been placed on these estimates recently to advocate emissions reduction targets that will limit global warming to 2°C.

It should be noted that there is a wide range of temperature increase indicated in Figure SPM.11 for each scenario, reflecting the upper and lower bounds of 'likely' temperature increases. Thus, even Category I allows for the possibility of an increase of up to almost 4° C. Also, a footnote to Table SPM.6 makes the important qualification that these estimated temperature increases will achieve an equilibrium level over a few centuries. Therefore, they cannot be assumed to be applicable over the next 50–100 years, and indeed the footnote adds that 'estimates for the evolution of temperature over the course of this century are not available . . . for the stabilization scenarios'. Nevertheless, as noted above, the IPCC report does canvass elsewhere a potential increase of 1.8–4.0° C by 2100, depending on the level of emissions produced until then.

These are important concessions in terms of the level of current scientific understanding of the critical link between particular stabilisation targets and related likely temperature increase over the next 40–50 years. The correlation between a stabilisation target of 450 ppm and a limit in increase of global average temperatures of 2–2.4° C above pre-industrial levels is far from strong, and there is an acknowledged possibility that, even within this target, temperature may increase by up to almost 4° C.

Setting targets for GHG emissions reductions

The most critical question in terms of targets and timetable is the level of reduction in emissions of GHGs required to achieve a desired stabilisation target. The difficulty with the principal GHG, CO_2, is that it has a relatively long life in the atmosphere, so that to the extent that emissions are not able to be sequestered naturally in oceans or forests they will accumulate for many years (CO_2 will reduce by about one-third after ten years in the atmosphere, by two thirds after 100 years and by four-fifths after 1,000 years). It is the momentum from historic emissions of CO_2 that renders the task of achieving a low stabilisation target extremely difficult. This also explains why the longer the delay in reducing CO_2 emissions, the greater the eventual challenge will be in terms of achieving a target such as 450 ppm CO_2-eq.

Table SPM.6 in the IPCC Summary for Policymakers provides a summary of the peaking years and reduction in CO_2 emissions required by 2050 (compared to 2000 emissions) for each of the six stabilisation scenarios. For Scenario 1, it indicates that emissions of CO_2 would need to peak by 2015 and to be reduced by 50–80 per cent by 2050. To achieve this significant level of reductions, it will be necessary for developed countries (identified by reference to Annex 1 to the UNFCCC) to reduce emissions even more substantially in order to accommodate unavoidable growth in emissions in developing countries. In the 2007 report of the IPCC's Working Group III, it is suggested that to achieve a stabilisation target of 450 ppm CO_2-eq, reductions would be required from Annex 1 countries in the range of 25–40 per cent by 2020 and 80–95 per cent by 2050, compared to 1990 values (IPCC, 2007c, p. 776 – Box 13.7).

It is important to note the 'baseline' that is used to measure cuts against. Whilst Table SPM.6 calls for a 50–85 per cent reduction from 2000 levels by 2050, the WG III report identifies reduction targets that are set against the 1990 baseline adopted by the Kyoto Protocol. Thus, a reduction of 25–40 per cent by 2020 against a 1990 baseline will be more substantial than the same reduction measured against a 2000 baseline, given the increase in emissions between these years. In the current negotiations, it appears that negotiators prefer to refer to a 2000 baseline for reduction targets, but this may then require even larger cuts by developed countries than those suggested by WG III for 2020 and 2050 respectively.

Although the suggested levels of cuts in emissions for the Annex 1 countries under the 450 ppm stabilisation target are very substantial, even in the period between now and 2020, the IPCC Summary for Policymakers concludes that 'all stabilization levels assessed can be achieved by deployment of a portfolio of technologies that are either currently available or expected to be commercialized in coming decades' (IPCC, 2007b, p. 20). The Summary also suggests that '60–80% of the reductions would come from energy supply and use, and industrial processes, with energy efficiency playing a key role in many scenarios', but acknowledges that 'non-CO_2 and CO_2 land-use and forestry mitigation options provide greater flexibility and cost-effectiveness' (IPCC, 2007b, p. 20). It will be suggested in section 3 of this chapter that there is a far greater potential for bio-sequestration strategies to contribute to the goal of stabilising GHG concentrations than has been acknowledged by either the IPCC or those involved in the negotiation of the post-Kyoto agreement.

The IPCC WG III Report also recognises the need for mitigation efforts by many developing countries between now and 2020 if the lowest mitigation target of 450 ppm CO_2-eq is to be achieved (IPCC, 2007c, p. 775):

> Under most of the considered regime designs for low and medium stabilization levels, the emissions from developing countries need to deviate – as soon as possible – from what we believe today would be their baseline emissions, even if developed countries make substantial reductions. For the advanced developing countries, this occurs by 2020 (mostly Latin America, Middle East and East Asia).

This is proving to be a particularly contentious issue for negotiators of the post-Kyoto agreement. As will be discussed further below, emissions from developing countries such as China, India, Indonesia and Brazil have increased significantly in recent years and, if not restrained in the near future, could largely counteract any reductions achieved by developed countries. If global targets for both GHG concentrations and emission reduction targets are to be effective, it may therefore be necessary to include all nations in a scheme of targets and timetables under the proposed new international regime (Hodas and Mumma, 2008). Such an approach could involve reductions on a lesser scale than those required of developed countries for the so-called 'advanced' developing countries and even moderate increases for other developing countries, thus still keeping faith with the principle of 'common but differentiated responsibility' that is intended to underpin the climate change regime.

Despite the emerging, strong scientific consensus concerning the extent of the reductions required to achieve a stabilisation target of 450 ppm CO_2-eq, it is far from clear that developed countries or the advanced developing countries are prepared to make the required commitments in Copenhagen. The European Commission (2009) has shown some leadership on this issue by expressing a commitment to a reduction of 30 per cent in CO_2 levels from a 1990 baseline by 2020 provided comparable reductions are agreed by other developed countries; it has also suggested that developing countries should accept reductions as a group of 15–30 per cent by 2020. Other countries have been much more cautious in indicating their proposed targets. For example, the Australian government (2008) appears likely to commit to a 2020 target of only 5–15 per cent reductions in CO_2 levels from 2000 levels, with the final figure within this range depending on whether a comprehensive agreement is reached in Copenhagen. This position was announced in December 2008, following an independent expert review by Professor Ross Garnaut (2008) in which it was concluded that whilst a 450 ppm target is desirable, it is more realistic to aim for a 550ppm target in Copenhagen.

Whilst it may be pragmatic to suggest that a 450 ppm CO_2-eq stabilisation target will prove to be unrealistic politically in Copenhagen, it is likely that the success of the negotiations in Copenhagen in December 2009 will be measured by many observers against such a standard. However, recent scientific opinion expressed since the publication of the IPCC's Fourth Assessment Report suggests that even the 450 ppm target may be insufficient to avoid dangerous climate change, and that a stricter stabilisation target of 300–350 ppm CO_2 may be necessary. This emerging viewpoint serves to illustrate vividly the substantial gap that exists between the rapidly emerging science of climate change and the relevant developing law and policy.

Recent Reassessments of Targets and Timetables Required to Avoid 'Dangerous Climate Change'

Since the release of the IPCC's Fourth Assessment Report, a new wave of scientific opinion has emerged which strongly suggests that the IPCC's lowest stabilisation scenario (Category I), upon which the target of 450 ppm CO_2-eq is based, will not be capable of achieving the objective of avoiding dangerous climate change. The implication of this recent science is that greater levels of emissions reduction will be required than were contemplated by the IPCC in even the most stringent scenario considered in its Fourth Assessment Report. The most recent papers and reports highlight four aspects of the IPCC assessment that are considered questionable or out-of-date.

Acceleration of CO_2 emissions

A report published in September 2007 indicates that the growth in CO_2 emissions, especially since 2000, has been much faster than anticipated by researchers (Canadell et al., 2007). The report notes that during the period 1990–1999, the growth rate in CO_2 emissions averaged 1.3 per cent per year; however, this rose to 3.3 per cent in the period 2000–2006, driven to a large extent by the rapid expansion in energy production and use in China and India. The International Energy Agency, in its World Energy Outlook 2007 report released in November 2007, notes that between 2005 and 2007 China's emissions expanded to the extent that it has now overtaken the United States (US) as the world's largest emitter of CO_2. The Agency also notes that India will be the world's third largest emitter of CO_2 by 2015 under current trends (International Energy Agency, 2007).

It is not appropriate, however, to lay the blame solely on the rapid increase in emissions in these particular developing countries, given that a number of Annex 1 countries appear likely to be unable to meet the relatively modest commitments they entered into under the Kyoto Protocol. Canada, for example, is almost certainly incapable of meeting its Kyoto target for the current compliance period of 2008–2012, whilst Australia, which enjoyed the dubious privilege under the Kyoto Protocol of being allowed to increase its emissions during this first compliance period, will only be able to meet its target as a result of a significant reduction in the clearance of native vegetation that will offset its substantially increased industrial emissions of CO_2.

It has been suggested that the recent high growth rate in CO_2 emissions 'exceeds that in the most fossil fuel intensive emissions scenarios used by the IPCC' (Climate Institute, 2007, p. 3). Thus, GHG concentrations may rise more rapidly than predicted by the IPCC to levels that will result in 'dangerous climate change'; correspondingly, the capacity to achieve the IPCC's most rigorous stabilisation target, as depicted in the Category 1 scenario, could be significantly less than that envisaged by the IPCC. In short, the trends in CO_2 emissions indicate that the need for urgent mitigation strategies is much greater than was suggested by the IPCC report.

Declining carbon sinks

Recent studies indicate that the capacity of both the oceans and terrestrial ecosystems to serve as 'sinks' for CO_2 is decreasing, thereby contributing further to an acceleration in CO_2 concentrations beyond the rate of increase in emissions (Canadell et al., 2007; Climate Institute, 2007, p. 7; Hengeveld, 2008). It is argued that the IPCC has not taken this trend sufficiently into account in its predictions: 'IPCC science expected this decrease but the observed changes are larger than estimated, suggesting that the carbon cycle is resulting in stronger and earlier warming than anticipated' (Climate Institute, 2007, p. 7).

Proposals to enhance ocean CO_2 sequestration by iron fertilisation have been found in a recent Canadian experiment not to be feasible (Hengeveld, 2008, p. 70), despite efforts by several corporations to pursue such projects in developing countries where there were lax environmental assessment standards.[5] However, as will be discussed more fully below, other means of bio-sequestration not currently being employed may be able to assist in reversing this trend.

Accelerated warming and an associated 'tipping point'

Since the IPCC report in 2007, there has been a common assumption by international negotiators and national policy-makers that a 450 ppm CO_2-eq stabilisation level would restrict the amount of atmospheric warming since the industrial revolution to around 2° C. However, this assumption is not justified even by the IPCC report which, as noted previously, allows for the possibility of a rise of up to 4° C over this century. The evidence is growing that a 450 ppm stabilisation level could result in substantially more than a 2° C increase in average global temperature.

The US climate change scientist, James Hansen, has recently suggested that the Earth's climate system is about twice as sensitive to CO_2 pollution as indicated in the IPCC's century-long projections. He argues that current levels of GHGs are sufficient to trigger significant impacts: '[n]o additional forcing is required to raise global temperature to at least the level of the Pliocene, 2–3 million years ago, a degree of warming that would surely yield 'dangerous' climate impacts (Hansen, et al., 2008, p. 6).

Hansen notes that a rise of 0.8° C has already occurred since 1860, that another 1° C increase is inevitable given current levels of CO_2, and concludes that a much sharper increase could result from 'feedback' events, including:

- release of methane from the thawing of the perma-frost regions;
- additional release of CO_2 from drying and burning vegetation;
- release of stored CO_2 from carbonates and frozen methane deposits in the deep ocean; and
- major albedo effects associated with melting ice sheets and reduced aerosol levels.

It is thought that any or all of these events could rapidly drive up GHG concentrations, producing a 'tipping point' sufficient to cause catastrophic climate change (Glickson, 2008). Hansen defines a 'tipping point' as 'the concept that climate can reach a point where, without additional forcing, rapid changes proceed practically out of our control' and concludes that:

> Earth's history shows that the positive feedbacks allow global warming to be relatively rapid, including sea level rise as fast as several meters per century. Thus if humans push the climate system sufficiently far into disequilibrium, positive climate feedbacks may set in motion dramatic climate change and climate impacts that cannot be controlled (Hansen, et al., 2008, p. 11).

These alarming analyses have not been factored into the IPCC's stabilisation scenarios, which effectively take for granted a rise of at least 2° C whilst at the same time suggesting that limiting global warming to this extent will provide a good chance of avoiding dangerous climate change. The recent work of Hansen and others suggests that it may be necessary to restrict warming to around 1.5° C to achieve this objective, which will mean avoiding some of the increase in temperature currently anticipated from existing emissions. Again, only urgent and substantial mitigation strategies could possibly deliver this goal.

Ice-melt and associated sea-level rise

The most dramatic indication that climate change is progressing at a much faster rate than any of the IPCC's 2007 predictions arises from observations of melting sea ice in the Arctic ocean, land-based ice melt in Greenland and collapsing ice shelves in West Antarctica.[6] Hansen is particularly critical of the IPCC's projections in relation to sea-level rise (Hansen, et al., 2007, p. 21):

> Our concern that BAU (business as usual) GHG scenarios would cause large sea-level rise this century differs from estimates of IPCC, which foresees little or no contribution to twenty-first century sea-level rise from Greenland and Antarctica. However, the IPCC analyses and projections do not well account for the nonlinear physics of wet ice sheet disintegration, ice streams and eroding ice shelves, nor are they consistent with the palaeoclimate evidence we have presented for the absence of discernible lag between ice-sheet forcing and sea-level rise.

Hansen's recent research departs from the use of climate models to predict future climate trends and instead utilises evidence from the Earth's history, based on the study of cores taken from the bottom of the ocean, to measure CO_2 levels millions of years ago. These indicate that the world began to glaciate at the start of the ice age, around 35 million years ago, when the concentration of CO_2 was about 450 ppm. Hansen believes de-glaciation is already commencing and argues that it is possible that substantial increases in sea-level could follow in as little as a few decades, rather than the centuries contemplated by the IPCC:

> Sea level changes of several meters per century occur in the paleoclimate record, in response to forcings slower and weaker than the present human-made forcing. It seems likely that large ice sheet response will occur within centuries, if human-made forcings continue to increase. Once ice sheet disintegration is underway, decadal changes of sea level may be substantial (Hansen, et al., 2008, p. 6).

In a paper published in *Science* in September 2008, the first assessment of how changes to the ice sheet may affect sea-level rise has been presented. It concludes that a rise of between 80 cms and 2 metres could occur this century 'under physically possible glaciological conditions' (Pfeffer, et al., 2008). This assessment reinforces the arguments advanced recently by Hansen on the basis of his paleontological research.

Conclusions on recent scientific reports

The overall effect of these recent scientific observations is to cast some serious doubt on the accuracy of the IPCC projections, including the adequacy of its most stringent scenario for the stabilisation of GHG concentrations. With CO_2 concentrations already at 384 ppm, and being driven upwards at an accelerating rate, the strategy of limiting warming to just 2° C by stabilising concentrations at about 450 ppm CO_2-eq is now highly questionable. Hansen notes that the forcing effect of GHGs other than CO_2 is currently being off-set to a large extent by the cooling effect of high albedo aerosols. He argues that even if these aerosols are eliminated in the future through enhanced pollution controls, their loss could be counteracted by significant reductions in another GHG, black soot, through the same controls. He therefore concludes that an initial stabilisation target should be focused on CO_2 levels and set at 350 ppm, but 'to be reassessed as effects on ice sheet mass balance are observed' (Hansen, et al., 2008, p. 11).

It appears that a 350 ppm target for CO_2 concentrations is well below the contemplation of the negotiators of the post-Kyoto agreement. Instead, as noted above, some countries such as Australia have indicated their willingness to accept an initial stabilisation target of 550 ppm CO_2-eq in Copenhagen, which would allow for a further increase in CO_2 concentrations above their current levels. The Hansen target, on the other hand, requires radical mitigation strategies, including the phase-out of coal by 2030. He suggests that this 'zero-carbon' strategy could reduce CO_2 concentrations to 400 ppm and that accompanying bio-sequestration of CO_2 (through improved forestry and agricultural practices) will be required in order to reduce concentrations by a further 50 ppm or more (Hansen, et al., 2008, p. 14 – figure 6).

Nevertheless, the idea of a 350 ppm target for CO_2 concentrations is gathering support in the non-government sector: see, for example, the campaign launched in June 2008 to call on international leaders to adopt such a target (available at www.350.org). George Monbiot, who has written about climate change for over 20 years (Monbiot, 2006), has endorsed the 350ppm CO_2 target and suggested 'we are talking at a minimum of a 100% cut [in CO_2 emissions], and it looks like it might have to go to 110% or 115%'.[7] This extends the mitigation strategy from a 'zero carbon' to a 'negative carbon' scenario, something that must boggle the minds of those charged with negotiating a post-Kyoto agreement. In a similar vein, both Al Gore[8] and Lester Brown (2008) have recently urged cuts in the order of 80 per cent by 2020 in the US in order to reduce CO_2 concentrations to a safe level.

The message behind the calls for a 350 ppm target is brutally simple but highly unpalatable for policy-makers: it is necessary for all developed nations to shift to a low carbon economy (Epstein, et al., 2008; Flavin, 2008), or possibly even a zero carbon economy (Makhijani, 2007) within the next 20–30 years, whilst in addition finding the means to 'soak up' additional CO_2 from the atmosphere in natural sinks (to create, in effect, a 'negative carbon' economy) if dangerous climate change is to be avoided.

The de-carbonisation of developed national economies will require radical innovations with respect to energy systems, with a focus initially on energy efficiency and ultimately on alternative, renewable sources. These strategies will also need to be steadily deployed in developing countries over the same period of the next 10–20 years, with the support of schemes involving technology transfer and financial assistance. Nuclear power, the other option often touted as an alternative to fossil fuels, is susceptible to serious risks in relation to nuclear proliferation, terrorism, serious accidents and management of wastes (Makhijani, 2007, p. 11). Furthermore, it is also questionable whether the additional nuclear energy capacity can be developed within the necessary time scale; also, any benefits derived from nuclear energy will be offset initially by the substantial GHG emissions involved in the construction of nuclear facilities.

The need to supplement these energy-related strategies with additional measures focused on forests and soils, as suggested by Hansen, is the subject of section 3 of this chapter. In particular, the potential for bio-sequestration to contribute to the attainment of the 350 ppm target will be considered. Given that there is evidence of a declining capacity of natural sinks to absorb CO_2, as noted above, the question that arises is how can bio-sequestration possibly contribute to a reduction of CO_2 concentrations in the order of 50 ppm in the future? It will be suggested that opportunities exist in this context that have not yet been adequately appraised by negotiators of the post-Kyoto agreement.

3. STRATEGIES TO ADDRESS CLIMATE CHANGE – THE ROLE OF BIO-SEQUESTRATION

Linking Targets to the Global Carbon Cycle

To examine the potential role of bio-sequestration as a mitigation strategy, it is necessary first to explain the global carbon cycle – that is, the process by which carbon is emitted and absorbed from natural sources, in particular the oceans and forests. This understanding is critical to the calculation of appropriate targets for the reduction of greenhouse gas emissions, since it enables negotiators to take into account the amount of CO_2 that is likely to be sequestered naturally.

A recent Canadian government publication that summarises 25 years of research on the carbon cycle provides an excellent overview of the global sources and sinks of GHGs (Canada, 2008). The introductory chapter indicates that whilst global, anthropogenic emissions of CO_2 from fossil fuels and land-use have averaged 8.6 gigatons (Gt C) per year from 2000–2005 (and, as noted before, have increased by 3 per cent per annum during this period), the net addition to the atmosphere is approximately 3.4–4.4 Gt C per annum after allowing for the operation of natural sinks (Hengeveld, 2008, pp. 4–5).

The report indicates that the oceans comprise the largest reservoir of CO_2 and have provided a net sink since 1990 of about 2.2 Gt C per year. However, as noted above, there are serious concerns about the declining capacity of the oceans to serve as a carbon sink, which suggest that it is not safe to assume that this level of uptake will continue in the future. Terrestrial ecosystems have provided an additional sink of 2–3 Gt C per year, though this reduces to 1Gt C/year if emissions from deforestation and other land use change are taken into account (Hengeveld, 2008, pp. 4–5). As these figures reflect, land-use change, largely associated with deforestation, accounts for approximately 20 per cent of global CO_2 emissions at present. Therefore, strategies to reduce emissions from deforestation and other forms of forest degradation could contribute significantly to the continued bio-sequestration function of these natural sinks. This is discussed further below.

It is important also to understand that there are substantial variations from year to year in the natural uptake of CO_2: '[t]he amount of excess carbon removed will vary considerably from one year to the next, so that some years experience less than 20% removal while others experience more than 70% removal. That is because the processes for removal are sensitive to changes and variations in global climate' (Hengeveld, 2008, p. 2). The figures cited above from this recent Canadian report serve to explain the basis of the emissions reduction targets suggested by the IPCC and which are currently under consideration by negotiators of the post-Kyoto agreement, as discussed above. If global emissions of CO_2 from fossil fuels and land-use change can be reduced to half of their current level of 8.6 Gt C per annum, this should achieve a situation where emissions no longer exceed the rate of natural uptake – assuming that the sequestration capacity of the biosphere does not decline significantly in the meantime. Indeed, the US-based Union of Concerned Scientists has proposed a mitigation strategy based on this calculation (Luers, et al., 2007). This would involve a peaking of the atmospheric concentration of CO_2 somewhere above the current level of 385 ppm, the exact point depending on when and to what extent the current trend of a net annual increase in CO_2 emissions is reversed. Meinhausen, for example, has suggested that CO_2 emissions should peak at 8 Gt C/ year (a figure slightly below the latest estimations of their current levels), then decline to 3 Gt C/year by 2050 and 1 Gt C/year by 2100 for a peak concentration of 470 ppm CO_2-eq, declining to below 400 ppm CO_2-eq eventually (Meinhausen, et al., 2005). It is this broad strategy that underpins the IPCC's most rigorous scenario for achieving stabilisation of GHGs at a level that will have a strong chance of limiting global warming to 2° C.

However, if account is taken of the most recent scientific advice to the effect that CO_2 concentrations should be reduced much more radically to about 350 ppm in order to avoid the risk of dangerous climate change, it will be necessary to:

(1) reduce global CO_2 emissions even further, for example by 75–80 per cent of current levels to around 2 Gt. C per annum, which could mean having to achieve virtually zero energy-related emissions in developed countries in order to allow for some increase in emissions in developing countries; and
(2) adopt strategies for increasing the current level of removal (sequestration) of CO_2 from the atmosphere, thereby achieving a 'negative-carbon' emissions situation.

Both of these approaches clearly are essential if the more radical stabilisation target is adopted. However, if the enhancement of natural sinks is used primarily to 'offset' ongoing emissions from industrial sources, this will detract considerably from the efforts to achieve such a target. In this sense, allowing carbon off-sets under the CDM for reforestation or afforestation projects may have short-term benefits by allowing more time for the transition to low or non-carbon technologies, but it will not contribute directly to the achievement of a long-term, radical stabilisation goal for CO_2.

Options for the Sequestration of CO_2

Carbon sequestration strategies can contribute both to the reduction of CO_2 emissions from various sources and to the increased uptake of emitted CO_2 via enhanced natural sinks. It is the latter aspect that is of particular interest in this chapter, but some brief attention will also be directed to the first function, particularly since this is attracting more attention currently from negotiators of the post-Kyoto regime than is enhanced bio-sequestration. Specifically, there is a strong focus on two sequestration-related emissions reduction strategies: carbon capture and storage, particularly with respect to coal-fired power plants; and avoidance of emissions from deforestation and forest degradation (referred to commonly as 'REDD').

Carbon capture and storage

Considerable research effort is being devoted to the feasibility of carbon capture and storage ('CCS') technologies as a means of achieving the long-term sequestration of CO_2 underground in geological formations or in the deep ocean environment, thereby reducing emissions to the atmosphere from sources such as coal-fired power stations. A substantial report on this subject by the IPCC in 2005 concluded that CCS has the 'economic potential' to reduce CO_2 emissions cumulatively by 60–600 Gt C by 2100, or 15–55 per cent of the cumulative mitigation effort required during this period (Metz, 2005). An MIT report in 2007 recommended that US government grants be provided to energy companies to fund CCS projects on a trial basis at several new power plants and also advocated US government funding for several, large-scale geologic injection projects (Deutch, et al., 2007). However, the latter strategy was dealt a serious blow when the US Department of Energy withdrew $950 million in funding for the proposed FutureGen 'clean coal' project in January 2008 (due to projected increases in the overall cost of the project to $1.8 billion) and instead called for proposals for a number of separate, smaller demonstration projects.[9]

Whilst there appears to be some consensus that carbon capture technology is already available for new coal plants at a relatively low end cost (Hawkins and Peridas, 2007, p. 5), it is the storage aspect that poses greater challenges in terms of both cost and capacity. In Australia, the Federal government has indicated that 'the cost of a GHG transport, injection and storage project would be in terms of hundreds of millions or some billions of dollars'.[10] Also in Australia, a joint project by Rio Tinto and British Petroleum to develop a deep-sea storage bed for CO_2 off the coast of Western Australia was abandoned because the area concerned did not have the capacity to hold the volume of CO_2 the companies wished to bury (Peating and Wilkinson, 2008). These factors may preclude the early adoption of CCS projects on a wide scale.

There is also a range of complex legal issues to be addressed in relation to CCS, both with respect to the development of CO_2 'performance measures' for carbon capture and in relation to the storage of CO_2. In the latter context, these include the definition of property rights, site licensing and monitoring, long-term responsibility for storage sites and liability/compensation rules (Deutch, et al., 2007). In Australia, the Federal government introduced the Offshore Petroleum Amendment (Greenhouse Gas Storage) Act 2008 to regulate CO_2 storage in offshore, Federal waters, but the draft legislation fails to make clear who is responsible in the long term for the proposed deep-sea storage sites (Peating and Wilkinson, 2008).

The interest in developing 'clean coal' methods of energy production is driven by the pressure, particularly from the coal industry, to continue the use of coal as an energy source. It appears inevitable that this pressure will result in CCS attracting significant levels of interest and funding from governments. However, it is arguable that funding priorities for research into mitigation strategies might be better directed to the development of alternative, renewable energy sources and innovative approaches to enhanced bio-sequestration (as discussed further below). It is interesting to note also that proposals to include CCS under the CDM in the proposed post-Kyoto international agreement appear likely to be rejected as a result of strong objections from countries such as Brazil, which argue that 'the CDM is supposed to be about clean development, not subsidizing fossil fuels' (Wilson, 2008). If CCS projects do not qualify under the CDM, there will be significantly less prospect of them becoming viable as a mitigation strategy.

REDD schemes

As noted above, deforestation currently accounts for approximately 20 per cent of global emissions of CO_2. One country, Indonesia, is now the world's third largest emitter of CO_2, primarily due to the substantial logging of its forests, much of which has occurred illegally. However, the current climate change legal framework fails to address this issue in any specific way. Article 4 of the UNFCCC requires parties to develop and update inventories of GHG emissions and to include therein emissions and removals from land use, land use change and forestry (LULUCF), but only the Annex I countries are obliged to report on LULUCF emissions and removals in terms of meeting their obligations with respect to emissions targets. Proposals several years ago to include 'avoided deforestation' as an eligible activity under the CDM were rejected due to methodological concerns at that time in relation to additionality, permanence, baseline setting and leakage (Fry, 2008, p. 166).

However, a revised concept of 'reducing emissions from deforestation and forest degradation', or 'REDD', has been given fresh attention since 2005, particularly following the UNFCCC COP 13 in Bali in December 2007. The concept has been advanced vigorously by developing countries such as Papua New Guinea, Costa Rica and Indonesia which have experienced substantial deforestation and are prepared to consider a scheme in which payments would be received from parties seeking carbon credits in return for the protection and sustainable management of their forests (Fry, 2008, p. 167).

The Bali Action Plan calls for: '[p]olicy approaches and positive incentives on issues relating to reducing emissions from deforestation and forest degradation in developing countries; and the role of conservation, sustainable management of forests and enhancement of forest carbon stocks in developing countries' (UNFCCC, 2008a). Methodological concerns previously raised are being addressed through the Subsidiary Body for Scientific and Technological Advice (SBSTA) established under the Kyoto Protocol. At a workshop in Japan in June 2008, the SBSTA identified a number of such issues, including estimating and monitoring changes in forest cover, reference emissions levels and displacement of emissions; it also identified 'cross-cutting' issues such as non-permanence, comparability and transparency, uncertainties in estimates and implications for indigenous peoples and local communities (Earth Negotiations Bulletin, 2008).

During a meeting in Accra in August 2008 of the Ad Hoc Working Group on Long-Term Cooperative Action (AWG-LCA) established under the UNFCCC, a workshop on REDD considered the SBSTA report. The workshop report called for further work by the Convention parties on the use of non-market financial resources and market-based mechanisms and also with respect to the issues of permanence, additionality and displacement of emissions (UNFCCC, 2008b). It is clear that considerable effort will be directed by negotiators of the post-Kyoto agreement to addressing these issues in the lead-up to COP 15 in Copenhagen in December 2009.

The effectiveness of REDD schemes in a global context as a means of reducing CO_2 emissions will depend in particular on whether there is a capacity to 'offset' emissions in other locations under such schemes, pursuant to the CDM or some equivalent mechanism. The report of the Accra workshop notes a divergence of views currently on this issue (UNFCCC, 2008b, p. 2):

> Some parties opposed the use of forest mitigation activities in developing countries to meet emissions reduction commitments of industrialized countries . . . efforts should be additional to the overall mitigation efforts by developed

countries, in order to attain the stabilization of GHG concentrations in the atmosphere.

This viewpoint is strongly supported by the author, on the basis of the analysis previously offered of the reduction and stabilisation targets that may need to be adopted in order to avoid dangerous climate change. It should also be noted that whilst REDD schemes are primarily intended as a strategy to reduce the rate of future emissions of CO_2, recent scientific studies have suggested that the rate of uptake of CO_2 in mature trees is far greater than previously thought (Science Daily, 2008), thus underlining the importance of this strategy in contributing also to the continued uptake of CO_2 through bio-sequestration.

Enhancement of existing carbon sinks

REDD schemes that are focused on the avoidance of future deforestation will not 'enhance' the existing capacity of forests as natural sinks, but it is interesting to note that the Bali Action Plan also calls for 'policy approaches and positive incentives on the role of conservation, sustainable management of forests and enhancement of forest carbon stocks in developing countries'. It appears that this is a separate and additional strategy option besides REDD schemes (Fry, 2008, p. 167).

The call for the enhancement of forest carbon stocks raises afresh the role of afforestation and reforestation as a mitigation strategy. Such activities may be allowed to count towards emissions reduction targets for Annex I parties under the Kyoto Protocol, subject to a cap on the number of credits able to be claimed. However, less than 1 per cent of CDM projects to date have related to afforestation and reforestation activities.[11] A concern is that any net increase in carbon stocks may be achieved through plantations of introduced species or monocultures whilst deforestation continues with respect to natural forests – with China being cited as an example (Fry, 2008, p. 181). As a result, there is resistance, particularly on the part of environmental NGOs, to the use of this mitigation strategy. Also, as with CCS, there is a limit to the capacity of afforestation and reforestation projects to provide for enhanced CO_2 uptake, particularly given that much deforested land has been converted to agricultural use – often for the purpose of oil palm production in countries such as Indonesia, and in conditions where the long-term viability of this land-use is unlikely.

The limitation in the Bali Action Plan of an enhanced bio-sequestration strategy to *forest* carbon stocks is understandable insofar as the context for the particular reference was forestry activities more generally. However, there may be other means of enhancing the capacity of the earth's natural sinks, in particular those provided by both the Earth's oceans and soils. How such approaches might fit within the legal framework of a post-Kyoto regime, if attention were to be directed to them by negotiators alongside the discussion of forestry options, is of particular interest. It is these possibilities that will be addressed in the final part of this chapter.

Enhancement of ocean sinks: the role of sea-grasses Oceans perform a major role in carbon sequestration, with some 39,000 Gt C being stored in the intermediate and deep oceans, including sediments on the ocean floor, whilst the upper surface waters store an additional 900 Gt C (Hengeveld, 2008, p. 2). It has been noted already that there are concerns that global warming may be reducing the capacity of the oceans to maintain their function as a major carbon sink, leading in turn to proposals to enhance the uptake of CO_2 in the oceans by artificial means such as iron fertilisation. However, these have been found not to be practical (Canada, 2008, p. 69).

Instead of engineered approaches, it may be preferable to consider options which, like afforestation/reforestation in the terrestrial environment, utilise the natural components and functions of the marine biosphere. For example, recent scientific studies (described below) suggest that sea-grass meadows serve a significant function as carbon sinks. Given the substantial destruction of sea-grass colonies in many parts of the world in recent decades, largely due to land-based pollution runoff and coastal development, an effort to 're-afforest' these colonies and to protect those that remain could be a significant parallel strategy to those contemplated for terrestrial forest systems.

A study undertaken for the South Australian Environment Protection Authority noted that over 25,000 hectares of sea-grass meadows have been subject to catastrophic loss in the southern Australian coastal environment (Balance Carbon, 2007). The study noted that the dominant species of seagrass in this region (Posidonia Australis, which is similar to the Posidonia Oceanica species found in the Mediterranean region) has a large proportion of its biomass (up to 70 per cent) stored in the sea-bed, most of which is likely to be stable provided the above ground plant material is not destroyed. The report suggests that bio-sequestration of CO_2 for decades to centuries could be achieved if environmental conditions can be maintained so that sea-grass plants can remain intact above their stored carbon. Based on another recent study (Mateo, et al., 2006), the South Australian report concludes that 'sea-grasses appear to be one of the most productive plant types with capacity to sequester carbon on the planet', with a global sequestration potential of almost 2.4 million tons of CO_2 per year (Balance Carbon, 2007).

This is a most interesting possibility, not only in terms of mitigation of climate change but also in relation to marine biodiversity protection, should it be possible to reverse the substantial losses of sea-grass colonies that have occurred in recent decades in many regions. In particular, it may offer a strategy for carbon sequestration in many developing countries, such as Indonesia, where substantial destruction of marine coastal environments has occurred in recent years. It could be possible, first, to extend the REDD concept to existing sea-grass colonies and, second, to allow carbon credits for 'reafforestation' of destroyed sea-grass colonies. In both instances, this would involve an extension of the legal mechanisms that may be developed in relation to terrestrial forestry initiatives under the post-Kyoto agreement. Whilst such an approach would involve the same issues with respect to measurement of carbon uptake and long-term storage as apply in relation to forests, there may be considerably less complexity in terms of land tenure and ongoing management of the relevant resources.

Considerable attention is now being focused by climate change negotiators on the role of terrestrial forests in contributing to climate mitigation strategies, but the potential contribution of 'marine forests' in the form of sea-grass meadows does not seem to have yet been recognised. It may be that there is a lack of awareness of the relevant science that indicates the sequestration potential of sea-grasses, or simply that forestry issues have a much higher profile politically than do marine conservation issues. Nevertheless, the protection of existing sea-grass colonies from further destruction and the regeneration of lost colonies through appropriate marine protection and conservation strategies could provide an important additional mitigation approach, particularly for developing countries, and deserves closer attention by negotiators.

Enhancing the role of soils as carbon sinks It is estimated that at the beginning of the industrial revolution, terrestrial ecosystems stored some 2,300 Gt C, with 80 per cent contained in soils and surface litter, the rest being in above ground vegetation (Hengeveld, 2008, p. 2). In Canada, the conversion of natural ecosystems to croplands has caused the loss of 15–30 per cent of the carbon originally present in the surface soil layer, mostly during the first two decades of cultivation. However, it is thought that 'soil carbon is reaching a new equilibrium, largely due to smaller amounts of land being converted to cropland, decreases in summerfallow, increases in no-till farming and increased fertilizer use' (Hengeveld, 2008, p. 8).

Of course, the situation is substantially different in other parts of the world, particularly in developing countries, where conversion to cropland (for example for the production of biofuels) is proceeding at a rapid pace.

It seems not to be fully appreciated that such practices lead not only to a loss of stored CO_2 from the surface vegetation that has been cleared, but also from the soils on which this vegetation depends. In this regard, measures to avoid further deforestation can have an additional effect with respect to carbon stored in the underlying soil. Similarly, measures to avoid soil degradation will serve to protect the carbon that remains stored therein, thus providing an additional rationale for their development and effective implementation.

As with the marine environment, scientists are now exploring ways of enhancing the capacity of soils to sequester CO_2. For example, an Australian team of scientists has suggested that by selecting particular strains of crops such as wheat and sorghum CO_2 can be locked up in 'plantstones' (otherwise referred to as 'phytoliths') which form around a plant's cells as they take up minerals from the soil (Nowak, 2009). These microscopic balls of silica are virtually indestructible, so that after a plant dies they remain behind in the soil where they can sequester CO_2 for thousands of years. Thus, as the report notes, it might be possible for farmers to earn income from carbon credits by changing to crops that store the most CO_2 – provided, once more, that reliable methods of measuring the increased carbon uptake can be developed. For developing countries, selection of agricultural crops that have a capacity to promote CO_2 uptake might provide another avenue for the generation of income from carbon credits.

A more radical and potentially far-reaching option that has been proposed recently is to turn biomass into 'bio-char' and thereby store very large amounts of carbon in soils for centuries or even thousands of years. A short paper released by the Institute for Governance and Sustainable Development (IGSD) in March 2008 provides a concise summary of the recent scientific literature concerning bio-char.[12] It notes that soils enriched with bio-char contain substantially more carbon (150 gC/kg) than their surrounding soils (20–30 gC/kg), and on average are more than twice as deep, thus enhancing further their carbon storage capacity. It also notes suggestions that up to 9.5 Gt C per year could be stored by 2100 using bio-char sequestration schemes, in combination with bio-fuel programmes. This exceeds the current total emissions of CO_2 annually.

This idea is based on an ancient Amazonian technique for the production of the fine-grained residue known as bio-char that involves covering biomass with soil and letting it smoulder. A modern chemical process known as pyrolosis can enable biomass to be converted to bio-char whilst at the same time yielding a liquid or gas bio-fuel. Thus, there may be a double benefit in terms of both enhanced CO_2 uptake and the generation of an alternative source of energy.

The IGSD paper claims that bio-char schemes 'have the potential to be implemented quickly and at scale in developing countries', and suggests that a conversion in such countries from 'slash and burn' methods of land clearance to 'slash and char' could reduce emissions from land-use change by 12 per cent per annum. Other benefits include reduced emissions of methane and nitrous oxides from soils, restoration of degraded soils and increased crop yields[13] (IGSD, 2008).

There may be serious questions to be resolved once more with respect to how to measure accurately the uptake of CO_2 though this soil sequestration technique, and also as to how to source the substantial volumes of biomass required for producing bio-char on a major scale.[14] It also remains to be seen whether the bio-char method is feasible across different soil types from those found in the Amazon Basin. Nevertheless, the potential to achieve a significant enhancement of soil sinks appears to exist, and it would seem appropriate for further research and policy appraisal of this option to be pursued.

Whilst governments and industry are prepared to invest very substantial sums in 'clean coal' via CCS, the use of bio-char to increase the bio-sequestration of CO_2 does not appear to have a high level of interest or support amongst negotiators of the post-Kyoto agreement. It is interesting to note, however, that the Secretariat of the UN Convention to Combat Desertification has proposed in recent submissions to the Ad Hoc Working Group on Long-term Cooperative Action under the UNFCCC that biochar be included in the CDM under the Kyoto Protocol as a way for developed countries to earn certified emissions reduction credits towards meeting their emissions targets.[15] This may reflect an increasing awareness of, and interest in, this option on the part of negotiators.

4. CONCLUSIONS

The underlying theme of this chapter is that the development of strategies to avoid dangerous climate change requires a strong linkage between science and law and policy, particularly in the negotiation of a post-Kyoto international agreement. For both developed and developing countries, the emerging science of climate change presents fresh challenges with respect to the positions they may adopt in this negotiation process.

In particular, the objective of stabilising CO_2 and other GHGs at concentrations in the atmosphere that are likely to limit global warming sufficiently for this purpose may require far greater reductions in emissions from current levels than was envisaged by the IPCC in the most stringent scenario presented in its Fourth Assessment Report in 2007. Thus, it may be necessary for firmer obligations to be assumed by developing countries than have been considered politically acceptable until now, whilst at the same time developed countries may need to aim for a zero or even a negative carbon economy within the next 20–30 years to accommodate some ongoing growth in emissions in the developing world.

These observations are based on very recent scientific reports that suggest that a stabilisation target of 300–350 ppm CO_2 may be required to avoid potentially disastrous feed-back loops that could, for example, trigger significant warming and related sea-level rise over the course of this current century. The challenge for the negotiators of a post-Kyoto agreement in this respect is to deal with potential risks that may be of a low likelihood but which could have catastrophic consequences – a task that is not readily accommodated by the complex and convoluted climate change international negotiation process in which traditional North/South divisions have prevailed. A means of moving beyond these divisions to adopt a precautionary approach, whilst still respecting the principle of common but differentiated responsibility, needs to be found by the time negotiators reach Copenhagen in December 2009 for COP15.

Given that a target of 300–350 ppm is below the current concentration of atmospheric CO_2 and that the momentum associated with the longevity of CO_2 in the atmosphere means that concentrations must inevitably continue to rise for some time yet, a goal of achieving near-zero fossil fuel emissions of CO_2 over the next 20–30 years in developed countries, coupled with strategies to achieve reductions from current levels in countries such as China, India, Brazil and Indonesia, and to limit the growth of emissions in other developing countries is extremely challenging. In addition to a focus on the reduction of fossil fuel emissions, it would be necessary simultaneously for negotiators to find new means of extracting at least a portion of the accumulated concentrations of CO_2 from the atmosphere.

In this latter respect, whilst 'engineered' approaches such as carbon capture and storage and the artificial stimulation of the ocean sink have been canvassed, the enhancement of the functions of natural sinks is only being recognised presently by negotiators of the proposed post-Kyoto regime in the context of forests. There may be considerable potential for much greater gains in the natural uptake of CO_2 by the oceans through the replenishment of seagrass colonies that have been severely depleted in coastal environments over recent decades. There appears also to be potential for a substantial increase in the sequestration of CO_2 in soils through the production of bio-char and its application to agricultural soils in particular. Additional reductions in emissions could also be achieved where liquid or gas by-products of the pyrolosis process are captured for use as bio-fuels. Strategies focused on enhanced bio-sequestration could therefore assist considerably in the delivery of a 350 ppm CO_2 stabilisation target if they were to be accommodated within the post-Kyoto legal framework.

There are two ways in which this outcome might be achieved. First, carbon uptake through these avenues could be taken into account in calculating overall emissions for the purpose of determining compliance with specified emissions targets. This may apply in the future not only to the industrialised Annex I Countries but also to emerging industrial economies such as China, India and Brazil. Second, it may be possible to extend the CDM beyond its current focus on the uptake of CO_2 by forest ecosystems, so as to allow for other sinks such as marine sea-grass 'forests' and soils. Once again, this may be of particular benefit to developing countries, not only in relation to the generation of income from carbon credits but also in enhancing the sustainable management of their coastal marine and soil environments. There is still time for negotiators to take on board such options, but a heightened level of awareness of the relevant science and a willingness to extend thinking about sequestration to other contexts than just forests and CCS are required.

NOTES

* Adjunct Professor in the Law School at the University of South Australia.

1. The term 'dangerous climate change' is adapted from Article 2 of the UNFCCC, which provides that:

> The ultimate objective of this Convention and any related legal instruments that the Conference of the Parties may adopt is to achieve, in accordance with the relevant provisions of the Convention, stabilization of greenhouse gas concentrations in the atmosphere at a level that would prevent dangerous anthropogenic interference with the climate system.

2. (1994) ILM 31, 849.
3. (2005) ILM 37, 32.
4. J. Russell (2008), 'Climate Change Accelerates', in *Vital Signs Online*, Worldwatch Institute, at www.worldwatch.org/node/5634 (verified 26 January 2009).
5. R. Warner, 'Preserving a Balanced Ocean: Environmental Regulation of Climate Change Mitigation Activities in the Offshore Marine Areas of Developing Countries', paper presented at IUCN Academy of Environmental Law Conference on *Climate Law in Developing Countries*, 26–28 September 2008, Ottawa, Canada, www.iucnael.org/content/view/93/30/lang,english/ (verified 26 January 2009).
6. Supra note 4.
7. G. Monbiot (2007), Speech at the Camp for Climate Change, London, 18 August 2007, www.indymedia.org.uk/en/2007/08/378866.html (verified 26 January 2009).
8. A. Gore (2008), 'A Generational Challenge to Repower America', speech delivered in Washington DC, 17 July 2008, http://dotearth.blogs.nytimes.com/2008/07/17/the-annotated-gore-climate-speech/ (verified 26 January, 2009).
9. M. LaMonica (2008), 'DOE Scraps FutureGen "Clean Coal" Project for New Tack', 30

January, http://news.cnet.com/8301-11128_3-9861473-54.html?tag=rtcol;relnews (verified 27 January 2009).
10. Commonwealth of Australia (2008), Explanatory Memorandum, *Offshore Petroleum Amendment (Greenhouse Gas Storage) Bill 2008,* www.austlii.edu.au/au/legis/cth/bill_em/opagsb2008520/memo_0.html (verified 27 January, 2009).
11. CDM (2008), Clean Development Mechanism web-site, http://cdm.unfccc.int/Statistics/Registration/RegisteredProjByScopePieChart.html, (verified 27 January 2009).
12. IGSD (2008), Institute for Governance & Sustainable Development, 'Significant Climate Mitigation from Bio-Char', 17 March 2008, www.igsd.org/docs/Biochar%20Note%2010July08.pdf (verified 27 January 2009).
13. Ibid.
14. For a vigorous criticism of biochar on these and other grounds see a recent declaration, 'Biochar, a new big threat to people, land, and ecosystems', signed by a large number of environmental non-government organisations and published by Friends of the Earth Australia, www.foe.org.au/news/2009/2018biochar2019-a-new-big-threat-to-people-land-and-ecosystems (verified 11 April 2009).
15. Secretariat, UN Convention to Combat Desertification, 'Required policy actions to Include carbon contained in soils including the use of biochar (charcoal) to replenish soil carbon pools, and restore fertility and sequester CO2', submission to the 5th Session of the Ad Hoc Working Group on Long-term Cooperative Action under the UNFCCC Convention (AWG–LCA 5), Bonn, Germany, April 2009, www.unccd.int/publicinfo/AWGLCA5/UNCCD_2nd_submission_land_soils_and_UNFCCC_process_05Feb.pdf (verified 11 April 2009).

REFERENCES

Australian Government (2008), *Carbon Pollution Reduction Scheme: Australia's Low Pollution Future,* White Paper (Department of Climate Change).

Balance Carbon (2007), *Australian Seagrass Meadows as Potential Carbon Sinks: Focus on Gulf St. Vincent, South Australia* (Report prepared for the Environment Protection Authority, South Australia) (on file with the author).

Brown, L.R., et al. (2008), *Time for Plan B: Cutting Carbon Emissions 80 Per Cent by 2020* (Earth Policy Institute).

Canada (2008), *Enhancement of Greenhouse Gas Sinks: A Canadian Science Assessment*, PERD POL 6.2.1 Final Report, Minister of Public Works and Government Services, Canada, available at Canadell et al. (2007).

Canadell, J. et al. (2007), 'Contributions to Accelerating Atmospheric CO_2 Growth from Economic Activity, Carbon Intensity, and Efficiency of Natural Sinks', *Proceedings of the National Academy of Sciences,* September 2007.

Climate Institute (2007), *Evidence of Accelerated Climate Change* (University of Melbourne).

Deutch, J., et al. (2007), *The Future of Coal: Options for a Carbon-Constrained World* (Massachusetts Institute of Technology).

Earth Negotiations Bulletin (2008), 'Summary of the UNFCCC Workshop on Methodological Issues Relating to Reducing Emissions from Deforestation and Forest Degradation in Developing Countries: 25–27 June 2008', *Earth Negotiations Bulletin* 12(376), 30 June.

Epstein, P.R., et al. (2008), *Healthy Solutions for the Low Carbon Economy – Guidelines for Investors, Insurers and Policy Makers* (Centre for Health and the Global Environment, Harvard Medical School).

European Commission (2009), Communication from the Commission to the European Parliament, The Council, the European Economic and Social Committee and the Committee of the Regions, *Towards a Comprehensive Climate Change Agreement in Copenhagen,* COM(2009)39, 28 January 2009.
Flavin, C. (2008), 'Low-Carbon Energy: A Roadmap', *Worldwatch Report* 178 (Worldwatch Institute).
Fry, I. (2008), 'Reducing Emissions from Deforestation and Forest Degradation: Opportunities and Pitfalls in Developing a New Legal Regime', *Review of European Community and International Environmental Law* 17(2), 166.
Garnaut, R. (2008), *Garnaut Climate Change Review. Final Report* (Cambridge University Press).
Glickson, A.Y. (2008), 'Milestones in the Evolution of the Atmosphere with Reference to Climate Change', *Australian Journal of Earth Sciences* 55, 365.
Hansen, J., et al. (2007), 'Climate Change and Trace Gases', *Philosophical Transactions of the Royal Society* 365.
Hansen, J., et al. (2008), 'Target Atmospheric CO2: Where Humanity Should Aim?', *Open Atmospheric Science Journal* 2, 217.
Hawkins, D. and G. Peridas (2007), 'No Time Like the Present: NRDC's Response to MIT's 'Future of Coal', *NRDC Brief* (Natural Resources Defense Council).
Hengeveld, H. (2008), 'An Overview of the Background Science and Goals of Canadian Research Activities Relevant to Enhancement of Greenhouse Sinks in Canada', in Environment Canada, *Enhancement of Greenhouse Gas Sinks: A Canadian Science Assessment* (Environment Canada).
Hodas, D. and A. Mumma, (2008), 'Designing a Global Post-Kyoto Climate Change Protocol that Advances Human Development', *Georgetown International Environmental Law Review* 20(4), 619.
International Energy Agency (IEA) (2007), *World Energy Outlook 2007 – China and India Insights* (IEA).
Intergovernmental Panel on Climate Change (IPCC) (2007a), *Fourth Assessment Report: Climate Change 2007* (Cambridge University Press).
Intergovernmental Panel on Climate Change (IPCC) (2007b), *Fourth Assessment Report: Climate Change 2007, Summary for Policymakers* (Cambridge University Press).
Intergovernmental Panel on Climate Change (IPCC) (2007c), *Mitigation of Climate Change,* Working Group III (Cambridge University Press).
Luers, A., et al. (2007), *How to Avoid Dangerous Climate Change: A Target for US Emissions Reductions* (Union of Concerned Scientists).
Makhijani, A. (2007), *Carbon-Free and Nuclear-Free* (Institute for Energy and Environmental Research).
Mateo, M.A., et al. (2006), 'Carbon Flux in Seagrass Ecosystems', in A.W. Larkum, et al. (eds), *Seagrasses: Biology, Ecology and Conservation* (Springer).
Meinhausen, M. et al. (2005), 'Multi-Gas Emissions Pathways to Meet Climate Targets', *Climatic Change* 573.
Metz, B. (2005), et al., *Carbon Dioxide Capture and Storage* (Cambridge University Press).
Monbiot, G. (2006), *Heat: How to Stop the Planet Burning* (Penguin Books).
Nowak, R. (2008), '. . . While Others Lock Carbon Away for Years', *New Scientist,* 5 January, 9.
Peating, S. and M. Wilkinson (2008), 'Hope Still High on Carbon Burial', *Sydney Morning Herald,* 19 May.

Pfeffer, W.T., et al. (2008), 'Kinematic Constraints on Glacier Contributions to 21st Century Sea-Level Rise', *Science* 321, 1340.

Science Daily (2008), 'Old Growth Forests are Valuable Carbon Sinks', *Science Daily*, 14 September.

UNFCCC (2008a), *Report of the Conference of the Parties on its Thirteenth Session, held in Bali from 3 to 15 December 2007*, FCCC/CP/2007/6/Add.1, 14 March.

UNFCCC (2008b), Ad Hoc Working Group on Long-Term Cooperative Action Under the Convention, *Report of the Workshop on Policy Approaches and Positive Incentives on Issues Relating to Reducing Emissions from Deforestation and Forest Degradation in Developing Countries; and the Role of Conservation, Sustainable Management of Forests and Enhancement of Forest Carbon Stocks*, FCC/AWGLCA/2008/CRP.5, 25 August.

Wilson, P. (2008), 'Brazilians Kill Off Aussie-led Proposal on Carbon Capture', *The Australian*, 11 December, 6.

9. Adaptation to climate change to save biodiversity: lessons learned from African and European experiences

Saja Erens, Jonathan Verschuuren and Kees Bastmeijer*

1. INTRODUCTION

Climate change is increasing the pressure on the dwindling biodiversity of the Earth. In 2004, the Conference of the Parties (COP) to the Convention on Biological Diversity (CBD)[1] requested all Parties to 'integrate climate change adaptation measures in protected area planning, management strategies, and in the design of protected area systems'.[2] IUCN experts have explained that certain organisms 'will move along altitudinal gradients in response to climate change' and that establishing networks of protected areas may in most parts of the world be crucial for species of plants and animals to adapt to climate change.[3]

In the current battle to minimize actual and projected climate change impacts on persistently declining biodiversity, a great deal of faith is placed in the creation of ecological networks and large protected areas. We examine the role played by these two approaches in current legislative efforts to address the negative effects of climate change on biodiversity. While recognizing the urgent need to designate and protect marine areas to conserve marine biodiversity, we focus here on the terrestrial environment.

The goal of comparing African and European experiences is not to decide which continent is doing a 'better job'. This would require development of an objective framework to make judgements, but this is not what is emphasized. Furthermore, such an approach is likely to disrespect the differences between the two regions, for instance in respect of ecological values, climate change sensitivity or the intensity and characteristics of human activities. The goal of comparing and contrasting is to find inspiration in the legal elaboration of adaptation measures to save biodiversity.

Before discussing experiences in Europe and Africa, we first consider the challenges for biodiversity due to a changing climate in general and the importance of large natural areas and ecological networks in particular. Next, we canvas relevant developments, first in international law, then in Europe and finally in Africa. We present our main conclusions in the final section.

2. BIODIVERSITY AND ADAPTATION TO CLIMATE CHANGE: SIZE AND CONNECTIVITY OF NATURAL AREAS

The potential consequences of climate change for biodiversity are receiving increasing attention worldwide. Changing temperatures and related changes such as decreasing availably of fresh water in certain areas and changing weather patterns (Hannah, et al., 2002a, p. 264) pose challenges to animal and plant species and various types of habitat. Research provides increasing information on these challenges, but there are still many gaps in our knowledge. What is certain is that the challenges for biodiversity are diverse. For instance, climate change may influence habitat conditions (e.g., availability of fresh water) in such a way that the area is no longer a suitable habitat for certain species. Climate change may also influence animal and plant biorhythms: plants may flower earlier in the year, young may be born earlier and animals may start to migrate earlier or later than they used to do. This may cause problems in the food chain. Ecological relationships may be disturbed by overpopulation of a particular species, caused by a warmer climate. A warmer climate may permit alien species to survive and spread in areas that were previously too cold for them. In addition, climate change may have very specific influences for certain species; for instance, '[s]ome reptile species exhibit temperature-dependent sex determination during egg incubation that could be influenced by changes and variability in global climates'.[4]

Policy documents and scholarly literature (for example Hannah, et al., 2002a, p. 264) stress that the response should be twofold: (a) limiting the human causes of climate change as much and as soon as possible and (b) taking management measures to help species, habitats and ecosystems adapt to climate change. The second category of responses is possibly more complex than the first. Climate change may require plant and animal species to migrate to other areas to avoid extinction. If there are barriers blocking species from finding their new 'climate space', populations or even entire species may become extinct. Habitat fragmentation (which typically leads to a loss of biodiversity), isolation (which leads to a loss of connectivity) and range shifts of species due to climate change are together seen as major threats to biodiversity. Further fragmentation of habitats will only exacerbate the problems many species face in adapting to climate change.

One strategy to tackle fragmentation is restoring natural interconnections by means of redeveloping ecological coherence through networks (Tyteca, et al., 2006, p. 10). The concept of ecological networks is relatively new. Previously, the focus was much more on protecting discrete areas. From the 1980s on, this gradually shifted to the point that ecological networks came into use as a practical conservation tool (Bischoff and Jongman, 1993; Bennett and Mulongoy, 2006). The cradle of the model of ecological networks was found in Central and Eastern Europe some 30 years ago, in particular in Estonia. Based on the polarized-landscape theory of the Russian geographer Boris Rodoman and the landscape-stabilization approach, the idea displayed an integrated approach to environmental management, focusing more on achievement of sustainable development than conservation of priority habitats. These initiatives were strongly government-influenced (which is not strange, given the political situation) (Bennett and Mulongoy, 2006, pp. 13 and 25). In Western Europe the network idea was more the outgrowth of new theories in the field of ecology, including MacArthur and Wilson's equilibrium theory of island biography and the concept of metapopulations (MacArthur and Wilson, 1967).

These new developments came down to the idea that larger islands contain larger populations of species which are less vulnerable to extinction. Besides, populations are dynamic and more disturbed in fragmented landscapes and habitats of varying quality, because of limited dispersal, migration and genetic change. This finding in ecology influenced ideas on nature conservation. Diamond, for example, argues that reserves should be as large as possible, as round as possible, as close as possible to each other, and connected with each other as far as possible (Diamond, 1975). The same idea is found in the IUCN's World Conservation Strategy (Bennett and Mulongoy, 2006, p. 27). Ecologists do not agree on the ecological value of large and small patches. The debate involves two basic questions: firstly, whether a large or small patch is better (the 'LOS' debate) and, secondly, whether it is better to have a single large patch or several small ones (the 'SLOSS' debate) (Forman, 1995).

Ecological networks have been often mentioned as a solution in recent discussions concerning climate change adaptation measures. Ecological networks typically have the same basic structure: core areas, corridors, stepping stones, restoration areas and buffer zones which are ecologically connected by landscape design. Ecological networks can serve several goals: to protect the most important habitats and species, develop a coherent structure of habitats, facilitate migration of species, and prevent further demise of habitats and species (Bischoff and Jongman, 1993, p. 25). The impact of climate change on species survival correlates with the degree of connectedness of the habitat of that species (Nijhof, Vos and van Strien, 2007, p. 14).

It is not a matter of 'the more connectivity, the merrier': connecting habitats carries certain ecological risks, and species differ in the way they make (or do not make) use of connective structures. Connectivity needs to be improved, but it also needs to be supplemented with other measures. Conservationists increasingly propose that effective responses to climate change also encompass regional reserve networks, landscape connectivity, and management of the matrix between core reserves (Hannah, et al., 2002a, p. 267).

While networks play an important role in fragmented ecosystems, conservation should also where still possible focus on protecting robust, large areas. Species in those areas will be less vulnerable to the impacts of climate change, since they may be able to migrate to new habitats within the same area (McNeely, 1994; Halpin, 1997; Hannah, et al., 2002a, 2007; Hannah, Midgley and Millar, 2002; Opdam and Wascher, 2004; Araújo, et al., 2004). These areas could be designated and protected in accordance with a strict wilderness protection policy. Experiences with this approach are available in various countries (for example the US and Finland: Kormos, 2008). For many regions (particularly the polar regions) this approach is – at least in theory – still an option (Bastmeijer, 2009).

Many protected areas are badly suited to overcoming climate change-induced changes in species' geographic ranges. Recent studies show that protected areas 'have not been designed for efficient (or even complete) representation of species' (Hannah, et al., 2007, p. 131), although 'protected areas are a useful conservation response to climate change', and that 'creation of new protected areas can substantially improve the likelihood of species conservation as climate changes' (ibid., p. 135). This calls for changing the scope of area conservation from fixity to adaptation. Since adapting to the impacts of climate change involves a lot of uncertainties, the need for flexibility is apparent:

> The objective in selecting an anticipatory adaptation policy should be to enhance the ability to meet stated objectives under a wide range of climatic conditions. As such a policy may be either robust, meaning it allows the system to continue functioning under a wider range of conditions, or resilient, meaning it allows to quickly adapt to changed conditions (Smith and Lenhart, 1996, p. 194).

This is also reflected in the Millennium Ecosystem Assessment (2005a, p. 70):

> [c]orridors and other habitat design aspects to give flexibility to protected areas are effective precautionary strategies. Improved management of habitat corridors and production ecosystems between protected areas will help biodiversity adapt to changing conditions.

A combination of several measures (enlarging areas, securing robust large areas, securing ecological connections between areas and establishing real ecological networks) therefore seems to be the best approach to maximize the ability of nature to cope with the pressure of climate change on biodiversity. This approach is also advocated under the name of 'landscape approach' (Millennium Ecosystem Assessment, 2005b, p. 145; Opdam and Wascher, 2004, p. 293).

3. CLIMATE CHANGE ADAPTATION AND INTERNATIONAL CONVENTIONS

During the last decades awareness of the global character of nature has gained a firm foothold in international nature protection law. Already in the older conventions, such as the Convention on Wetlands of International Importance Especially as Waterfowl Habitat,[5] the Convention on Migratory Species of Wild Animals[6] and the Convention on the Conservation of European Wildlife and Natural Habitats,[7] attention is paid to the designation of transboundary protected areas or the creation of ecological networks. Nevertheless, climate change was not yet an issue in those times. Even in the 1992 CBD, climate change only gained attention as a biodiversity issue through later decisions of the COP. We now discuss the extent to which those conventions offer starting points for a more global approach to biodiversity conservation in the light of climate change.

Convention on Wetlands of International Importance Especially as Waterfowl Habitat

The Convention on Wetlands of International Importance Especially as Waterfowl Habitat, better known as the Ramsar Convention for the city where it was concluded, is an example of an international convention that pays attention to the designation of transboundary protected areas.[8] For transboundary wetlands ('shared wetlands' or 'international wetlands'), the Convention provides in Article 5(1) that parties must consult with each other about implementing obligations arising from the Convention. They must also endeavor to coordinate and support present and future policies and regulations concerning the preservation of wetlands and their flora and fauna.

A vast quantity of resolutions, handbooks and guidelines has been adopted since 1971 which further define the provisions of the Convention. The 'Ramsar Toolkit' is a set of no fewer than 17 Handbooks about the wise use of wetlands, including one on transboundary wetlands (Ramsar Convention Secretariat, 2007). The toolkit gives detailed advice on how to pursue international cooperation on the management of such areas. Referring to the 1992 Helsinki Convention on the Protection and Use of Transboundary Watercourses and International Lakes,[9] one of the Handbooks indicates that multi-state management commissions should be established to promote international cooperation, and urges states to harmonize wetland management with the obligations arising from watercourse agreements (Ramsar Convention Secretariat 2007, p. 13). Africa has many transboundary protected areas, including transboundary wetlands that have been designated under the Ramsar Convention.[10]

More generally, it can be observed that over the last few years wetland management has been integrated into river basin management, recognizing the fact that wetlands usually are only a part of a bigger catchment area and, for their conservation, largely depend on the quality of the entire catchment.[11] To achieve this integration, the Ramsar Convention Bureau and the Secretariat of the Convention on Biodiversity have joined forces in a River Basin Initiative. In 2005, the ninth COP to the CBD adopted a resolution that laid down practical guidelines for the integration of wetland management into river basin management.[12] The guidelines focus, among other things, on upgrading wetlands management to the river basin level.

The Convention on Migratory Species of Wild Animals

Because of the nature of migratory species, the Convention on Migratory Species of Wild Animals (CMS) focuses on establishing international networks to conserve migratory routes.[13] The Convention promotes international cooperation to protect animals that cross one or more national boundaries while migrating. For species listed in Appendix I to the Convention, states must prevent, remove, compensate for or minimize the adverse effects of activities or obstacles that seriously impede or prevent the migration of the species, and conserve and restore the habitats of endangered species.[14] For species listed in Appendix II, the range states of migratory species (i.e., states on or over whose territory the species range) are required to conclude agreements that may provide for, among other things, 'maintenance of a network of suitable habitats appropriately disposed in relation to the migration routes'.[15]

A series of such agreements has been concluded, some of which aim to create a network. Thirty-six Eurasian and 24 African states have signed the African-Eurasian Waterbird Agreement (AEWA), concluded in 1995, which focuses on protecting migratory water birds that breed in Europe and Asia, and winter in Africa.[16] The Parties to the AEWA agree to 'coordinate their efforts to ensure that a network of suitable habitats is maintained or, where appropriate, re-established throughout the entire range of each migratory waterbird species concerned, in particular where wetlands extend over the area of more than one Party to [the] Agreement'.[17]

Another agreement under the CMS is the Agreement on the Conservation of Gorillas and their Habitat (Gorilla Agreement, 2007).[18] A similar provision is included in this agreement, under which Parties agree to 'coordinate their efforts to ensure that a network of suitable habitats is maintained or re-established throughout the entire range of all species and sub-species, in particular where habitats extend over the area of more than one Party to [the] Agreement'.[19]

For elephants in West Africa there is a less legally binding Memorandum of Understanding concerning Conservation Measures for the West African Populations of the African Elephant (West African Elephant MOU).[20] It does not contain the strong wording found in the Gorilla and Waterbird Agreements, nor does it in any way recognize the need to establish a network of protected areas across the region for this highly migratory species. Although the preamble states that the signatories 'understand' that 'this species is essentially migratory, which makes the survival of the individuals dependent upon the conservation of habitats, including corridors for movements between or among Range States', the text of the MoU itself only generally requires the states involved to 'take steps to conserve and, when and where appropriate, to strictly protect the African Elephant and to conserve and sustainably use the habitats essential for its survival'.[21] Several transboundary migration corridors for elephants are being established, however: two in Burkina Faso and Ghana (a western and an eastern corridor), and one in Guinea and Liberia.[22]

Convention on the Conservation of European Wildlife and Natural Habitats

The 1979 Convention on the Conservation of European Wildlife and Natural Habitats (better known as the Bern Convention for the city where it was concluded) is a regional treaty to which, among others, the European Union and several North African states are parties.[23] The involvement of the African countries is important, given that the continent hosts migratory species that breed in Europe.

The Bern Convention aims to preserve species of great importance for the biological diversity of both Europe and the countries involved. The Convention promotes conservation of natural habitats in transboundary areas as a whole.[24] It also stresses the importance of specific measures for migratory species that cross political boundaries. In 1989, the Council of Europe launched the Emerald Network as an optional initiative under the Bern Convention.[25] Its purpose was to create an ecological network to conserve wild flora and fauna and their natural habitats in Europe. The network was to consist of Areas of Special Conservation Interest (ASCIs) created by parties to the Convention. The network was introduced by a recommendation under the Convention. As such it is an optional 'soft law' initiative, but the obligations under the Convention itself have a 'hard' international law status. The Emerald Network is still in the implementation phase (identification of specics and habitats, collection of data, selection of potential ASCIs, creation of databases, and designation of lists). The real work of management, maintenance and restoration of species and habitats will come afterwards. For the African area, three pilot projects are underway (in Burkina Faso and Senegal). In 2007 the Standing Committee of the Bern Convention considered extending the Emerald Network methodology to any African countries that so wished, to enable broader protection of species and contribute to the commitments made at the 2002 World Summit on Sustainable Development to achieve a significant reduction in the current rate of biodiversity loss by 2010 and to set up global networks of protected areas.[26]

Convention on Biological Diversity

One of the main goals of the CBD is 'to anticipate, prevent and attack the causes of significant reduction or loss of biological diversity at source'.[27] Although it is a hard law instrument, the CBD is as a 'framework convention' with relatively weak provisions. It offers 'guiding principles' to support parties during their efforts to implement national law and policy concerning biodiversity matters (Birnie and Boyle, 2002, p. 571). A jumble of later decisions have been adopted to fill in this framework further, but they generally lack legal force.

The CBD's embrace of the 'ecosystem approach' is a good example. The CBD defines 'ecosystem approach' in Article 2, but its meaning has been worked out by means of further COP decisions. The COP declared the ecosystem approach to be 'the primary framework for action under the Convention'[28] and proposed an adaptive management strategy with a strong focus on integrating other management and conservation approaches such as Biosphere Reserves, protected areas, single-species conservation programmes.[29] It also declared that the ecosystem approach is 'a tool that contributes to the implementation of various issues addressed under the Convention, including the work on, inter alia, protected areas and ecological networks'.[30] Recently it has adopted further decisions aimed at fuller integration of the ecosystem approach into climate change adaptation and mitigation activities.[31]

The governing organs of the CBD increasingly recognize the link between climate change and biodiversity.[32] They have called for integration of biodiversity considerations into climate change mitigation and adaptation plans to strengthen ecosystems' resilience (CBD Secretariat, 2003, p. 76).[33] In this connection they have proposed, among other things, the establishment of networks of terrestrial, freshwater and marine protected areas that take into account projected changes in climate (ibid., p. 77). A technical research report published by the CBD Secretariat suggested linking currently fragmented reserves and landscapes through corridors or habitat matrices as an option to provide the potential for climate change-induced migration (ibid.). The report defined 'corridor' as 'habitat areas sufficiently close to each other (i.e. functionally linked) to enable dispersal' (ibid.). The report underlines the value of corridors for animals but stresses that their utility for vegetation is less certain and that in some cases corridors will not be enough to deal with climate change impacts, making other adaptation measures necessary. The COP has encouraged parties and other governments 'to cooperate regionally in activities aimed at enhancing habitat connectivity across ecological gradients, with the aim of enhancing ecosystem resilience and to facilitate the migration and dispersal of species with limited tolerance to altered climatic conditions'.[34]

In addition to protected area networks, the COP has encouraged Parties to take management measures to make ecosystems capable of coping with extreme climate impacts,[35] and to promote the integration of these considerations into national policies, programs and plans.[36] It has also urged the CBD secretariat, UNFCCC secretariat, their Parties and other relevant organizations to coordinate their activities in the area of biodiversity conservation, climate change mitigation and adaptation, and combating land degradation.[37]

The COP has also encouraged Parties and invited relevant organizations to 'enhance research and awareness of the role that protected areas and the connectivity of networks of protected areas play in addressing climate change'.[38] It has drawn attention to the need for funding, especially for developing countries such as those in Africa, to allow the establishment of ecological networks and improve management of existing protected areas.[39]

In 2007, the COP adopted a Programme of Work on Protected Areas (PoWPA).[40] One of its targets is that '[b]y 2015, all protected areas and protected area systems are integrated into the wider land- and seascape, and relevant sectors, by applying the ecosystem approach and taking into account ecological connectivity and the concept, where appropriate, of ecological networks'.[41] To achieve this, the COP suggested activities concerning integration of network systems, development of tools to realize ecological connectivity and rehabilitation and restoration of habitats and degraded ecosystems to (re)shape ecological networks, corridors and buffer zones.[42] The maintenance of structural and functional viability of ecosystems via ecological networks was confirmed during the recent COP in Bonn (2008), which promoted the application of appropriate tools and policy measures better to integrate protected areas into the broader context.[43] In the same decision, the COP proposed concrete tools to implement the PoWPA and establish transboundary protected areas and ecological networks. The CoP also recognized a lack of implementation and constraints on capacity building in developing countries in relation to, among other things, integration of protected areas into wider landscapes and seascapes, evaluation of the effectiveness of protected areas management, protection of large intact areas. The COP urged Parties to address those gaps and constraints as a matter of priority.[44]

The CBD itself does not refer explicitly to the transboundary aspects of biodiversity conservation. The transboundary dimension is mentioned explicitly for the first time in the PoWPA. Goal 1.3 of the Programme is 'to establish and strengthen regional networks, transboundary protected areas (TBPAs) and collaboration between neighbouring protected areas across national boundaries'.[45] The PoWPA encourages Parties to realize this goal through, among other things, collaboration and coordination in the establishment and management of regional networks.[46] In particular, it suggests the establishment of new TBPAs with adjacent Parties and countries, enhancement of effective collaborative management of TBPAs, and collaboration between protected areas across national boundaries.[47]

The PoWPA also envisages a supporting role for the Executive Secretary, for example in the development of guidelines for establishing TBPAs and collaborative management, and compilation and dissemination of information on regional networks of protected areas and TBPAs.[48] It also urges Parties to 'review the potential for regional cooperation under the Convention on Migratory Species with a view to linking of protected area networks across international boundaries and potentially beyond national jurisdiction through the establishment of migratory corridors for key species'.[49] Notwithstanding the fact that the CBD lacks an explicit focus on transboundary issues, the approach of the PoWPA is without doubt transboundary.

The COP has devoted explicit attention to Southern states. The latest COP decision '*invites* Parties, other Governments, regional and international organizations to support South-South Cooperation by facilitating projects and programmes aimed at joint conservation and sustainable use of cross border ecosystems to further contribute towards halting biodiversity loss'.[50]

4. CLIMATE CHANGE ADAPTATION AND EUROPEAN LAWS AND REGULATIONS

European climate change policy is shifting only very slowly from a focus on mitigation toward a more proactive emphasis on adaptation. Its silence on adaptation is especially noticeable in the area of biodiversity. In theory, most European nature conservation is based on the ecosystem approach, addressing the whole system of habitats, species, organisms, their environment and their interactions.[51] In recent years several initiatives to create ecological networks have been introduced in Europe, of which the most significant is Natura 2000, which aims to create an ecological network throughout the 27 EU Member States. This is presented as the panacea in combating the adverse effects of climate change on European biodiversity (European Environment Agency, 2006). In this section we discuss what role ecological networks, connectivity and cross-border coordination and cooperation play in European policy on climate change and biodiversity.

Law and Policy Development in 'Wider' Europe

The European Community implemented the Bern Convention by means of the Habitats Directive and the Natura 2000 network, which we discuss below. For Bern Convention Parties that are Member States of the European Union, 'Emerald Network sites are those of the Natura 2000-network'.[52] The Emerald Network is a joint effort. The EU is responsible for sites in EU Member States and the Council of Europe for sites outside the EU. The areas belonging to the Emerald Network (including Natura 2000 sites) are considered core areas of the Pan-European Ecological Network (PEEN), which is a soft law initiative of the Council of Europe.

PEEN is an overarching framework of ecological networks in Europe: '[i]t acted as both a physical network through which ecosystems, habitats, species, landscapes and other natural features of European importance are conserved, and a co-ordinating mechanism through which the partners in the Strategy could develop and implement co-operative actions'.[53] Besides Natura 2000 and the Emerald Network, PEEN incorporates the UNESCO World Network of Biosphere Reserves, the European Network of Biogenetic Reserves, Ramsar Convention sites, UNESCO World Heritage Sites, protected areas under Europarc Federation management, European Diploma of protected areas, the Bonn Convention, and regional and national networks.

Law and Policy Development in the European Union

The two focal points of EU nature protection policy are the Wild Birds Directive of 1979[54] and the Habitats Directive of 1992.[55] Although the Wild Birds Directive illustrates the tendency at that time to address a single issue, it shows awareness of the transboundary character of the subject of conservation of migratory birds. The Habitats Directive took things a step further, introducing the idea of cross-border conservation by proposing a European ecological network: Natura 2000. Natura 2000 is a network of protected areas across the EU, consisting of Special Areas of Conservation (SACs) designated under the Habitats Directive and Special Protection Areas (SPAs) designated under the Wild Birds Directive. Natura 2000 is the major instrument to achieve the European target of halting biodiversity decline by 2010. The aim of the Natura 2000 network is to ensure that fauna and flora of Community interest receive adequate protection to ensure their long-term viability. It is seen, moreover, as *the* measure in EU nature conservation legislation to maintain and restore ecological connectivity.

The Habitats Directive speaks of establishing a 'coherent ecological network'.[56] Article 3 proposes the network as a way to maintain and restore the favourable conservation status of sites hosting listed habitats and species in their natural range. Article 10 proposes maintaining and developing features of the landscape of major importance for wild fauna and flora as a possible way to improve the ecological coherence of Natura 2000. Both provisions leave a lot of questions unanswered, however. Both speak of 'ecological coherence' without defining it. The European Commission issued a guidance document on Article 10 which, although not legally binding, aims to 'help develop and implement integrated ecological connectivity related measures' to maintain and restore connectivity and to respond to the impacts of climate change (Kettunen, et al., 2007, p. 10). The guidance document ('the EU Guidance') makes a big step forward by acknowledging that climate change requires flexibility in protected area management instead of only aiming for preservation within specific fixed locations (ibid., p. 47). It does not, however, really provide guidance in the interpretation of Article 10. It only provides recommendations to the Member States for the implementation of Article 10, which is discretionary.

The EU Guidance notes that the EU Water Framework Directive of 2000 ('WFD')[57] 'provides a good opportunity to manage river basins at transnational scale' (ibid., p. 83). The goal of the framework is to prevent European waters and their ecosystems from (further) deterioration and to promote sustainable water use. A further goal is to soften the effects of floods and droughts. To achieve this, Member States are obliged to designate river basin districts and draw up a River Basin Management Plan (RBMP) for each district.[58] Where necessary, basins must be designated internationally. EU Member States are to ensure coordination of these international river basins together. In this respect, the WFD calls for transboundary cooperation.[59]

The EU Guidance sheds some light on the relationship between the WFD and the Habitats Directive. Although the WFD does not explicitly mention obligations to implement the provisions of the Habitats Directive, it 'has been seen to provide important support to the management and monitoring of the Natura 2000 network in the future' (Kettunen, et al., 2007, p. 82). Since river basins often cross borders, Member States should explore ways to use 'the framework provided by the WFD to prevent fragmentation and enhance connectivity between Member States' (ibid., p. 83). Further integration between the WFD and Habitats Directive could be achieved by integrating connectivity issues into the RBMPs, as the Guidance advises. The WFD itself does not mention climate change. The EU Guidance, however, discusses climate change in relation to the WFD. Since the WFD is still in the process of being implemented, Member States are advised 'actively [to] support capacity building in relation to the importance and value of inland water ecosystem biodiversity, including issues related to the maintenance of ecosystems services and climate change' (ibid., p. 83).

Joint Initiatives by European States

We turn now to the case of the Danube river basin. This is not meant to be a comprehensive case study. It does, however, provide a good example of an area covered by a patchwork of legal regimes and forms of (cross-border) cooperation. It illustrates the challenges of concentrating protection around a river basin.

The Danube river basin

Flowing through ten countries, the Danube crosses political and conceptual borders. The floodplains at the upper part of the river are mainly cultivated, while those downstream are largely untouched and of high environmental value. The source of the river is in the rich western part of Germany, while the estuary on the Black Sea is located in relatively poor former communist countries such as Moldova and Romania. Among the ten Danube countries, some are 'old' EU members, some are 'new', and some are not members of the EU. Along the approximately 2,800 kilometer length of the Danube, national parks, Ramsar sites, UNESCO World Heritage Sites and Biosphere Reserves, and potential Natura 2000 sites can be found. Many protection initiatives involve cross border cooperation but not many involve the entire course of the second longest river in Europe (Turnock, 2001, p. 659).

In 1994, 11 countries and the European Community signed the Danube River Protection Convention (DRPC).[60] The Convention entered into force in 1998. The Preamble refers to the Parties 'striving at a lasting improvement and protection of Danube River and of the waters within its catchment area in particular in the transboundary context'.[61] The Parties agree explicitly to cooperate in this field and to take 'all appropriate legal, administrative and technical measures'[62] to achieve the Convention's goals and prevent or limit transboundary impacts on water. In this respect, the Convention is the overarching legal instrument for co-operation and transboundary water management in the Danube basin. Although climate change is not mentioned (it was hardly an issue at that time), the DRPC's broad legal framework leaves enough room for the development of a climate change policy.

Things are somewhat different with the Declaration on the Cooperation for the Creation of a Lower Danube Green Corridor (LDGC Declaration), signed in 2000 by Romania, Bulgaria, Ukraine and Moldova.[63] In contrast to the Danube River Convention, the soft-law LDGC Declaration starts by mentioning the natural and ecological uniqueness of the Lower Danube basin, with its usefulness to humans relegated to second place (healthy floodplains and wetlands help to maintain water quality and provide economic opportunities). The Declaration is clear about how to reach its goals. A 'green corridor' is to be established, consisting of more than one million hectares of protected wetlands. This is to include over 770,000 hectares of existing protected areas, more than 160,000 hectares of new protected areas and more than 223,000 hectares to be restored to natural floodplains. The LDGC emphasizes flexibility, since it consists of three types of areas: those with a strict protection regime, buffer zones with a differentiated protection regime in which human activity can be permitted and degraded areas restored, and areas in which there is room for 'sustainable economic activities'.[64] The LDGC is by definition a transboundary initiative, aimed at creating a corridor of interconnected protected areas along the final 1,000 kilometers of the Danube, including the globally important Danube Delta. The LDGC initiative was facilitated by the WWF and presented to the WWF as a Gift to the Earth, as part of the WWF's Living Planet campaign.[65] By early 2008 the goal of one million hectares of protected wetlands was achieved and more than 50,000 hectares had been restored (roughly one quarter of the goal).[66] The LDGC parties recognize their obligations under the Ramsar and Bern Conventions and their national commitments to the protection of biodiversity, while 'taking into account' the DRPC.[67] The Declaration emphasizes partnership with local, national and international organizations and governments, 'the crucial role of environmental Non-Governmental Organisations' as the mouthpiece of public interests and ideas, and the importance of offering citizens and environmental NGOs an active role in decision making processes.[68]

A good example of the central role of environmental NGOs can be found in the WWF, which is active along the Danube (Turnock, 2001) and spearheaded the LDGC initiative. It recently issued a report that 'shows that practical adaptations to climate change impacts on freshwaters may have immediate benefits for peoples' livelihoods and to conserve ecosystems' (WWF, 2008). In the case of the Lower Danube river basin this means for example that by dismantling dykes and polders and restoring floodplains and wetlands, the capacity of the river basin to cope with floods would be enhanced, economic activity (e.g., tourism and fishery) would grow, and biodiversity conservation would get a boost (as was shown by growing numbers of bird species in pilot projects) (ibid., pp. 12–14). Even though climate change might not always be mentioned explicitly, work on waterways that starts with a different motivation may also be beneficial to issues concerning climate change.

Of a very substantive nature is the European Green Belt initiative, formed in 2004. This joint venture of government agencies, NGOs and other stakeholders emerged from the German Green Belt initiative. It strives to create an ecological network that runs from the Barents Sea to the Black Sea like a European 'backbone' (Geidezis and Kreutz, 2004). The Green Belt follows the trail of the former Iron Curtain, and therefore also partly the course of the Danube. With its route having been left in 'peace' for over 40 years, this corridor through Europe developed into a relatively 'wild' piece of nature. The initiative takes an explicit crossborder stance, as it runs along state borders and connects areas on both sides (ibid., p. 137). The Green Belt could be of help in realizing international ecological networks such as the Emerald Network, Natura 2000 and PEEN. Amazingly enough, this relatively new initiative makes no direct or indirect reference to climate change.

5. LARGE NATURAL AREAS AND ECOLOGICAL NETWORKS IN AFRICA

The global initiatives described in section 3 also apply to the African continent. In addition, there are some continent-specific instruments, including initiatives of the African Union and sub-regional organizations such as the Southern African Development Community (SADC). The latter have extensive transfrontier management and conservation programs and pay explicit attention to cooperation around water.

The Convention on the Conservation of Nature and Natural Resources

The African Convention on the Conservation of Nature and Natural Resources of 2003, signed under the auspices of the African Union (AU), provides rules regarding the establishment, maintenance and extension of conservation areas to ensure the long-term conservation of biological diversity.[69] The Convention marked a decades-long process of revising the earlier 1968 Convention of the same name,[70] and when it comes into force it will replace the earlier Convention. The Convention states that conservation areas should in particular conserve those ecosystems that are most representative of and peculiar to areas under the jurisdiction of that state or are characterized by a high degree of biological diversity, and ensure the conservation of all species and the habitats that are critical to their survival, or which are threatened or of special scientific or aesthetic value.[71] Explicit attention is paid to transboundary cooperation with regard to protected areas: parties agree to cooperate in the management of transboundary ecosystems as the need arises, and set up interstate commissions for their conservation and sustainable use.[72] The latter is also true for transboundary wetlands.[73] Besides this, Article XII(4) obliges states to establish buffer zones around the borders of conservation areas to control activities outside these areas that are detrimental to the achievement of the conservation goals of the protected areas.[74]

Law and Policy Development in African Regions

Numerous regional nature protection initiatives are found in Africa. The most advanced of these are found in Southern Africa. Before turning to that region, it is useful to survey briefly developments elsewhere in Africa. The 1999 Treaty establishing the East African Community (EAC) has several provisions that promote co-ordination, the adoption of common policies, and the harmonization of policies and regulations concerning shared natural resources and ecosystems.[75] It also states that the Partner

States 'encourage the joint use of training and research facilities and develop common management plans for trans-border protected areas'.[76] In addition, it calls on Partner States to 'develop common management plans for trans-border protected areas'.[77] There are however no explicit references to the creation of ecological networks in this Treaty or in other legal documents of the EAC. The Action Plan for the Environment Initiative of the AU's 'New Partnership for Africa's Development' (NEPAD) program refers to an initiative in the North and West African regions to create a network of wetlands (NEPAD, 2003, p. 48). However, there are no legally binding documents to support this initiative, except for the general provision on international cooperation of the Ramsar Convention.

Law and Policy Development in the SADC

Harmonization of laws is evolving rapidly within the SADC. As in the EU, harmonization efforts within SADC are being made in the field of nature conservation. The 1999 Protocol on Wildlife Conservation and Law Enforcement to the SADC Treaty obliges the Member States to establish management programs for the conservation and sustainable use of wildlife, integrate such programs into national development plans, and assess and control activities which may significantly affect the conservation and sustainable use of wildlife so as to avoid or minimize negative impacts.[78] It obliges them to promote cooperative management of shared wildlife resources and wildlife habitats across international borders, and to develop transfrontier conservation and management programs.[79] In particular, the Protocol regulates the establishment of Transfrontier Conservation Areas (TFCAs).[80]

In 2006, the SADC TFCA Office was established in Gaborone, Botswana. This office promotes the establishment and development of TFCAs in the SADC region and generally promotes conservation and sustainable management of ecosystems that transcend international boundaries within the SADC region. It also develops and facilitates the implementation of guidelines, standards and best practices for the establishment and development of TFCAs, and maintains a network with other TFCA practitioners working in the SADC region.

TFCAs are also known as peace parks. The idea to create transboundary protected areas for conservation and tourism purposes originally came from WWF South Africa. Since 1997, a Peace Parks Foundation has facilitated the process of TFCA establishment and funding. It currently co-funds conservation measures in seven TFCAs.[81] TFCAs are not automatically protected in the countries involved. In South Africa for example a TFCA has to be designated under one of the categories of protected areas according to the Protected Areas Act, for instance as a National Park (Tanner, et al., 2004, p. 169; Mramba, 2004, p. 214). We will return to this point below.

Within SADC there also exists a Shared Watercourse Systems Protocol (2000),[82] which is based on both the UNECE Convention on the Protection and Use of Transboundary Watercourses and International Lakes[83] and the UN Convention on the Law of the Non-navigational Uses of International Watercourses.[84] The Protocol is an important instrument for the topic of this chapter, as the protection of watercourses is usually essential for biodiversity conservation. Large wetlands like river estuaries, often located in more than one country, are the lifeline of many species. The same is true of international watercourses. The Shared Watercourse Systems Protocol requires the parties to enter into consultations on the management of shared watercourses, which may include the establishment of a joint management mechanism if one of the states requests it. A shared watercourse agreement providing for the establishment of a shared watercourse institution such as a shared water commission will be the likely outcome of such consultations.[85]

By 2008, 18 TFCAs had either been established or were in the process of establishment. Once an agreement has been signed, a Memorandum of Understanding (MOU) is drafted on the basis of which joint management decisions are made and a joint management plan is prepared (Mramba, 2004, pp. 216–25).

Examples of multilateral agreements to establish TFCAs are agreements signed by Mozambique, South Africa and Zimbabwe on the Great Limpopo Transfrontier Park, and the agreement signed by Angola, Botswana, Namibia, Zambia and Zimbabwe on the Kavango Zambezi TFCA, for which a MOU was also signed in 2006. The latter area is huge, situated in the Okavango and Zambezi river basins. It covers parts of five countries and a total area of 287,132 km^2 (more than 28 million hectares). Conservation status varies within the TFCA. It consists of 36 national parks, game reserves, community conservancies and game management areas, for each of which specific conservation requirements have been set in national law. Another example of a TFCA is the |Ai-| Ais/Richtersveld Transfrontier Park in South Africa and Namibia. This TFCA is connected by the Orange River to the Orange River Mouth Ramsar site, a transboundary wetland designated by both countries under the Ramsar convention. There are plans to extend the Park further north into Namibia and Angola, bringing it to a total protected area of more than 19 million hectares.

The fact that so many huge areas have been designated as TFCAs in a relatively short time must be considered a big success. The key reason for this success is the broad political support the initiative has gained throughout the region. Political leaders all embraced the idea of creating TFCAs as a means of peaceful cooperation with neighboring countries. The fact that the areas are considered to be a potential source of revenue for the tourist industry also helped to create political momentum.[86]

The limited harmonization of national law in the southern African region is one of the shortcomings of the TFCA agreements. Although they should be aimed at harmonizing law, thus facilitating transfrontier conservation, the agreements are vague and abstract and do not add much to the texts of international law documents, such as the SADC Protocol on Wildlife Conservation and Law Enforcement (Lubbe, 2008, p. 144). Instruments to facilitate transfrontier biodiversity conservation are basically lacking. The actual conservation measures taken depend largely on the existing national legal frameworks of the countries involved. Colonial legacies and insistence on state sovereignty are the main reasons for this fragmented and nationally oriented approach to transfrontier biodiversity conservation (ibid., p. 148). Resource and capacity constraints aggravate this situation.[87] It must be observed however that for some transboundary areas cooperative management plans are being drafted and a cooperative governance approach is being pursued in which all stakeholders from the various countries are involved. Sometimes international NGOs such as the Peace Parks Foundation, WWF and Wetlands International provide resources for specific conservation projects. These are all relevant factors for the conservation of transboundary protected areas. In other words, despite the absence of a solid legal framework, local governance initiatives can be successful (Verschuuren, 2008).

As to river basin management, throughout Africa there are many bi- or multilateral management systems in place on joint watercourses. A multilateral example in southern Africa is the 2000 treaty by which all Orange River riparian states (Botswana, Lesotho, Namibia and South Africa) established the Orange–Senqu River Commission (ORASECOM).[88] The Council of this Commission provides technical advice to the authorities of the states involved on matters relating to the development, utilization and conservation of the water resources of the river system.[89] The Parties to the treaty agree, among other things, '[i]ndividually and jointly [to] take all measures necessary to protect and preserve the river system from its sources and headwaters to its common terminus',[90] 'including the estuary of the river system and the marine environment taking into account generally accepted international rules and standards'.[91] The vastness of this river basin however makes it difficult, if not impossible, to develop a precise and effective policy on common management of the entire area. Research shows that local governance initiatives are more successful (Verschuuren, 2008).

6. CONCLUSIONS

We started this contribution by asking what Europe and Africa could learn from each other with regard to the legal elaboration of climate change adaptation measures to save biodiversity. First of all, it is important to note a difference between the two continents. Africa has more experience with creation of large and robust natural areas and transboundary protected areas, while in Europe, at least on paper, the focus is on creating ecological networks. At the international level, there are some cautious steps to promote connectivity, create ecological networks, and protect transboundary areas, but they are generally not the subject of clear 'hard law' obligations for state governments. It is remarkable that neither in Europe, in Africa nor at the international or regional level do initiatives show real awareness of the importance of flexibility to cope with hard-to-predict future impacts of climate change on biodiversity. That climate change is hardly mentioned in the relevant instruments could be due to the fact that most initiatives were adopted when climate change was simply not yet a (big) issue. Only the CBD seems to offer possibilities to put more emphasis on climate change adaptation activities. Here, the embrace of the ecosystem approach and the stress on networks, connections and transfrontier protection could offer a flexible global framework to address climate change adaptation activities, but only if it gains more judicial power.

Many nature protection initiatives in the wider Europe stress the need for ecological networks. Laws, policies, practical tools, ecological knowledge, political feasibility and states of mind however sometimes seem stuck in the traditional way of thinking about nature conservation in terms of insular protected areas. In general, nature conservation legislation pays enough attention to 'core areas', but ignores the urgent importance of corridors, especially in relation to climate change adaptation. Even the EU initiative to create an ecologically coherent network, Natura 2000, suffers from the same malady. Although the Natura 2000 network is promoted as the main instrument to protect the EU's wild flora and fauna, the Habitats Directive does not distinguish among core areas, corridors and buffer zones, but seems to stick to the idea of 'just' protecting core areas. The Habitats Directive leaves open what a coherent ecological network is, although a cautious impulse toward an integrated landscape approach can be found in Article 10. Besides this, Natura 2000 lacks transboundary coherence and there are still a lot of physical barriers within and outside protected areas, which will stand in the way of making the network 'climate proof'.

In southern Africa, governments have opted to combine bigger and smaller areas into large transboundary protected areas (TFCAs). Co-operation appears to be the main objective, since these areas are almost always much larger than in Europe and span the territories of several states. TFCAs are created through bi- or multilateral agreements, but the parts are then designated under national law, so that all parts of a TFCA do not have the same conservation status. Conservation measures thus depend largely on national legal frameworks. There are however also examples of areas with truly co-operative management across borders. TFCAs cover areas as big as Italy, while in the EU mostly smaller areas have been designated. In a way however, the distinction between networks and large areas is relative, since the formation of large areas out of numerous smaller ones can also be seen as the transformation of a network into a single protected area. This differs fundamentally from Europe, where the practice is to connect central Natura 2000 areas by means of often narrow corridors.

In both regions, the integrated water basin approach offers an additional stimulus to transboundary cooperation and the creation of ecological networks. Both regions also show the relevance of the involvement of NGOs and other stakeholders. Climate changes poses new and as yet unknown challenges to nature conservation efforts. Given this uncertainty, we argue for a protection regime that offers space and flexibility to give legal status to those areas that potentially qualify in the future. In this respect, Europe can learn from Africa about the benefits of designating larger, more robust areas. We conclude further that the European Union should put more emphasis on flexibility in its system of designating protected areas. The southern African region on its part could benefit from putting more emphasis on connectivity.

NOTES

* Saja Erens is a PhD researcher at the Tilburg Graduate Law School, Tilburg University, The Netherlands. Kees Bastmeijer is Professor of Nature Conservation and Water Law at Tilburg Law School. Jonathan Verschuuren is a Professor of International and European Environmental Law at Tilburg University.

1. 5 June 1992, (1992) ILM 31, 818 ('CBD').
2. Conference of the Parties to the Convention on Biological Diversity, 'Protected Areas (Articles 8 (a) to (e))', Decision COP VII/28 (Kuala Lumpur, 9–20 February 2004), para. 1.4.5.
3. J. Willot and C. Thomas (2001), 'Implications of Climate Change for Species Conservation' (Briefing Paper, IUCN Workshop, Gland, Switzerland, 19–21 February), http://cmsdata.iucn.org/downloads/ccspeciessurvival_2.pdf, p. 14 (visited 25 February 2009).
4. US Forest Service, Climate Change Resource Center, 'Amphibians and Reptiles and Climate Change', www.fs.fed.us/ccrc/topics/amphibians-reptiles.shtml (visited 27 January 2009).

5. 2 February 1971, (1972) ILM 11, 963 ('Ramsar Convention').
6. 23 June 1979, (1980) ILM 19, 15 ('CMS').
7. 19 September 1979, Eur. TS 104 ('Bern Convention').
8. Ramsar Convention, supra note 5.
9. 17 March 1992, (1992) ILM 31, 1312.
10. Some examples of transboundary Ramsar wetlands in Africa are the Parc National du W in Burkina Faso, Benin and Niger, the Niumu/Saloum wetland in Gambia and Senegal, and the Orange River Mouth in South Africa and Namibia.
11. Conference of the Parties to the Ramsar Convention, 'Guidelines for integrating wetland conservation and wise use into river basin management', Resolution COP VII.18 (San José, 10–18 May 1999).
12. Conference of the Parties to the Ramsar Convention, 'Additional scientific and technical guidance for implementing the Ramsar wise use concept', Resolution COP IX.1 (Kampala, 8–15 November 2005).
13. CMS, supra note 6.
14. Ibid., Art. III(4)(a) and (b).
15. Ibid., Art. V(5)(f).
16. 16 June 1995, 5 *Y.B. Int'l Envtl. L.* 907 ('AEWA').
17. Ibid., Art. III(2)(d).
18. 26 October 2007, UN Doc. No. UNEP/CMS/GOR-MOP1/Inf.1.
19. Ibid., Art. III(2)(c).
20. Memorandum of Understanding concerning Conservation Measures for the West African Populations of the African Elephant, 22 November 2005, available at the CMS website www.cms.int/species/elephants/index.htm (visited 3 February 2009) ('West African Elephants MOU').
21. Ibid., Preamble and Art. 1.
22. CMS Secretariat, 'Memorandum of Understanding concerning Conservation Measures for the West African Populations of the African Elephant', www.cms.int/species/elephants (visited 20 January 2009).
23. Bern Convention, note 7.
24. Ibid., Art. 4.
25. Standing Committee of the Bern Convention, Resolution No. 1 (1989).
26. Plan of Implementation of the World Summit on Sustainable Development, in *Report of the World Summit on Sustainable Development*, Johannesburg, 26 August–4 September 2002, UN Doc. A/CONF.199/20, p. 7, at pp. 33–34.
27. CBD, supra note 1, Preamble.
28. Conference of the Parties to the Convention on Biological Diversity, 'Ecosystem Approach', Decision COP V/6 (Nairobi, 15–26 May 2000) and 'Ecosystem Approach', Decision COP VII/11 (Kuala Lumpur, 9–20 February 2004).
29. Decision COP V/6, supra note 28, para. A.4–5.
30. Decision COP VII/11, supra note 28, para. 4.
31. Conference of the Parties to the Convention on Biological Diversity, 'Ecosystem Approach', Decision COP IX/7 (Bonn, 19–30 May 2008).
32. Conference of the Parties to the Convention on Biological Diversity, 'Progress report on the implementation of the programme of work on marine and coastal biological diversity (implementation of decision IV/5)', Decision COP V/3, and 'Progress report on the implementation of the programme of work for forest biological diversity', Decision V/4 (both Nairobi, 15–26 May 2000), and 'Biodiversity and climate change: guidance to promote synergy among activities for biodiversity conservation, mitigating or adapting to climate change and combating land degradation', Decision COP VIII/30 (Curitiba, 20–31 March 2006); see also CBD Secretariat 2003.
33. See also Conference of the Parties to the Convention on Biological Diversity, 'Biodiversity and climate change', Decision COP IX/16 (Bonn, 19–30 May 2008), para. B.2.
34. Decision COP VIII/30, supra note 32, para. 4.

35. Conference of the Parties to the Convention on Biological Diversity, 'Biodiversity and climate change', Decision COP VII/15 (Kuala Lumpur, 9–20 February 2004).
36. Decision COP VIII/30, supra note 32.
37. See for instance Conference of the Parties to the Convention on Biological Diversity, 'Biological Diversity of Inland Water Ecosystems', Decision COP VII/4 (Kuala Lumpur, 9–20 February 2004) and Conference of the Parties to the Convention on Biological Diversity, 'Biological Diversity of Dry and Sub-humid Lands', Decision COP VIII/2 (Curitiba, 20–31 March 2006).
38. Conference of the Parties to the Convention on Biological Diversity, 'Protected areas ', Decision COP IX/18 (Bonn, 19–30 May 2008), para. A.23.
39. Ibid., para. B.6a; CBD, supra note 1, Art. 8(m).
40. Decision COP VII/28, supra note 2, Annex: Programme of work on protected areas ('PoWPA').
41. Ibid., Ch. II, goal 1.2.
42. Ibid., Ch. II, s. 1.2.3–1.2.5.
43. Decision COP IX/18, supra note 38, para. A.4a.
44. Conference of the Parties to the Convention on Biological Diversity, 'Protected Areas', Decision COP VIII/24 (Curitiba, 20–31 March 2006), para. 9.
45. PoWPA, supra note 40, Ch. II, Goal 1.3.
46. Ibid., Ch. II, s. 1.3.1.
47. Ibid., Ch. II, s. 1.3.3 and 1.3.4.
48. Ibid., Ch. II, s. 1.3.5 and 1.3.6.
49. Ibid., Ch. II, s. 1.3.7.
50. Conference of the Parties to the Convention on Biological Diversity, 'South-South Cooperation on Biodiversity for Development', Decision COP IX/25 (Bonn, 19–30 May 2008), para. 5.
51. European Commission, *Communication from the Commission: Halting the Loss of Biodiversity by 2010 – and Beyond: Sustaining Ecosystem Services for Human Well-being*, 22 May 2006, COM(2006)216 final.
52. Standing Committee of the Bern Convention, Resolution No. 5 (1998).
53. Ibid., p. 10.
54. Directive 79/409/EEC of the European Council of 2 April 1979 on the conservation of wild birds, 1979 OJL103 (Wild Birds Directive).
55. Directive 92/43/EEC of the European Council of 21 May 1992 on the conservation of natural habitats and of wild fauna and flora, 1992 OJL206 (Habitats Directive).
56. Ibid., Art. 3(1).
57. Directive 2000/60/EC of the European Parliament and the Council of 23 October 2000 establishing a framework for community action in the field of water policy, 2000 OJL327 (Water Framework Directive).
58. Ibid., Art. 4(1).
59. Ibid., Art.3(4).
60. Convention on Co-operation for the Protection and Sustainable use of the Danube River, 29 June 1994, 5 *Y.B. Int'l Envtl. L.* Doc. 16.
61. Ibid., Preamble.
62. Ibid., Art. 2(2).
63. Declaration on the Cooperation for the Creation of a Lower Danube Green Corridor, Bulgaria–Moldova–Romania–Ukraine, 5 June 2000, www.ramsar.org/key_danube_corridor.htm (visited 26 January 2009) (LDGC Declaration).
64. Ibid., Art. 2.
65. Ibid., Art. 9.
66. WWF, 'Lower Danube Green Corridor: Europe's Most Ambitious Wetland Project', www.panda.org/about_wwf/where_we_work/europe/what_we_do/danube_carpathian/our_solutions/freshwater/floodplains/lower_danube_and_danube_delta/index.cfm (visited 26 January 2009).
67. LDGC Declaration, supra note 63, Preamble.

68. Ibid., Arts 6 and 7.
69. 11 July 2003, www.africa-union.org/root/au/Documents/Treaties/treaties.htm (visited 3 February 2009) (not in force) (New African Nature Convention).
70. 15 September 1968, Organization of African Unity Doc. No. CAB/LEG/24.1, www.africa-union.org/root/au/Documents/Treaties/treaties.htm (visited 3 February 2009).
71. New African Nature Convention, note 69, Art. XII.
72. Ibid., Art. XXII(2)(e).
73. Ibid., Arts VII(3) and XXII(2)(e).
74. Ibid. Art. XII(4).
75. 30 November 1999, www.eac.int (visited 3 February 2009), Arts 112(2)(j), 114(1)(c), and 114(2)(b)(ii).
76. Ibid., Art. 116(d).
77. Ibid.
78. SADC Protocol on Wildlife Conservation and Law Enforcement, 18 August 1999, www.sadc.int/index/browse/page/164 (visited 25 February 2009), Art. 7(1) and (2) (SADC Wildlife Protocol).
79. Ibid., Art. 7(5)(a) and (9).
80. Ibid., Art. 4(2)(f).
81. Peace Parks Foundation, 'Facilitating Peace Parks', www.peaceparks.org (visited 3 February 2009).
82. 7 August 2000, www.sadc.int/index/browse/page/158 (visited 3 February 2009) (SADC Shared Watercourses Protocol).
83. 17 March 1992, (1992) ILM 31, 1312.
84. 21 May 1997, (1997) ILM 36, 700.
85. SADC Shared Watercourses Protocol, supra note 83, Arts 5(3) and 6.
86. Louis Kotze, North West University (Potchefstroom, South Africa) and Alexander Paterson, University of Cape Town (Cape Town, South Africa), personal communications with the authors, September 2008.
87. Ibid.
88. Agreement on the Establishment of the Orange–Senqu River Commission, 3 November 2000 (unpublished; copy on file with authors).
89. Ibid., Art. 4.
90. Ibid., Art. 7(12).
91. Ibid., Art. 7(14).

REFERENCES

Araújo, M.B., et al. (2004), 'Would Climate Change Drive Species Out of Reserves? An Assessment of Existing Reserve-selection Methods', *Global Change Biology* 10, 1618.

Bastmeijer, K. (2009), 'Protecting Polar Wilderness: Just a Western Philosophical Idea or a Useful Concept for Regulating Human Activities in the Polar Regions?', *Yearbook of Polar Law*, 1.

Bennett, G. and K.J. Mulongoy (2006), *Review of Experience with Ecological Networks, Corridors and Buffer Zones* (CBD Technical Series No. 23) (Secretariat of the Convention on Biological Diversity).

Birnie, P.W. and A.E. Boyle (2002), *International Law and the Environment* (Oxford University Press).

Bischoff, N.T. and R.H.G. Jongman (1993), *Development of Rural Areas in Europe: The Claim for Nature, Preliminary and Background Studies* (Report No. V79/1993) (Netherlands Scientific Council for Government Policy).

CBD Secretariat (2003), *Interlinkages Between Biological Diversity and Climate Change: Advice on the Integration of Biodiversity Considerations into the Implementation of the United Nations Framework Convention on Climate Change and its Kyoto Protocol* (CBD Technical Series No. 10) (Secretariat of the Convention on Biological Diversity).
CBD Secretariat (2006), *Guidance for Promoting Synergy Among Activities Addressing Biological Diversity, Desertification, Land Degradation and Climate Change* (CBD Technical Series No. 25) (Secretariat of the Convention on Biological Diversity).
Diamond, J. (1975), 'The Island Dilemma: Lessons of Modern Biogeographic Studies for the Design of Nature Preserves', *Biological Conservation* 7, 129.
European Environment Agency (2006), *Vulnerability and Adaptation to Climate Change in Europe* (Technical Report No. 7/2005) (European Environment Agency).
Forman, R.T.T. (1995), *Land Mosaics: The Ecology of Landscapes and Regions* (Cambridge University Press).
Geidezis, L. and M. Kreutz (2004), 'Green Belt Europe – Nature Knows No Boundaries: From Iron Curtain to Europe's Lifeline', *Urbani Izziv,* 15(2), 135.
Halpin, P.N. (1997), 'Global Climate Change and Natural-area Protection: Management Responses and Research Directions', *Ecological Applications* 7(3), 828.
Hannah, L. et al. (2002a), 'Conservation of Biodiversity in a Changing Climate', *Conservation Biology* 16(1), 264.
Hannah, L. et al. (2007), 'Protected Area Needs in a Changing Climate', *Frontiers in Ecology and the Environment* 5(3), 131.
Hannah, L., G.F. Midgley and D. Millar (2002), 'Climate Change-integrated Conservation Strategies', *Global Ecology & Biogeography* II, 485
Kettunen, M., et al. (2007), *Guidance on the Maintenance of Landscape Features of Major Importance for Wild Flora and Fauna – Guidance on the Implementation of Article 3 of the Birds Directive (79/409/EEC) and Article 10 of the Habitats Directive (92/43/EEC)* (Institute for European Environmental Policy).
Kormos, C.F. (ed.) (2008), *A Handbook on International Wilderness Law and Policy* (Fulcrum Publishing).
Lubbe, W.D. (2008), 'Straddling Borders and Legal Regimes: The Case for Cooperative Transfrontier Biodiversity Conservation in the Southern African Development Community', *Yearbook of International Environmental Law* 18, 126.
MacArthur, R.H. and E.O. Wilson (1967), *The Theory of Island Biogeography* (Princeton University Press).
McNeely, J.A. (1994), 'Protected Areas for the 21st century: Working to Provide Benefits to Society', *Biodiversity and Conservation* 3, 390.
Millennium Ecosystem Assessment (2005a), *Ecosystems and Human Well-being: Biodiversity Synthesis* (World Resources Institute).
Millennium Ecosystem Assessment (2005b), *Ecosystems and Human Well-being: Policy Responses* (Vol. 3) (World Resources Institute).
Mramba, S.J. (2004), 'The Peace Parks Initiative: A Breakthrough Towards Sustainable Natural Resource Management in Southern Africa?', *South African Journal of Environmental Law and Policy* 11, 213.
NEPAD (2003), *Action Plan for the Environment Initiative* (NEPAD).
Nijhof, B.S.J., C.C. Vos and A.J. van Strien (2007), *Indicators for the "Convention*

on Biodiversity 2010" – Influence of Climate Change on Biodiversity (Working Paper No. 53.7a) (WOT Natuur and Milieu).

Opdam, P. and D. Wascher (2004), 'Climate Change Meets Habitat Fragmentation: Linking Landscape and Biogeographical Scale Levels in Research and Conservation', *Biological Conservation* 117, 285.

Ramsar Convention Secretariat (2007), *Ramsar Handbooks for the Wise Use of Wetlands* (3d edn, Ramsar Convention Secretariat).

Smith, J.B. and S.S. Lenhart (1996), 'Climate Change Adaptation Policy Options', *Climate Research* 6, 193.

Tanner, R.J., et al. (2004), 'Transfrontier Conservation Areas of Southern Africa and International Law in the Context of Indigenous Community Involvement', *South African Journal of Environmental Law and Policy* 11, 167.

Turnock, D. (2001), 'Cross-border Conservation in East Central Europe: The Danube-Carpathian Complex and the Contribution of the Worldwide Fund for Nature', *GeoJournal* 54, 655.

Tyteca, D., et al. (2006), *Feasibility of Ecological Networks: Ecological, Economic, Social and Legal Aspects (ECONET)* (Final Report) (Belgian Science Policy).

Verschuuren, J. (2008), 'The Case of Transboundary Wetlands under the Ramsar Convention: Keep the Lawyers Out!', *Colorado Journal of International Environmental Law and Policy* 19(1), 49.

Worldwide Fund for Nature (WWF) (2008), *Water for Life: Lessons for Climate Change Adaptation from Better Management of Rivers for People and Nature* (WWF-UK).

PART 3

The Clean Development Mechanism and mitigation strategies

10. The deadlock of the Clean Development Mechanism: caught between sustainability, environmental integrity and economic efficiency

Christina Voigt*

1. INTRODUCTION

At first sight, the Clean Development Mechanism (CDM)[1] tells a story of success. It has rapidly been developing as an important element in international climate policy by providing to industrialized nations a cost-effective means of complying with their greenhouse gas (GHG) reduction commitments in the Kyoto Protocol. By early 2009, there were more than 1,300 registered CDM projects in 50 developing countries, and approximately another 4,500 projects in the project registration pipeline.[2] As of early 2009, the registered projects have resulted in some 140 million issued certified emission reductions (CERs). The CDM is expected to generate more than 1.4 billion CERs by the time the first commitment period of the Kyoto Protocol ends in 2012, each equivalent to one tonne of carbon dioxide.[3]

Despite these encouraging responses so far, it has become clear that the CDM is burdened with structural flaws, which must be dealt with if the mechanism is to achieve long-term climate benefits. These flaws can be linked to the conflicting interests informing the mechanism. The protection of the CDM's contribution to sustainable development and its environmental integrity, on the one hand, are counterweighted by demands of procedural efficiency and economic feasibility on the other. Because most of the participants in the CDM cycle, such as investors, project developers, host countries and Annex I governments, have a strong interest in realizing low-cost emission reductions and receiving many CERs, this means that ensuring the environmental integrity and contribution to sustainable development of the CDM rests entirely on its regulatory framework and legal safeguards. In the absence of checks and balances, legal rules and careful administration are the only means to enable the CDM to achieve its three often conflicting purposes. If the rules and administration of this mechanism are not strong enough, the CDM could actually lead to increases in global GHG emissions. So far, the intentions behind the CDM mechanism are good, but its rules and their enforcement are not.

This chapter first briefly describes the current crisis of the mechanism before analysing the (often) colliding objectives that inform the CDM. It then suggests various proposals for political and legal reform. It will be shown that the economic efficiency required to attract investment in CDM projects often outweighs the existing legal and institutional safeguards designed to ensure that CDM projects achieve sustainable development and environmental integrity. It is argued that unless the global carbon market in general, and the CDM in particular, are grounded in a sound and effective commitment to sustainability, no real long-term climate benefit can be expected from these instruments.

2. THE CDM IN CRISIS

The CDM serves a temporary role in the structure of the international climate regime. For as long as developing countries remain not bound by quantified emission reduction targets, the CDM is a means for reaching out to these countries by providing a financial incentive for climate investments to, and climate mitigation action in, developing countries. Despite (or because of) its temporary character as a bridging mechanism to involve developing countries during the term of the Kyoto Protocol until 2012, with no assurance of its extension beyond, the market strongly embraced the mechanism.[4] The high demand for CERs and the large number of projects spurred the CDM machinery. As a result, the CDM institutions have been strained by the market surge.[5] Yet, the effectiveness of the CDM in delivering on its three main goals depends upon its methodological, institutional and legal safeguards. In the face of strong market forces, it has become evident that the CDM in its current design is flawed in several ways. I will focus on only a few of these, namely additionality; carbon leakage and the rebound effect; environmental impacts other than GHG emissions; adverse policy incentives; contribution to sustainable development; the credibility of the CDM; and the ethics underpinning the CDM.

Non-additionality of Projects

Article 12 of the Kyoto Protocol requires the CDM to generate GHG emission reductions that are real, measurable, long-term and additional to any that would have occurred in the absence of the project.[6] An emission reduction project is additional when the project goes beyond 'business as usual' and thus deserves to be rewarded with tradable credits. If a CDM project is not additional, but nevertheless leads to the issuance of CERs, these credits result in an increase in global GHG emissions as they enable Annex I countries or companies to emit more GHGs than they would have been permitted to, whereas the emission reductions from the project in the non-Annex I country would have occurred anyway. Several recent surveys show that for a high number of registered projects the additionality criterion was not fulfilled. Between 40 and 70 per cent of the analysed projects would have been carried out anyway, even in the absence of 'carbon finance', as CDM project financing is often described.[7]

Demonstration of additionality is not straightforward. The controversial issue is that the question whether emissions reductions would have occurred in the absence of the project is hypothetical and needs to be measured against a counter-factual baseline. A baseline for a CDM project activity is a hypothetical reference case, representing the volume of GHGs that would be emitted if the project were not implemented.[8] In other words, additionality can never be proven with absolute certainty. Moreover, the claim of financial additionality (i.e., providing an investment that otherwise would not have occurred) is often based on subjective or company internal criteria, for example, that management is not willing to invest without carbon finance (Harthan, 2008; Michaelowa and Purohit, 2007).

The CDM Executive Board (EB), the United Nations (UN) organ charged with registering CDM projects,[9] has made some improvements in terms of setting up benchmarks and barriers for additionality analysis. In order to demonstrate additionality the so-called 'additionality tool' established by the EB or the 'combined tool for baseline selection and demonstration of additionality' (combined tool) can be used.[10] These tools involve an analysis of various barriers facing the implementation of projects to show that were it not for the revenue from the sale of CERs, a project activity would not be implemented. It must also be shown that implementation of the project was not required under any mandatory law or policy that is widely enforced. Other improvements are pending.[11] Still, the high number of projects and the pressure on the CDM institutions lead to registration of projects the additionality of which is questionable. This problem is enhanced by the strong market demand for CERs and inaccuracies within project applications (project design documents (PDDs)) as well as occasional non-compliance with the original project proposal (Ramachandran, 2009). These issues continue to haunt the CDM and affect its credibility.

Carbon Leakage and Rebound Effect

Another critical aspect of the CDM is carbon-leakage, meaning the possible reallocation of polluting industries within the CDM host country or to other countries. Because the CDM in its current form only validates and verifies reductions in GHG emissions on a project basis, it does not take account of increased emissions elsewhere in the host country, or outside its borders, caused by the implementation of the project (Kallbekken, et al., 2007; Boer, et al., 2007, pp. 1169–88). Suggestions have been made to meet this shortcoming. These include policy crediting (i.e., crediting for emissions reduction policies rather than single projects), and sectoral or target-based crediting (i.e., crediting emissions reductions below the benchmark or target set for an entire industrial sector) (for an early suggestion see Sterk and Wittneben, 2007; for an overview see Sterk, 2008; Americano, 2008). Both types of mechanisms are not possible under the current CDM.

Similar to the issue of carbon-leakage is the concern that energy reduction generated by a valid CDM project can lead to increased energy consumption elsewhere. This energy saving-rebound[12] effect was recently observed in an extensive study of Dutch CDM projects (IOB Evaluations, 2008). These are systemic flaws which can be addressed only by design changes.

Negative Impacts on Biological Diversity, Local Communities and Water Quality

The current design of the CDM focuses its approval of projects only on their ability to achieve GHG reductions. Other effects of the projects on local populations or other environmental media are neither credited nor discredited. This is particularly controversial with regard to forestry projects that could potentially lead to a decline in freshwater quality and loss of biological diversity. Similar concerns relating to indirect damages exist for large hydro dam projects that impact on river ecosystems and water quality and require relocation of local populations.[13] In order for the CDM to gain international acceptance and to provide a credible means for addressing global (climate) inadequacies and inequities, these 'collateral' damages urgently need to be addressed. Given that Article 12(2) of the Kyoto Protocol proclaims that 'achieving sustainable development' is one of the core goals of the CDM, it is crucial that such collateral ecological impacts be addressed in project planning.

Adverse Policy Incentives

The CDM as a mechanism for attracting financial flows for climate change mitigation into developing countries holds the danger of dis-incentivizing these states from committing to binding international climate targets. Because some developing countries experience an influx of green investment and technology transfer via the CDM, this reduces the need or willingness to exchange this CDM benefit for an obligation to cap GHG emissions in an *international* agreement. A cap on national emissions would not fit well with the current design of the CDM as a baseline-and-trade mechanism. Though the CDM is generally adaptable to such change, the possibility of losing CDM-related investment and revenue acts as a 'political chill' in climate negotiations on the issue of capping developing countries' emissions.

Moreover, the additionality criterion in its present state can create adverse policy incentives to *national* climate change mitigation and clean energy regulation. The potential of CDM projects to generate much-needed investment flows into a host country has led some developing countries to back off from more progressive energy or climate policies and legislation. These policies and laws, if integrated into the baseline, would disqualify CDM projects that aim at meeting these new standards or thresholds because they would make it impossible to prove additionality (for examples in Ecuador, Mexico and Colombia: see Figueres, 2006).

Also, the idea of a mandatory environmental and sustainability impact assessment for all CDM projects has been seen as an infringement of the sovereignty of potential host states. As a result, the final language of the CDM in Kyoto is weak, requiring nothing more than an 'analysis of environmental impacts only if the host country makes it mandatory for the project to be approved'.[14] The CDM Modalities and Procedures do not provide for the situation in which the host country does not have any laws on environmental impact assessment. If stakeholders have concerns about the local environmental or social impacts of a CDM project, it should be evaluated according to the highest international environmental and social assessment procedures and standards (UNEP, 2004, pp. 61–2).[15]

However, the more stringent the rules on environmental and sustainability impact assessment are, the more costly CDM projects may become. Since a host country benefits from a CDM project, the absence of harmonized international rules may create an incentive for the host country to refrain from insisting on a thorough impact assessment, in order to make its own market attractive for CDM projects. 'The CDM's geographical flexibility', warn Meijer and Werksman, 'should not become a means of channeling projects to host countries with the lowest environmental standard' (Meijer and Werksman, 2005, p. 210). Internationally harmonized rules on environmental and sustainability impact assessment of CDM projects would counterbalance such adverse incentives.

Contribution to Sustainable Development?

Another critical aspect of the CDM is its expected contribution to the sustainable development of CDM host countries. Some sustainable development benefits could be directly related to GHG reductions, such as technology transfer needed for GHG reductions, energy supply diversification and improved energy efficiency. These benefits can also be indirectly linked to GHG abatement, for example by job creation, local community support, poverty alleviation, improved water quality and quantity. Yet no common indicators or international standards have been accepted for measuring sustainable development benefits. Experience has shown that host countries have been quite willing to approve CDM projects with little or no added 'sustainability' value. Host countries often define the contribution of CDM projects to sustainable development in line with existing national development strategies, which more often than not are based on economic development rather than sustainability (Michaelowa, 2005). Although some countries, such as China, India and Brazil, have (different) sets of criteria for sustainable CDM projects, they fail to include verifiable indicators to measure the outcome, or do not have the means for monitoring or enforcing the sustainability benefits (Castro and Michaelowa, 2008, p. 8). Two recent reports also highlight the fact that local stakeholders, and partly also host countries' Designated National Authorities (DNAs) which are in charge of writing a letter of approval for each project application, are not sufficiently informed about the additional economic and social benefits that could 'trickle down' from sustainable CDM projects (Castro and Michaelowa, 2008; IOB Evaluations, 2008).

The steep increase in international demand for CERs has led Annex I governments to engage in projects which have a strong benefit in terms of GHG reduction, such as projects reducing HFC23 and other fugitive industrial emissions, such as SF6, but which contribute little in terms of sustainable development (IOB Evaluations, 2008).[16] Despite the clear goal in Article 12 of the Kyoto Protocol, achieving sustainable development benefits remains uncertain. The CDM institutions, such as the EB and accredited Designated Operational Entities (DOEs), do not assess or review projects for sustainable development benefits; neither is monitoring or enforcing of these benefits structurally integrated into the CDM.[17]

Another critical issue is how to balance two of the purposes of the CDM as written into the Kyoto Protocol: cost effectiveness and promotion of sustainable development. Critics have claimed that basing GHG emission reductions on a market mechanism will necessarily lead to pursuing emission reductions at the lowest possible cost and to purchasing CERs at the lowest possible price. Yet this search for least-cost carbon credits sidelines sustainable CDM projects, such as renewable energy projects, by not rewarding the additional benefits they provide (Voigt, 2008; Streck and Chagas, 2007). Ultimately, as one critic says, the danger is that 'the CDM's first mandate to help reduce Kyoto Protocol compliance costs is making it impossible to fulfill its second mandate to promote sustainable development' (Pearson, 2004, p 1). However, this should not be perceived as a question of 'either-or'. In order for the CDM to play a role in any future climate agreement, a balance between cost effectiveness and promotion of sustainable development has to be found.

Sinking Credibility

In the wake of substantive criticism leveled at how CDM projects are conducted, the credibility of the CDM has suffered. The fact that many projects turned out not to have been in need of carbon financing through the CDM, and would have gone ahead anyway, led to general skepticism in relation to this regime. There have consequently been calls raised to abolish the CDM altogether (see for example Hagem and Holtsmark, 2008, pp. 10–18).

A successful CDM project requires legal, technological and financial expertise, knowledge of the carbon market and a fruitful connection between project developers, host governments and investors. Because of this intricate and complex relationship, the CDM has turned out to be a 'cash-machine' for lawyers, accountants, economic counsellors, brokers and intermediaries. Moreover, in some cases the CDM has also generated substantive revenue and investment returns (Schneider, 2007), which has led to some questioning the market rationale of the mechanism: is it still an economic means serving an environmental end or have economic rationales taken over entirely?

After all, the CDM functions as an offset mechanism, allowing capped countries to emit higher amounts of GHGs than they are permitted to under their Kyoto Protocol commitment, and justifying inaction within their own territories. In other words, via the CDM, rich polluting countries are able to buy their way out of their 'climate responsibility'.

Further, the influx of high numbers of CERs into the market has the effect of diluting the emission caps set by Annex I States. The European Union's (EU) '2008 Climate and Energy Package' recognizes this concern:

> there is a risk that too generous a use of CDMs can dilute the effectiveness of the ETS [EU Emissions Trading Scheme] by increasing the supply of credits and thereby cutting the demand for allowances, and reducing the incentive for governments and companies to promote emissions reductions at home. This can also limit the ETS' capacity to act as the key driver to realize the target for renewable energy (Commission of the European Union, 2008).

The EU proposal is a clear indication of the diminishing credibility of the CDM.[18]

In this context, suggestions that aim at restructuring the CDM away from an off-set mechanism to a crediting mechanism that would truly combine cuts in GHG emissions with sustainable climate benefits in developing countries gain force (Zapfel, 2008; see also Cames, 2008, and Schneider, 2008).

Outstanding Issues

The CDM is not a finished product, yet the regulatory tools are not developing as fast as the market is driving the mechanism. Regulation of the CDM happens on a 'learning while doing' basis and is marked by 'trial and error'. A number of critical issues remain unresolved and contentious. In terms of methodologies for present or future project areas, the following remain under constant criticism: HFC 23 projects (see for example Wara and Victor, 2008),[19] Carbon Capture and Storage (CCS) projects (CAN-International, 2006), biofuel projects,[20] forest projects (especially when including genetically modified plant species or mono-cultural tree plantations) (Forests and the European Union Resource Network and SinksWatch, 2003; and World Rainforest Movement, 2007), and large hydro dam projects (International Rivers Network and CDM Watch, 2004; International Rivers Network and CDM Watch, 2002; Haya and McCully, 2007).

Another outstanding issue regarding the implementation of the CDM is the question of supplementarity, namely the quantification of what part of their emissions reduction commitment Annex I Parties can cover by using CERs. In other words, should climate targets be met where they are cheapest or are there any other restrictions on where climate mitigation measures are to be implemented (Luhmann and Sterk, 2008)? This issue relates to the deeper dilemma of how to balance global cost-effectiveness in climate mitigation with demands of global equity and responsibility. Although emission reductions in developing countries can (still) be realized at lower costs than in industrialized countries, this should not be an excuse for inaction and 'business-as-usual' at home.

In this context, quantitative and, possibly, qualitative limitations which buyers, especially governments as compliance buyers, can impose on the purchase of CERs play an important role. On 17 December 2008, the European Parliament adopted a legislative resolution on the effort of Member States to reduce their GHG emissions to meet the Community's GHG emission reduction commitments up to 2020. Here it noted that '[i]t is important that credits from projects used by Member States represent real, verifiable, additional and permanent emission reductions and have clear sustainable development benefits and no significant negative environmental or social impacts. Member States should also report on the qualitative criteria they are applying for the use of such credits'.[21]

From 2013, only CERs which comply with these criteria will be accepted in the European Community scheme. No such qualitative criteria have yet been adopted by the CDM EB or at the Conference of the Parties (COP) to the UNFCCC. If environmental integrity is to be a central element of the CDM, it is necessary to define it in a wider sense than exclusively in the context of additionality, including environmental benefits – or, at least, the absence of environmental damage.[22]

Finally, there is also uncertainty concerning *retroactive crediting* of projects, for example the crediting of projects which started years ago but requested carbon credits only years later. In order to ensure that the CDM delivers additional GHG reductions, retroactive crediting of projects should be stopped, since in such cases it is not convincing that the CDM was seriously considered in the investment decision.[23]

These are only a few of the important issues and questions which still need answers.

What are the Ethics Behind the CDM?

The final point in this non-exhaustive list of the weaknesses of the CDM is the question of which ethics actually inform the CDM (Livermann, 2008, pp. 47–58). Is the CDM aimed at GHG abatement with high environmental integrity and, if so, what does this actually entail? Is every CDM project 'right' as long as it delivers (additional) GHG reductions, in line with a deontological ethics? Or are these projects supposed to satisfy market demands of increased economic efficiency and cost-effective climate mitigation measures, in line with utilitarian values? Or is the CDM based on a consequentialist ethics of bringing about the sustainable development of the 'South'? The answer probably lies somewhere in the middle, or is simply 'all of the above'. However, by trying to serve 'different masters', different interests are pursued which often compete against each other. The ethical challenge is how to balance global cost-effectiveness in climate mitigation with demands of global equity and responsibility.

3. THE CORE OF THE PROBLEM: COMPETING INTERESTS IN THE CDM

At the heart of the CDM crisis lies a tridimensional problem: the pursuit of at least three different and often competing interests: environmental integrity, sustainable development and economic efficiency (see Figure 10.1).

First, the CDM is supposed to deliver real, measurable and lasting climate benefits; this capacity is often referred to as its environmental integrity. So far, the environmental integrity of the CDM rests largely on the 'integrity of the process' carried out by the UN institutions and organs involved in the CDM[24] and corresponds to the technical and administrative capacity of the EB to develop and apply methodologies that validate projects and certify emissions that are additional to those which would have occurred in the absence of the CDM. The integrity of the CDM thus depends on applying conservative methods on accuracy and transparency (allowing for a safe margin of error). Moreover, it depends on the confidence of the EB to not certify registered projects that fail to meet agreed criteria. It also requires decision making that is not politically biased or influenced or under the threat of legal claims for compensation of financial losses of project participants (Werksman, 2008, pp. 95–104; Flues, Michaelowa and Michaelowa, 2008).

Second, in addition to environmental integrity, the CDM must promote the sustainable development of host countries. As mentioned above, a number of uncertainties exist in this respect. Yet, the requirement of

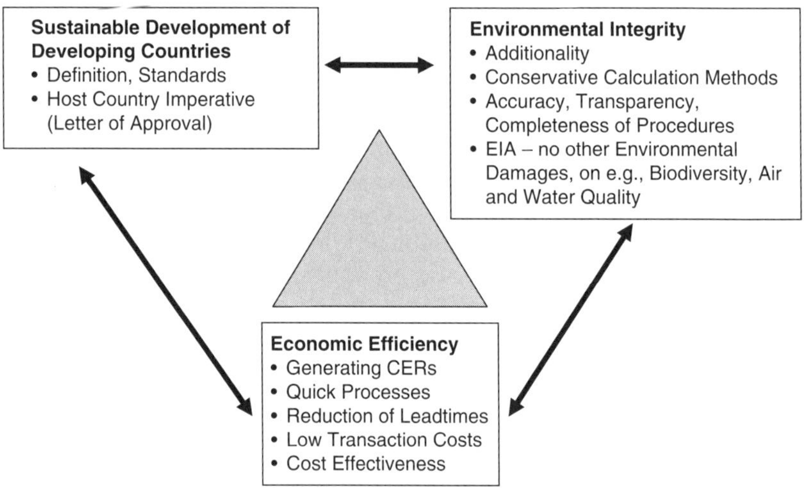

Figure 10.1 Competing interests

additionality is intricately related to the sustainable development goal. In practice, projects that are clearly additional have proven to contribute very little to the sustainable development of host countries. Projects that are undoubtedly additional are those that would not be economically feasible without the CDM, such as HFC23 and projects involving other industrial gases. Yet these projects use end-of-pipe technology that does not bring about any technological changes in terms of avoiding the generation of GHGs or any other sustainable benefits. Some of those projects even give incentives for producing more GHG gases.[25] On the other hand, projects that do contribute to sustainable development (e.g., renewable energy projects, fuel-switch projects) are often not additional. The CDM does not play any, or only an insignificant, role in the investment decision. Investment in these kinds of projects is often based on 'business as usual' economic calculations.

Third, the CDM needs to be economically interesting. In order for the CDM to play a role in the climate market while being a significant response to climate change, it must succeed in attracting a 'critical mass' of participants, especially project developers and investors, which are willing to participate in 'good' CDM projects, namely projects that are both additional and contribute to sustainable development. This will require transparency, consistency, certainty and predictability of the process, reduction of lead times (especially the duration of review) and transaction and administrative costs, and increased overall efficiency and cost-effectiveness (Streck, 2007, pp. 91–100). The dilemma, however, is that projects that both are additional and contribute to sustainable development are extremely rare and often have to go through a long, opaque and very bureaucratic process before registration, which affects their economic efficiency.

In this context, Werksman (2008, p. 99) notes that '[a]t issue is the tension between the care required to ensure the environmental quality of projects, and the bureaucratic efficiency and technocratic precision required by the demands of the market'. The crisis described above indicates an imbalance of interests tilting towards market demands and surrendering to market imperatives. It also indicates that the care required to ensure environmental integrity and to deliver on all three goals of the CDM is not systematically built into its design. This situation demands a re-evaluation of the CDM and the role of its governing organ, the EB.

4. WHICH WAY TO GO?

The CDM has reached a watershed point: the next step will be decisive for its future. Is the CDM fatally flawed and a failure? Will its credibility sink to the point of being entirely rejected? Or is it worthwhile to embark on a journey of reform? Given the current slow pace of international climate negotiations, in my opinion it is worthwhile to undertake substantial reform of the Mechanism. There are several reasons for preferring reform to rejection.

To start with, the CDM is not dead. It is of no use rejecting it altogether and thereby 'throwing the baby out with the bath water'. Trying to get rid of the CDM as a bad thing might succeed in destroying whatever good there was as well. The CDM is an innovative mechanism that builds a bridge over the 'North/South' gap in the Kyoto scheme, and it brings together private economic interests and public climate policy by helping to channel private sector investment toward climate friendly projects that otherwise might not have taken place (Holm and Fenham, 2008).

At present, there exists no alternative to the CDM. In international climate negotiations developing countries show a strong reluctance to committing to quantified emission targets. Moreover, there are strong expectations that, for reasons of historical responsibility and current capability, climate mitigation action conducted in the 'South' should be financed by the 'North'. Such expectations are legitimate and in line with the principle of common but differentiated responsibilities and respective capabilities in Article 3.1 of the UNFCCC. Yet developed countries are demanding that developing countries with the strongest economies commit to emission cuts (Planet Ark, 2008). The EU, Japan and Australia are among those nations that claim it is unfair to expect 'the rich group from almost two decades ago to keep on taking the lead' (Planet Ark, 2008). This 'Catch-22' might be a defining aspect of international climate negotiations for a long time to come.

Climate action is urgently necessary in both the developed and the developing world and the CDM is so far *the only* international mechanism financially and politically to reach out to developing countries and implement much needed technology changes.

Moreover, the problems with the CDM described above are not intrinsic problems. Rather, they are mostly linked to lacking political willingness to set up a strong and secure framework for the CDM, ensuring its environmental integrity. The market will always attempt to find the cheapest emission cuts. As a result, the effectiveness of the CDM at additionally reducing emissions and creating sustainability benefits depends on its strict regulation, monitoring and enforcement (Streck 2007, pp. 91–100).

Therefore, the task should be to make the CDM *better, safer and stronger* and to make it deliver on the goals it has ascribed to: long term, measurable, additional emission reductions that contribute to the sustainable development of 'uncapped' developing countries. There is ample room for improvement, though little time. In my opinion, reform, including *legal* reform, is not only possible; it is necessary to make the CDM survive the current crisis. These changes need to be systematically implemented into the design of the CDM, and they have to happen quickly.

5. LEGAL REFORM: MAKING THE CDM BETTER, SAFER AND STRONGER

The reform needed to revitalize the CDM has to happen on different levels and in different arenas. Political will and leadership are important, as is financial support. But the legal framework of the CDM, its administration, administrative regulation, and the possibility for judicial review of decisions made by the organs governing the CDM are also defining elements that must be part of the course of reform. In what follows I will highlight some legal changes and proposals that aim to ensure the CDM delivers on its goals.

The Rule of Law

In order for the CDM to achieve its goals, it has to be based on the rule of law. In general, the rule of law entails imposing restraint on the actions of political leaders and providing predictability for the citizenry. The World Bank (2004, p. 4) explains, 'the rule of law requires transparent legislation, fair laws, predictable enforcement, and accountable governments to maintain order, promote private sector growth, fight poverty and have legitimacy'. In terms of the CDM, the rule of law translates into (i) setting up clear and comprehensive rules, (ii) transparency, consistency and predictability in decision making (due process), participation and information, and (iii) the possibility for judicial review and appeal.

To start with, ambiguous and broad wording in COP/MOP documents leaves much room for clarification and interpretation by the EB, which leads to uncertainty and unpredictability. Reform of the CDM would need to ensure that new rules and decisions are phrased clearly and precisely in their content, leaving little room for circumvention. With regard to the existing framework, it would be advisable to ask the UNFCCC Secretariat for authoritative interpretations if controversial debate arises.

The existing regulatory framework of the CDM is an intricate system involving a large number of documents containing COP/MOP decisions and decisions of the EB.[26] Many of the rules governing the CDM came about in an ad hoc manner and have led to a fragmented system, where single, technical issues are dealt with individually. Providing an official compilation of the documents in an ordered, systematic and regularly updated manner would promote the legal certainty of the CDM.

Problems of non-transparency, inconsistency and unpredictability of decision-making partly relate to communication deficiencies among the main CDM institutions, bodies and sub-bodies and project participants (Streck, 2007; Streck and Chagas, 2007, and references therein). The number of institutions and bodies involved during the project cycle is increasing. For example, the recently created Registration and Issuance Team (RIT) is endowed with the task of assisting the EB when considering requests for registration of project activities and requests for issuance of CERs submitted to the Executive Board by DOEs. The decisions of the RIT are highly influential, yet neither its documents nor its meetings and decisions are open to the public. Clear rules and guidance through the administrative procedures and competences of such organs would help increase transparency.

This decision-making remains highly unpredictable in part because recent objections and criticism directed at the EB from nongovernmental organizations (NGOs) and the private sectors have led to a sudden increase in reviewed and rejected projects (De Jonge, 2008). While this, of course, is a laudable development, it shows that decision-making can become politically influenced and thus less predictable. Furthermore, analysts have listed a number of factors that influence the consistency, and thus predictability, of EB decisions: lack of collective or institutional memory in the EB due to rotation of its members; insufficient technical, administrative and/or financial expertise; and conflicts of interests of EB members due to concurrent exercise of other functions, such as acting as negotiators for their country or as representatives of DNAs for the CDM (Streck, 2007; Streck and Chagas, 2007). The last point in particular often leads to political friction and bias and hinders efficient and credible decision-making (Streck and Lin, 2008, p. 417; Castro and Michaelowa, 2008; Flues, Michaelowa and Michaelowa, 2008).

Further, strengthening the role of the UNFCCC Secretariat in terms of providing binding authoritative interpretations of the COP/MOP legal texts to the EB would bring collective memory (the Secretariat is staffed with long-term specialized experts and lawyers) into the decision-making process of the EB. Finally, in order to secure working capacity and impartiality of EB members, they would need to be employed on a full-time basis, selected for competence (technical, administrative, financial and legal) and salaried via the UN system. Moreover, support by sufficient and skilled staff would ease the pressure to deal with important work in an effective manner.

Standards for Sustainable Development

The absence of measurable standards or indicators is the main impediment to ensuring the contribution of CDM projects toward sustainable development. Despite the resistance of host countries to commit to binding standards for sustainable development, reform of the CDM needs to include objective, international quality criteria for local and global environmental sustainability; social sustainability and development; and economic and technological development.[27] Here is it important to make host countries aware of the additional benefits that could be generated via carbon-financed projects. A post-Kyoto agreement and future decisions of COP/MOP ought to entail clearer formulation of the requirements of sustainability, both substantial and procedural, the latter including Environmental Impact Assessment (EIA) and/or Sustainability Impact Assessment (SIA).

Where agreement on sustainability standards proves impossible, non-binding guidelines for developing countries based on national criteria could be a second best alternative. Yet, the advantage of binding, objective sustainability criteria is that they facilitate approval, monitoring and enforcement by the EB and DOEs of projects in accordance with these requirements. In this regard, the role of the EB and of the DOEs warrants revision and extension. The contribution of CDM projects to sustainable development ought to be continuously assessed during the entire project cycle. Validation of the project should require a sustainability check alongside proof of additionality. Similarly, the registration, verification and issuance of credits need to be based on proof that the project contributes to sustainable development. The requirement of contribution to sustainable development needs to be systematically built into the CDM legal framework and to be systematically checked. This includes monitoring of contribution to sustainable development which needs to be structurally incorporated into the general design of the CDM, and into each project.

As concerns CER buyers, no restrictions currently apply as to the choice of projects from which credits are acquired. However, buyers in general and Annex I governments in particular can, via 'sustainability conscious' investments and purchases, pull the development of CDM projects in a more sustainable direction. Yet the demand for CERs is great, and buyers still prefer cheap credits to 'good credits'. Thus, the use of quantitative and/or qualitative limitations for CER buyers could bundle purchasing power in order strongly to guide the CDM in a sustainable direction.[28] Alternatively, for those projects which do not directly contribute to sustainable development, it could be a project requirement to provide for financing measures or other means that support the contribution to the sustainable development of host countries in other ways. Moreover, buyers and hosts of CDM projects stand in a contractual relationship. This contract can much more strongly and clearly demand sustainable development benefits and envisage procedures to enforce such benefits by host countries. Contractual consequences for non-sustainable performance of projects, such as devaluation of CERs, penalties, taxation and so on can in themselves provide incentives for a more sustainable CDM.

Definition of Environmental Integrity

Similar challenges exist with regard to ensuring the environmental integrity of the CDM. First and foremost, there is no clear definition of environmental integrity. Although the Modalities and Procedures for the CDM demand that 'environmental integrity is to be achieved through sound modalities, rules and guidelines for the mechanisms, sound and strong principles and rules governing land use, land use change and forestry activities, and a strong compliance regime',[29] it remains unclear what is meant by the term. Is environmental integrity understood solely in terms of additionality or does it require *other* environmental benefits, or at least the absence of other environmental damages, such as to water quality or biodiversity? The relationship of the Kyoto Protocol to other MEAs also plays a role in this respect, and requires further legal elaboration (Wolfrum and Matz, 2004).

When it comes to ensuring *additionality*, a number of legal options for reform exist. The assessment of additionality should be based on more transparent and objective criteria, such as ambitious benchmarks. This should involve increased use of technology benchmarks for baseline setting and additionality testing. The use of subjective barriers (e.g., such as 'the own management is not willing to invest without the CDM') should be abandoned.

For enhancing compliance with the legal framework, the review of DOE decisions by the EB could be made mandatory and sanctions could be strengthened, the latter entailing not only the replacement of missing CERs for projects which are not additional, but penalty fines or CER premiums. However, ultimately it is the EB which registers projects as valid and issues CERs for generated reductions below a hypothetical baseline. Yet no independent judicial review mechanism exists for the decisions and acts of the EB. The establishment of an independent panel and appeal body – similar to the WTO Dispute Settlement Body – could function as a corrective to faulty decisions of the EB (Voigt, 2008).

Another reason for imperfections in the CDM system is strong differences in what the project developers state in the Project Design Document

(PDD) as the expected potential of a CDM project regarding emissions reductions, and the actual reductions achieved during the performance of the project. Project developers, as well as host countries and DOEs, have a pecuniary interest in receiving high amounts of CERs out of a CDM project. CER numbers are often exaggerated, barriers overstated, financial decisions are non-transparent or documentation does not provide all necessary information (e.g., on external financing). One incentive to avoid and detect fraud is the inclusion of a so-called 'truth clause' in the PDD and contracts; a statement from the project developer that all information is correct and complete. Such a statement opens the possibility of criminal proceedings against a company that is found to have provided incorrect information in the PDD. Assuming that such a clause can function as a general fraud prevention tool, its inclusion in the PDD should be made mandatory.

It is suggested that stronger financial barriers for profitable projects[30] and environmental quality requirements come into place. On the European level, for example, the EU Parliament Environment Committee recently voted for the adoption of quality criteria for CERs. From 2013, only high quality CERs from third countries that have ratified the future international agreement on climate change will be accepted in the Community scheme. Such high quality CERs are credits that represent real, verifiable, additional and permanent emission reductions from projects with clear sustainable development benefits and no significant negative environmental or social impacts.[31]

No such qualitative criteria have yet been adopted at the UN level. If environmental integrity is to be a central element of the CDM, it is necessary to define it in a wider sense than exclusively in the context of additionality, including environmental benefits or, at a minimum, the absence of environmental damage. As a starting point, environmental integrity of the CDM ought to require emission reductions that are real, measurable, long term, and additional, *and* which do not lead to environmental harm directly or indirectly caused by the project activity. Yet at the same time the challenge remains to keep the validation, registration and verification processes simple and transparent.

Stakeholder Involvement

Much criticism has been leveled at the democratic deficit in the CDM, which falls short of the best practice standards established by the Aarhus Convention on Access to Information, Public Participation in Decision-making and Access to Justice in Environmental Matters.[32] Significant parts of meetings of the EB have largely been carried out behind closed doors. The public is generally excluded from meetings of expert panels, teams and working groups (Streck, 2007, pp. 58–59). Likewise, DOEs do not involve the public in the course of their assessments (Streck, 2007, p. 59). Although the reports of EB and panel meetings are publicly available on the UNFCCC website, the minutes of the meetings are not.

Apart from these procedural issues, stakeholder involvement and public participation are also requirements for project validity. However, the rules remain unclear, as it is within the host countries' discretion to set up the framework for local/stakeholder involvement. Anxious about interference with their sovereign authority, host countries have been unwilling to accept binding standards for public participation. In many instances stakeholder involvement has been illusory. Either the public was unable to submit comments or 'stakeholders' were persons closely connected to, or held a pecuniary interest in, the project itself (IOB Evaluations, 2008). In order to increase the legitimacy of the CDM, stakeholder rights need to be strengthened, or established where they are absent. This warrants a wide definition of 'stakeholder', including local and indigenous communities, and due process rules for participation (especially regarding form, time, hearing, information, consent, etc.) as well as possibly also national stakeholders in the host country, stakeholders in the buyer's home state, or those from third party states affected by the environmental consequences of the project (Voigt, 2009). The mechanisms of prior informed consent or an advance informed agreement procedure, including a mandatory environmental impact and sustainability impact assessment, could provide a workable approach as a starting point.

Moreover, these rights need to include the right to environmental information as a prerequisite to public participation in decision-making. The right to information also corresponds to the duty of the host state to have and to provide information on the CDM project. This also means that publicly accessible lists, registers and files, and contact points ought to be in place, which inform the public about planned and implemented CDM projects.

Further, stakeholder involvement includes public participation in decisions on specific CDM activities. In order to participate effectively, the public must receive adequate information as to the proposed activity, the nature of the decision, draft decisions, the proposed site, physical and technical characteristics of the project, the effects on the environment, measures envisioned to reduce and prevent these effects, an outline of alternatives, and a non-technical summary and notification (sometimes even individual notification).[33] Stakeholders or the public should also have access to a review procedure before a national court or an independent and impartial body, if they can claim sufficient interest or impaired rights.

Such access to justice should provide adequate and effective remedies – including injunctive relief – that must be equitable and timely.

Finally, the procedural inclusion of public comments and concerns is one thing, but the substantive impact of these comments quite another. The fact that, for example, local communities strongly oppose a certain project currently does not lead to the rejection or review of the project. In this respect, judicial review could provide an adequate means to prevent damage to affected persons, or compensation for resulting harms.

Judicial Review

The issue of establishment of a competent judicial panel under the CDM to review decisions was already mentioned above. This is an important aspect of securing the functioning of the CDM, while delivering environmentally sound results and projects that contribute to sustainable development. The CDM is an innovative legal instrument which is designed to attract the participation of the private sector and is largely dominated by its interests. In this sense, the CDM is at odds with the traditional view of only sovereign states being directly affected by international regulation (Meijer, 2007, pp. 873–928). The CDM creates much closer contact between international institutions and private entities, diminishing the role of states. EB decisions can have a direct bearing on private project participants. It is therefore not surprising that in several cases project participants, whose project registration was rejected by the EB, have threatened to sue the Board in order to recover alleged financial losses.[34] In this situation, it is important to uphold the rule of law by creating conditions for due process, stakeholder participation and access to justice, especially by guaranteeing fair and transparent review of decisions.

While the rules and decisions on environmental integrity, especially on additionality, are already constantly improving,[35] it is their implementation which gives rise to concern. In other words, some decisions simply should not have been taken. Decisions to register projects under the CDM which turn out not to be additional can partly be linked to the systematic flaws of the CDM detailed in this chapter. However, they are partly caused by mal-governance of the CDM, such as in cases where the EB, faced with high complexity, limited expertise, conflicting interests and/or time constraints, has acted outside its delegated authority or without legal foundation, or where decisions are based on factually incorrect technical or scientific conclusions. In particular, in cases where determinations of the EB infringe on the rights of private or public legal entities, and the Board has failed to take all measures as required by decisions of the COP/MOP in order to avoid such injury; these entities must be granted the right to challenge a decision. Further, relevant procedural rules and rights need to be elaborated, such as public proceedings, right to petition and appeal, and allocation of costs.

A review mechanism would first and foremost give project participants and other entities with rights under the CDM a right legally to challenge decisions.[36] In this context, questions of invocation and legal standing arise. Would a claim require the violation of a right under the CDM or would a legal interest in the environmental integrity of a particular project or an interest in its contribution to sustainable development be sufficient? Could, for example, an NGO or an academic in Norway demand the review of a decision of the EB because a registered project in China has a detrimental impact on the quality of freshwater and the ecological integrity of a river? What would the consequences of such review be? What kind of sanctions would be advisable – and effective – to ensure the efficiency of the CDM?

Review needs to include decisions of the EB, such as registration of projects and issuance of credits and the accreditation of DOEs, as well as decisions of DOEs such as validation and verification of projects for their contribution to sustainable development and environmental integrity. The problem which arises in this context is the question of the justiciability of sustainable development. In order to meet this challenge, objective criteria, standards or/and indicators for sustainable development must be created, as argued above.

6. CONCLUSIONS

This chapter has demonstrated that the current framework of the CDM is poorly equipped for its tripartite goals of balancing market forces with environmental integrity and contribution to sustainable development, and falls short of succeeding in practice. No strong regulatory framework exists yet, and is only very slowly developing.

Such 'face-lifting' has to happen quickly, before credibility is lost. The carbon market in general and the CDM in particular are not a panacea: they will need to be carefully regulated and flanked with other financial incentives to drive low carbon technologies in developing countries. As shown in this chapter, the CDM suffers from difficulties linked to additionality, to problematic project types (HFC23 and others), to validation and verification (especially the unclear regime for DOEs and the lack of verification and validation standards), quality control, standards for sustainable development, and outstanding governance issues of the CDM by the EB.

The time is now to strengthen the CDM, in the context of negotiations for the Copenhagen meeting. Crucially, real and additional emission reductions from CDM projects must be ensured. This involves increased use of technology benchmarks for baseline setting and additionality testing. Moreover, possibly perverse incentives resulting from (low cost) CDM projects have to be addressed. The EB has to be assigned a more executive and supervisory role, including delegation of decision-making and strengthened professional support staff. Revision of CDM decision-making procedures, including due process, appeals system and sanctions, is a valid means to secure 'good' projects. Reform also includes an assessment of the roles and responsibilities of DOEs. Due process, increased consistency of decisions and rules, and securing environmental integrity while establishing a balance between sustainable development of host countries and economic profitability for project developers – all the while reducing GHGs – would considerably strengthen the legitimacy of the CDM.

Its central role in the carbon market makes the CDM an important post-2012 climate agreement option.[37] Discussions on the improvement and reform of the CDM were central at COP 14 in Poznan[38] and are very likely to be high on the agenda in Copenhagen in 2009. An environmentally more effective CDM should continue to play a role, but offsetting alone is not enough. The CDM offers promising potential but must develop new tools that build on the sustainable development of developing countries.

NOTES

* Dr Christina Voigt, LLM, PhD, is a postdoctoral research fellow in international environmental law at the University of Oslo, Department of Public and International Law. The author is grateful for constructive comments provided by Professors Heather McLeod-Kilmurray, Benjamin J. Richardson and Hans Christian Bugge, and by the participants at the IUCN Academy's conference on *Climate Law and Developing Countries*, where this chapter was initially presented as a paper. All mistakes remain mine.

1. Defined in Art. 12 of the Kyoto Protocol, the CDM provides for Annex I Parties to implement project activities that reduce emissions of greenhouse gases (GHGs) in developing countries, in return for certified emission reductions (CERs). The CERs generated by such project activities can be used by Annex I governments to help meet their emission reduction targets under the Kyoto Protocol. They can also freely be traded on an emissions trading scheme or used for investment purposes. Article 12 also stresses that CDM projects are to assist the developing country host Parties in achieving sustainable development and in contributing to the ultimate objective of the UNFCCC.
2. UNEP Risø Centre for Energy, Climate Change and Sustainable Development, http://cdmpipeline.org/publications/CDMpipeline.xls (visited 13 January 2009).
3. See http://cdm.unfccc.int/Statistics/index.html (visited 13 January 2009).

4. The projected volume of the global CDM market in 2008 was 1,200 $MtCO_2$ e, corresponding to a projected value of US$22 billion: Point Carbon, Press release, 26 February 2008, www.pointcarbon.com/aboutus/pressroom/pressreleases/1.266588 (visited 1 April 2009).
5. Institutional reform of the CDM to meet these bottlenecks was a central topic of the recent CMP 4 in Poznan, Poland, December 2008, though no final decision was taken. See Decision -/CMP.4, *Further Guidance Relating to the Clean Development Mechanism*, http://unfccc.int/files/meetings/cop_14/application/pdf/cmp_cdm.pdf; also Reuters, *Carbon trade in U.N. climate spotlight*, Press release, 9 December 2008, www.reuters.com/article/topNews/idUSTRE4B82RJ20081209 (visited 1 April 2009).
6. Article 12(5) of the Kyoto Protocol defines a CDM project as additional if 'anthropogenic GHG emissions are reduced below those that would have occurred in the absence of the registered project activity'. It is similarly defined in 3/CMP.1, Annex, para. 43 as follows: '[a] CDM project activity is additional if anthropogenic emissions of greenhouse gases by sources are reduced below those that would have occurred in the absence of the registered CDM project activity' (3/CMP.1, Annex, para. 43).
7. In a survey carried out by the Öko-Institut, 71 per cent of the participants agreed with the statement that 'many CDM projects would also be implemented without registration under the CDM', and as many as 86 per cent of the participants affirmed that 'in many cases, carbon revenues are the icing on the cake, but are not decisive for the investment decision' (Schneider, 2007). The mass media have also covered the issue: N. Davies, 'Truth about Kyoto: Huge Profits, Little Carbon Saved', *The Guardian* (2 June 2007); N. Davies, 'The Inconvenient Truth about the Carbon Offset Industry', *The Guardian* (16 June 2007); and M. Gregory, 'The Great Carbon Bazaar', *BBC News* (4 June 2008).
8. The baseline derived from a 'baseline scenario' for a CDM project activity is defined in 3/CMP.1, Annex, para. 44 as follows: '[t]he baseline for a CDM project activity is the scenario that reasonably represents the anthropogenic emissions by sources of greenhouse gases that would occur in the absence of the proposed project activity (3/CMP.1, Annex, paragraph 44)'. The baseline can be used to determine whether a CDM project activity is additional; and also the volume of additional GHG emission reductions achieved by a project activity.
9. The EB supervises the CDM under the authority and guidance of the COP/MOP, and is fully accountable to the COP/MOP. It is the institutional entity which administers the CDM (3/CMP.1, Annex, para. 5).
10. The additionality tool includes the following steps for proving additionality: (1) identification of alternatives to the project activity; (2) investment analysis to determine that the proposed project activity is either: (i) not the most economically or financially attractive, or (ii) not economically or financially feasible; (3) barriers analysis; and (4) common practice analysis (See EB 39, Annex 10, at 1). Each of these elements comes with its own complex requirements. (See, for example, the guidance on the assessment of investment analysis adopted by the EB [EB 41, para. 65]). The use of the additionality tool is not mandatory, except for cases where the additionality tool is included in an approved methodology. Moreover, the use of this tool to assess and determine additionality does not replace the need for the baseline methodology to provide for a step-by-step approach to identifying the baseline scenario. Project participants proposing new baseline methodologies shall ensure consistency between the determination of additionality of a project activity and the determination of a baseline scenario (EB 39, Annex 10, para. 7).
11. See EB, Call for Public Comments on the 'Proposal for an enhanced barrier test for project activities that have a potentially high profitability without CER revenues', CDM – Meth Panel, Thirty-third meeting, Report, Annex 11 http://cdm.unfccc.int/public_inputs/Panels/meth/033/mp_033_an11.pdf (visited 1 April 2009).
12. The rebound effect occurs when energy savings lead also to saving of money and the money saved is then spent on additional energy-consuming activities that would not

otherwise have been undertaken. In this situation, some or all of the energy savings may be eliminated.

13. NGOs have criticized the inclusion of large hydropower projects, which they consider unsustainable, as CDM projects. In order to qualify for the EU's Emissions Trading Scheme, hydropower projects larger than 20MW must now document that they follow the relevant international criteria and guidelines, including those contained in the World Commission on Dams year 2000 Final Report.
14. Annex, Decision 17/CP.7, para. 37(c) and Appendix B, art. 2(e).
15. However, the identification of international standards provides additional challenges. It has been suggested that existing international or regional standards, such as the Convention on Environmental Impact Assessment in a Transboundary Context ((1991) ILM 30, 800; Espoo, 28 February 1991) or the World Bank's operational procedures on environmental assessment (World Bank Operational Policy/Bank Procedures 4.01) could be useful in this context. See Meijer and Werksman, 2005, pp. 191–211.
16. By 1 January 2009, credits from HFC23 projects stand for more than 50 per cent of all issued CERs: see UNEP Risø, http://cdmpipeline.org/publications/CDMpipeline.xls (visited 1 April 2009).
17. *Joint Implementation Quarterly*, 14(2) (July 2008), notes that '[c]onsequently, when a project generates CO_2 eq. emission reductions but fails to deliver all envisaged sustainable development benefits, this will generally have little contractual consequences for the project partners. Even in cases where host country governments intend to enforce the compliance of the project to its design, they may lack the means to do so', at 5.
18. Also, more substantial criticism of the CDM has become commonplace in the US: see Wara and Victor (2008). In November 2008, the Chairman of the Senate Energy and Natural Resources Committee, Jeff Bingaman, stated at a conference in Washington, DC, hosted by Point Carbon and the Pew Center on Global Climate Change: 'I think this whole issue of offsets, the more I've read about this issue, both international and domestic offsets, is fraught with opportunity for game playing, which will be fully exploited, I'm sure. We have a lot of creative people who can find ways to find offsets and to verify offsets if we open that door to occur' (reference on file with the author).
19. See also 'Kyoto Protocol "Loophole" has Cost $6 Billion', *New Scientist* (February 2007).
20. For an overview of several critical issues with regard to biofuels and the CDM see *Biofuels and the CDM*, CDM Investment Newsletter 3/2007 (Positiv, 2006).
21. P6_TC1-COD(2008)0014, para. 11.
22. Not just for ensuring environmental integrity, but also in order for the CDM to gain international acceptance and to provide a credible means of addressing global (climate) inadequacies and inequities, such 'collateral damage' needs to be addressed.
23. The CDM EB and investors have become concerned about hydro projects due to their potential lack of additionality. One reason was that many of these projects had started well before applying for CDM status. In June 2008, third party validator TÜV SÜD Group rejected a hydropower project in China because the project proponents could not document that they had seriously considered CDM at the time the project was started.
24. For an overview of the project cycle and mandate to all involved UN institutions see CDM Rulebook, at http://cdmrulebook.org (visited 10 March 2009).
25. See '*Carbon Credits Linked to Product Dumping*', Point Carbon, 6 *CDM & JI Monitor* 23, 26 November 2008, indicating that revenues from selling carbon credits from projects reducing nitrous oxide have been used to subsidize and increase the production of adipic acid in some Asian countries.
26. See, for an overview, www.cdmrulebook.org (visited 10 March 2009).
27. These criteria could be formulated similar to those established by the Gold Standard Foundation, a non-profit foundation based in Switzerland. The Gold Standard Foundation offers a quality label to CDM projects with sustainable development benefits (www.cdmgoldstandard.org).

28. See, for example, EU Parliament position, supra note 20.
29. FCCC/KP/CMP/2005/8/Add.1.
30. A first step in this direction is the EB's 'Proposal for an enhanced barrier test for project activities that have a potentially high profitability without CER revenues', CDM – Meth Panel, Thirty-third meeting, Report, Annex 11. http://cdm.unfccc.int/public_inputs/Panels/meth/033/mp_033_an11.pdf (visited 1 April 2009).
31. European Parliament (Committee on the Environment, Public Health and Food Safety) 'Proposal for a Directive of the European Parliament and of the Council Amending Directive 2003/87/EC so as to Improve and Extend Greenhouse Gas Emission Allowance Trading System of the Community' (11 June 2008) COM(2008)0016 – C6-0043/2008 – 2008/0013(COD).
32. UN Doc. ECE/CEP/43, 25 June 1998, entered into force on 30 October 2001.
33. Participatory rights and the right to access to justice in environmental issues are basically included in the 1998 Aarhus Convention on Access to Information, Public Participation in Decision-Making and Access to Justice in Environmental Matters, adopted at Aarhus, Denmark, on 25 June 1998 and entered into force on 30 October 2001, www.unece.org/env/pp/welcome.html. Most sovereign compliance buyers (Annex I States) are Parties to the Convention, while almost none of the host countries have ratified it. The problematic issue is that the inclusion of these rights in the CDM legal framework amounts to a 'creeping extension' of the Aarhus Convention to countries which have not ratified it. Yet, the involvement of sovereign states which have obligations under the Convention could possibly be used as a justification for the 'extraterritorial application' of those rights.
34. In 2007, the UNFCCC Secretariat noted 12 threats of legal proceedings in various legal forums against the EB (FCCC/KP/CMP/2007/2): see Wilder and Millar (2008).
35. For example the EB decided at its 43rd meeting in October 2008 to review 41 project activities for additionality, out of 66 registration requests. See http://cdm.unfccc.int/EB/043/eb43rep.pdf (visited 1 April 2009).
36. For a more detailed discussion of a possible review mechanism see Voigt (2009).
37. See Mueller and Ghosh (2008). These authors suggest the CDM as an implementing tool for measurable, reportable and verifiable developing country mitigation actions according to the Bali Action Plan, para. 1.b.ii (decisions 1/CMP.3). The authors suggest that the generated CERs be retired from the developed countries' accounts and not used as offsets for developed country compliance.
38. 'Poznan Conference to Debate key CDM Revisions', *Point Carbon* (5 November 2008).

REFERENCES

Americano, B. (2008), 'Towards Structural Change for Sustainable Development in Some Sectors', in K. Holm and J. Fenham (eds), *A Reformed CDM – Including New Mechanisms for Sustainable Development* (UNEP Risø Perspectives Series).

Blakker, S.J.A. (2006), *CDM and Biofuels – Can the CDM Assist Biofuel Production and Development?* (Energy Research Center of the Netherlands).

Boer, B., et al. (2007), 'Assessment of Carbon Leakage in Multiple Carbon-sink Projects: A Case Study in Jambi Province, Indonesia', *Mitigation and Adaptation Strategies for Global Change* 12(6), 1169.

Cames, M. (2008), 'Discounting CDM in a Global Carbon Market – Component of a Post-2012 Agreement?', Side Event at the Climate Change Talks, Accra, Ghana, www.unfcccc.org (visited 13 February 2009).

CAN-International (2006), *CCS in CDM: First Things First* (CAN-International).
Castro, P. and A. Michaelowa (2008), *Empirical Analysis of Performance of CDM Projects* (Climate Strategies Report).
Commission of the European Union (2008), *20 20 by 2020 – Europe's Climate Change Opportunity*, Communication from the Commission to the European Parliament, the Council, the European Economic and Social Committee and the Committee of the Regions (Commission of the European Union).
Figueres, C. (2006), 'Sectoral CDM: Opening the CDM to the Yet Unrealized Goal of Sustainable Development', 2 *International Journal of Sustainable Development Law and Policy* 1, 5.
Flues, F., A. Michaelowa and K. Michaelowa (2008), *UN Approval of Greenhouse Gas Emissions Reduction Projects in Developing Countries: The Political Economy of the CDM Executive Board*, CIS Working Paper No. 35 (Center for Comparative and International Studies).
Forests and the European Union Resource Network (FERN) and SinksWatch (2003), *Forest Fraud: Say No to Fake Carbon Credits* (FERN).
Hagem, C. and B. Holtsmark (2008), *Er det noen fremtid for CDM-ordningen?* Samfunnsøkonomen No. 5, 10.
Harthan, R.O. (2008), 'Is the CDM Fulfilling its Environmental and Sustainable Development Objectives? An Evaluation of the CDM and Options for Improvement' (presentation given at a seminar on the CDM, Ministry of Finance, Oslo, 22 May 2008).
Haya, B. and P. McCully (2007), *Failed Mechanism: Hundreds of Hydros Expose Serious Flaws in the CDM Mechanism* (International Rivers Network).
Holm, K. and J. Fenham (eds) (2008), *A Reformed CDM – Including New Mechanisms for Sustainable Development* (UNEP Risø, Perspectives Series).
International Rivers Network and CDM Watch (2002), *Damming the CDM: Why Big Hydro Is Ruining the Clean Development* (International Rivers Network and CDM Watch).
International Rivers Network and CDM Watch (2004), *CDM Large Hydro Status Note for the World Bank/IETA Carbon Expo* (International Rivers Network and CDM Watch).
IOB Evaluations (2008), *Clean and Sustainable? An Evaluation of the Contribution of the Clean Development Mechanism to Sustainable Development in Host Countries* (Ministry of Foreign Affairs, Netherlands).
de Jonge, L. (2008), 'Clean Development Mechanism – Results and Recent Developments (presentation given at a seminar on the CDM, Ministry of Finance, Oslo, Norway, 22 May 2008).
Kallbekken, S., L. Flottorp and N. Rive (2007), 'CDM Baseline Approaches and Carbon Leakage', *Energy Policy* 15, 4154.
Livermann, D. (2008), 'The CDM, Ethics and Development', in K. Holm and J. Fenham (eds), *A Reformed CDM – Including New Mechanisms for Sustainable Development* (UNEP Risø, Perspectives Series).
Luhmann, H.-J. and W. Sterk (2008), *Climate Targets – Should They be Met at Home or Where it is Cheapest?* (Friedrich Ebert Stiftung).
Meijer, E. (2007), 'The International Institutions of the Clean Development Mechanism Brought before National Courts: Limiting Jurisdictional Immunity to Achieve Access to Justice', *New York University Journal of International Law and Policy* 39, 873.

Meijer, E. and J. Werksman (2005), 'Keeping it Clean – Safeguarding the Environmental Integrity of the Clean Development Mechanism', in D. Freestone and C. Streck (eds), *Legal Aspects of Implementing the Kyoto Protocol Mechanisms: Making Kyoto Work* (Oxford University Press).
Michaelowa, A. (2005), 'Creating the Foundations for Host Country Participation in the CDM – Experiences and Challenges in CDM Capacity Building', in Y. Farhana (ed.), *Climate Change and Carbon Markets. A Handbook of Emission Reduction Mechanisms* (Earthscan).
Michaelowa, A. and P. Purohit (2007), *Additionality Determination of Indian CDM Projects: Can Indian CDM Project Developers Outwit the CDM Executive Board?* (Climate Strategies).
Mueller, B. and P. Ghosh (2008), *Implementing the Bali Action Plan: What Role for the CDM?* (Oxford Institute for Energy Studies and Climate Strategies).
Pearson, B. (2004), 'The Clean Development Mechanism and Sustainable Development', *Tiempo Climate Newswatch*, 1.
Planet Ark (2008), 'Rich or Poor? New Faultline in UN Climate Talks', *World Environment News*, 29 August.
Ramachandran, R. (2009), *Success in CDM Projects by Avoiding Pitfalls* (DNV Certification, India).
Schneider, L. (2007), *Is the CDM Fulfilling its Environmental and Sustainable Development Objectives? An Evaluation of the CDM and Options for Improvement* (Öko-Institute, report prepared for Worldwide Fund for Nature).
Schneider, L. (2008), *Key Lessons Learned on the CDM and its Use in the EU ETS. Briefing Paper for a Workshop by Ecologic at the EU Parliament* (3 June), www.ecologic-events.de/cdm-workshop/documents/schneider_briefing.pdf (visited 1 April 2009).
Sterk, W. (2008), *From Clean Development Mechanism to Sectoral Crediting Approaches – Way Forward or Wrong Turn?*, JIKO Policy Paper 1/2008 (Wuppertal Institute).
Sterk, W. and B. Wittneben (2007), 'Enhancing the Clean Development Mechanism through Sectoral Approaches: Definitions, Applications and Ways Forward', *International Environmental Agreements: Politics, Law and Economics* 6(3), 271.
Streck, C. (2007), 'The Governance of the Clean Development Mechanism: The Case for Strength and Stability', *Environmental Liability* 15(2), 91.
Streck, C. and T.B. Chagas (2007), 'The Future of the CDM in a Post-Kyoto World', *Carbon and Climate Law Review* 1, 53.
Streck, C. and J. Lin (2008), 'Making Markets Work: A Review of CDM Performance and the Need for Reform', *European Journal of International Law* 19(2), 409.
United Nations Environment Program (UNEP) (2004), *CDM Information and Guidebook* (2nd edn, UNEP).
Voigt, C. (2008), 'Is the Clean Development Mechanism Sustainable? Some Critical Aspects', *Sustainable Development Law and Policy* 8(2), 15.
Voigt, C. (2009), 'Responsibility for Environmental Integrity of the CDM: Judicial Review of EB Decisions', in D. Freestone and C. Streck (eds), *Legal Aspects of Carbon Trading: Kyoto, Copenhagen and Beyond* (Oxford University Press).
Wara, M. and D.G. Victor (2008), *A Realistic Policy on International Carbon Offsets*, Program on Energy and Sustainable Development Working Paper, No. 74 (Stanford University).
Werksman, J. (2008), 'The "Legitimate Expectations" of Investors and the CDM:

Balancing Public Goods and Private Rights under the Climate Change Regime', *Carbon and Climate Law Review* 1, 95.
Wilder, M. and I. Millar (2008), 'Is Suing the Executive Board the last Resort?', *Point Carbon CDM&JI Monitor*, 28 May.
Wolfrum, R. and N. Matz (2004), *Conflicts in International Environmental Law* (Springer).
World Bank (2004), *Initiatives in Legal and Judicial Reform* (World Bank).
World Rainforest Movement (WRM) (2007), *Proposed Changes to CDM Rules will Give Incentive to Clear Forest and Provide Subsidy for Plantations Industry*, WRM Bulletin N° 119 (June).
Zapfel, P. (2008), DG Environment, *From the EU ETS Review to the Global Carbon Market* (European Climate Change Law Conference, Berlin, 25 July 2008).

11. Beautifying Africa for the Clean Development Mechanism: legal and institutional issues considered

Damilola S. Olawuyi*

1. INTRODUCTION

The Kyoto Protocol is regarded as the most important global agreement of the late twentieth century, not only for setting greenhouse gas (GHG) emission limits for industrialized nations to achieve by 2012, but also for providing them with several flexible mechanisms through which they can achieve those targets. One such mechanism is the Clean Development Mechanism (CDM). Contained in Article 12 of the Kyoto Protocol, the CDM aims to enable developed nations to attain their emission targets by investing in projects that reduce GHG emissions in developing countries.

The CDM was designed to provide an incentive for governments and companies in industrialized countries to invest in GHG reduction projects in developing countries and be credited for GHG reductions achieved through these projects by means of the issuance of Certified Emission Reductions (CERs). Thus, the CDM allows developed countries investing in such projects to achieve their emission reduction commitments with much flexibility and at much lower costs, as it promotes sustainable development in developing countries hosting such projects. It is proposed that the CDM can act as a basis for such nations to achieve progress in environmental issues such as cleaner air and water, reduced deforestation, soil conservation, and biodiversity protection; to realize social benefits such as rural development, employment and poverty alleviation; and to encourage private investment and public–private partnerships in economic development (UNEP, 2005, p. 3). The CDM market has grown substantially in recent years, increasing from a mere five projects in 2003 to 676 by 2006 (Lecocq and Ambrosi, 2007, p. 140).

The CDM, being a market-based mechanism, allows governments of industrialized countries to decide in which developing country they wish to pursue their emission reduction activities. Consequently, like prudent investors, they often favour investment locations which guarantee high emission reductions at the least cost and with the least investment risks. Thus, a country with high mitigation potential, a safe and conducive investment climate and an appropriate legal framework on CDM implementation is considered as a very attractive locale for CDM investments (Jung, 2006, p. 2174).

Unfortunately, due to the distinct economic, social and administrative conditions among developing countries, 67 per cent of them have so far been unable to meet these requirements and have consequently been unable to attract prospective CDM investors (Silayan, 2006, p. 1). Project investors in pursuit of optimal profit have continued to support the same cluster of developing countries which offer the appropriate regulatory frameworks for CDM implementation. Statistics show that over 80 per cent of the current CDM projects are clustered in Asian countries, namely: India, China and Indonesia, collectively referred to by experts as CDM giants.[1] African countries comparatively have not made meaningful progress in terms of the CDM. Statistics show that Africa accounts for only a meagre 2 per cent of current CDM project investments.[2] This trend is attributable to the current perception of African countries as unattractive CDM investment locations (Jung, 2006, p. 2183).

This chapter analyses why African countries have enjoyed little patronage in the CDM market. It shows that the absence of sound legal frameworks governing CDM investments, inadequate institutional capacity, and the high rate of political insecurity in African countries have been the main reasons they have remained unattractive locations for CDM investments. The next section undertakes an overview of the CDM portfolio and a description of the current position of most African countries in the CDM market. Section 3 examines the barriers to the successful implementation of the CDM in most African countries. It shows that the absence of CDM laws, coupled with the archaic and unfriendly provisions in existing laws, is impeding the effective implementation of the CDM in Africa. The final part makes recommendations on the relevant legal and institutional restructuring needed in African countries to enable them to benefit optimally from the sustainable development gains of the CDM.

2. THE CLEAN DEVELOPMENT MECHANISM: AN OVERVIEW

The CDM is widely regarded as one of the most innovative mechanisms of the Kyoto Protocol. This is because it is the only flexible mechanism in the Protocol that allows the direct participation of developing countries in emission reduction activities. The CDM is designed to enable developed countries to transfer and acquire emission reductions by investing in projects which lower emissions in developing countries (United Nations Conference on Trade and Development (UNCTAD) and the Earth Carbon Market Program, 2006, p. 1). This partnership is designed to benefit both developed and developing countries.

Through the CDM, developed countries have the option of meeting their emission reduction commitments in a more cost-effective manner in developing countries. Studies show that it is cheaper to reduce a tonne of GHGs in a developing country than in a developed country.[3] While it takes, on average, US$50 to mitigate one tonne of CO_2 in developed countries, the same can typically be done in developing countries at the cheaper rate of US$15 per tonne of CO_2. Thus, through the CDM, developed countries can achieve GHG mitigation in poorer countries at costs three times cheaper than what would be expended in achieving the same results domestically. Similarly, it was expected that through such 'clean' projects rapidly industrializing developing countries would also be able to solve their environmental, social and economic problems while at the same time generating CERs.[4]

Despite these benefits, the CDM has been criticized for not guaranteeing the same level of participation by all developing countries. Large developing countries such as India and China have enjoyed the patronage of developed countries to the detriment of other smaller developing countries, especially those in Africa. Due to the competitive nature of the CDM market, the mechanism is not offering sustainable development to all the developing countries, as promised by the negotiators of the Protocol. Gupta and others have argued that:

> in all this, the basic criterion – the CDM must 'assist' developing countries for sustainable development – has got lost. Poor countries, with financially-strapped governments, are forced into a mindless competition to facilitate the selling of credits, cheaply and as fast as possible (Gupta, Kazi and Cheatle, 2005, p. 2).

Secondly, the CDM has been criticized for allowing 'business as usual projects' to be re-packaged as CDM projects (CDM Watch, 2004). Although the CDM rules provide that a project must meet the 'additionality' requirement, namely that anthropogenic emissions of GHGs must be reduced beyond a level that would have occurred in the absence of the registered CDM project activity, the governing regime does not provide any standard method for assessing the additionality of a CDM project. Industrialized countries have capitalized on this ambiguity to propose non-additional and 'free rider' projects that would have taken place anyway outside the CDM portfolio.

These criticisms notwithstanding, the CDM remains a proactive mechanism that has a lot to offer developing countries. It opens the door for huge investment flows to the South, providing solutions to social, economic and environmental problems. The CDM is expected to provide developing countries with the essential resources for cleaner economic growth. According to the United Nations (UN), a developing country can, through CDM project investments: attract capital for projects that assist in the shift to a more prosperous but less carbon-intensive economy; encourage the active participation of both its private and public sectors; provide a tool for technology transfer through projects that replace old, dirty and inefficient fossil fuel technology with cleaner; create new industries using environmentally sustainable technologies; and help define investment priorities in projects that meet sustainable development goals (United Nations Environment Program (UNEP), 2004, p. 14). Figueres summarized the benefits of the CDM to developing countries:

> The funding channelled through the CDM should assist developing countries in reaching some of their economic, social, environmental and sustainable development objectives, such as cleaner air and water, improved land-use, accompanied by social benefits such as rural development, employment, and poverty alleviation and in many cases, reduced dependence on imported fossil fuels. In addition to catalyzing green investment priorities in developing countries, the CDM offers an opportunity to make progress simultaneously on climate, development, and local environmental issues. For developing countries that might otherwise be preoccupied with immediate economic and social needs, the prospect of such benefits should provide a strong incentive to participate in the CDM (Figueres, 2004, p. 21).

Clearly, developing countries stand the chance of facilitating sustainable development, economic gains and solutions to challenging environmental issues through the CDM. However, the competitive nature of the market makes it imperative for them to create the necessary climate for such investments if they wish to secure the confidence of CDM investors.

3. HOST COUNTRY ATTRACTIVENESS FOR CDM INVESTMENTS

International Comparisons

Three factors that primarily determine the attractiveness of a CDM host country are its climate mitigation potential; the general investment climate; and the legal and institutional capacity of a country to host CDM projects (Jung, 2006, p. 2174). In essence, any host country with a high mitigation potential (an ability to reduce emissions at the lowest cost), a clear and effective legal framework and adequate institutional capacity for CDM projects, and a conducive and safe investment climate will be considered an attractive spot for CDM investments by developed countries. According to Michaelowa, any developing country that wishes to reap the benefits of the CDM must make itself as attractive as possible (Michaelowa, 2003, p. 201).

These requirements account for why most African countries have not secured significant investment under the CDM portfolio. African nations tend to be perceived internationally as an unattractive milieu for CDM investments. As such, developed countries have chosen to invest in Asia and Latin America where conditions appear more conducive for CDM projects. A cluster analysis of all the developing countries, based on their level of attractiveness for CDM projects, has been developed by Jung (2006, p. 2175). In the 'very attractive' cluster of eight nations, Jung lists only one from Africa, namely South Africa. In the 'attractive' cluster are 13 nations, of which again only one is from Africa (Mauritius). African countries are much more highly represented in the less attractive clusters. Of the 27 countries deemed to be 'attractive to a limited extent', eight are from Africa (i.e., Algeria, Equatorial Guinea, Madagascar, Mali, Morocco, Niger, Uganda, Zimbabwe). Finally, of the some 52 countries worldwide considered to be 'attractive', half are African (i.e., Botswana, Benin, Congo, Cameroon, Central African Republic, Cape Verde, Ethiopia, Gambia, Gabon, Ghana, Guinea, Ivory Coast, Kenya, Liberia, Lesotho, Libya, Malawi, Mozambique, Nigeria, Guinea-Bissau, Rwanda, Seychelles, Senegal, Sierra Leone, Sudan, Togo).

Thus, on Jung's analysis, South Africa is the only African country listed as 'very attractive' for the CDM. Most of its African peers occupy the very unattractive portion of the table. Why is this so? This chapter examines each of the requirements for host country attractiveness, with a view to ascertaining the main reasons for the unappealing position of most African countries for CDM projects. The first two requirements are addressed only briefly, while the last requirement, the bedrock of this chapter: legal and institutional capacity, is canvassed in depth.

Mitigation Potential

Mitigation potential refers to the level of emission reduction that can be achieved in a country and the cost of achieving such reduction. A developing country that offers high emission reduction at the very least cost is said to have a high mitigation potential. The mitigation potential of a country is assessed by the GHG emission intensity, the use of dirty technologies, the use of energy in efficient technologies, and the level of industrial activities carried out in that nation relative to others (Hanh and Michaelowa, 2006, p. 9). Thus, CDM investments will flow to countries that can generate cheap CERs at high volumes. Developing countries that have heavy industries with high GHG emissions tend to offer the best emission reduction prospects. Given the high transaction costs of most CDM projects and the present low value of CERs in the CDM market, a CDM project must be quite large to be economically viable. Thus a large CDM project that offers more CERs will enable an investor to accrue more CERs at about the same price of achieving fewer CERs through smaller projects. This explains why only countries with high GHG emission rates are considered particularly suitable for emission reduction activities (World Bank, 2004, p. 2). Michaelowa and Jotzo (2005, pp. 511–22) make this point when they observe that a minimum volume of 20,000 tonnes of CO_2 is required for a project to be considered cost effective, and that any project delivering less than 100,000 CERs per year is unlikely to entice prospective CDM investors.

Notably, 70 per cent of all CDM projects have been diverted to India and China for this reason, as they offer such economies of scale. For example, South Africa is reputed to be the highest CO_2 emitter in Africa, emitting 1.6 per cent of the total global CO_2 emission. This is followed by Algeria (0.7 per cent), Egypt (0.6 per cent), Nigeria (0.4 per cent) and Libya (0.2 per cent). These cannot be compared to the mega-emission rates of countries like China (18.4 per cent) and India (4.9 per cent).

Nigeria, however flares natural gases in huge proportions, and offers a very promising opportunity for developed countries significantly to reduce GHGs in the process of investing in gas flaring reduction projects. Nigeria currently flares more natural gas associated with oil extraction than any other country on the planet. Estimates suggest that of the 3.5 billion cubic feet of associated gas (AG) produced annually, 2.5 billion cubic feet, or about 70 per cent, is wasted via flaring. This equates to about 25 per cent of the UK's total natural gas consumption, and the equivalent to 40 per cent of the entire African continent's gas consumption (Friends of the Earth, 2004). This gives Nigeria a high rating when it comes to CDM mitigation potential, because it can generate large credits for CDM activities.

Numerous other CDM opportunities also exist in Africa, ranging from small scale biomass and renewable energy projects to larger projects such as electricity generation, fuel switching from coal or oil to gas, biomass in sectors like cement, pulp and paper, mining and refining of mineral resources amongst others.[5] For example, Mozambique offers a high potential for large energy efficiency improving projects (e.g., upgrading of industrial units) and medium potentials for windmill projects, and for projects related to better use of wood resources (improved stoves and efficient charcoal making processes). Botswana also offers high potentials for efficiency improvement in coal based thermo-electrical facilities, energy efficiency improving projects (industrial units revamp/technology transfer), and landfill gas-to-energy projects. Zambia is also reputed to have high potential for energy efficiency improving projects (industrial units revamp/technology transfer), landfill gas-to-energy projects and some potential for windmills. These attributes suggest that, subject to satisfactory performance of African countries in the other indicators for host country attractiveness, these areas should be considered by developed countries as good candidates for CDM investments.

Investment Climate

The second requirement – a positive and secure investment climate – concerns the level of investment risks associated with locating a CDM project in a particular country. The investment climate in a country is the collective set of incentives which establish the 'rules of the game' to which economic actors must adhere.[6] Set by a wide variety of sources, including government policies, the culture of public administration, as well as institutional, social and physical infrastructure, the investment climate determines the level and certainty of returns expected by CDM investors on their investments. According to Hanh and Michaelowa (2006), the CDM investment climate is usually considered to be based on the attractiveness of the general investment climate and associated level of financial and investment risks. Scholars have attempted to establish a link between the performance of a nation in terms of foreign direct investments (FDIs) and its suitability for hosting CDM projects. They take the view that the total FDI in a country offers a broader measure that best illustrates the general investment climate in that country. According to Ellis and others (2004, p. 2), countries expecting to attract the most CDM projects are countries that are recipients of a significant proportion of FDIs. The general investment climate of a country can be broken down into the following three main areas:

(i) *Macroeconomic and Trade Policy* – This factor covers the capacity of domestic institutions to reduce the costs of international trade and finance, and to ensure a consistently safe atmosphere for investments (e.g., fiscal, monetary, trade, and exchange rate policies, administration of customs and ports, security of lives and property, strength of the rule of law, and political stability). Hence, developing countries bedevilled by war, political instability, dictatorship, the absence of the rule of law, and heavy taxation burdens score very low on this point since they are considered too unsafe and risky for investments.

(ii) *Microeconomic Framework* – This focuses on the existence of trade-friendly regulations, predictable government policies and the availability of efficient enforcing agencies devoid of unnecessary bureaucracy and administrative bottlenecks. Thus, a country with flexible and less cumbersome rules on market entry and exit, macro-economic stability, comprehensive legal frameworks on contractual relations, proven enforcement capabilities, and an available pool of skilled workers and other sources of human capital will be perceived as an attractive location for CDM investments (Ellis, et al., 2004, p. 2).

(iii) *Enabling Infrastructure* – This covers the availability of key public infrastructures necessary for production activities and investments, including electricity, land, efficient security service systems, skilled employees, efficient transportation systems and the availability of basic infrastructural facilities. Since these basic infrastructures are pre-requisites for doing business in a country, developing countries that cannot guarantee them are often considered to be unattractive investment locations. Most African countries including Nigeria often score very low when it comes to security issues and investment climate due to the absence of these basic infrastructural facilities and the ineptitude of security agencies in most of these countries.[7]

These three indicators go a long way in shaping the direction of CDM project investments. They are interdependent and do not exist in isolation. As such, a change in policy in one facet affects the general perception of a country in the opinion of investors. Therefore, developing countries wishing to attract CDM investments must ensure good performance on each of these indicators.

Most African countries often score very low points when it comes to security issues relevant to their investment climate. This is due to the persistence of dictatorships, the fragile rule of law, the menace of rampant corruption, the prevalence of ethnic and religious tensions, and the ineptitude of law enforcement agencies in most of these countries (Alao, 1999; Reno, 1998; Sylvester, 1994). For example, Nigeria is increasingly viewed as a 'failed state', due to the high incidence of hostage taking, kidnapping, rape and other violence and civil unrest (Hebst, 1994). This is a major deterrent for CDM investors and indeed most investors of any kind. Other African countries such as Sudan, Somalia, Zimbabwe, Chad, Cote d'Ivoire, Democratic Republic of Congo, Guinea, Central African Republic, Uganda, Ethiopia and Burundi rank prominently on the international lists of unsafe countries and failed states. This is why most developed countries prefer to situate their investment projects in more hospitable developing states such as India and China.

Legal and Institutional Framework

The legal and institutional context has been described as the most important indicator of host country attractiveness (Point Carbon, 2006, p. 2). On this factor, a country is assessed based on its ability to lay down laws which enhance CDM implementation. A legal framework on CDM implementation is vital because it determines how CDM project investments are to be carried out in that country, the nature of projects considered compatible with the sustainable development goals of the country and the investment protection offered to prospective CDM investors. Thus, according to Michaelowa and Jotzo, 2005, p. 511):

> even if a host country has many attractive CDM project opportunities, it will not necessarily mean that many projects will actually be implemented. An effective national institutional structure is necessary to harness the CDM potential and attract investors. At the outset, a country should develop a clear understanding about its approval criteria and sectoral as well as technological priorities. In this process, the competitive nature of the CDM should be kept in mind.

For a developing country to be considered an attractive destination for CDM projects on these criteria, it must at a minimum:

- ratify the Kyoto Protocol;
- enact national laws which domesticate the Kyoto Protocol;
- establish a Domestic National Authority (DNA) to oversee CDM projects;
- identify eligibility requirements for CDM projects such as sustainable development criteria and highlighting specific projects that meet these criteria; and
- enact a comprehensive CDM law which specifies the procedure for proposing CDM projects and for obtaining national approval. This law should also establish the DNA, stating the scope of its authority, its objectives, organizational structure, functions, priorities and mode of operation.

Most African countries score very low in terms of creating the appropriate legal and institutional framework on CDM implementation. Emmanuel Kasimbazi's chapter in this book on implementation of CDM projects in the Ugandan forestry sector illustrates the formidable legal and institutional problems many African states face in creating an appropriate legal context for hosting CDM projects. These barriers will now be explored in detail.

4. LEGAL AND INSTITUTIONAL BARRIERS TO CDM IMPLEMENTATION IN AFRICAN COUNTRIES

Even though a large number of developing countries in Africa have ratified the Kyoto Protocol, most of them have yet to put in place the necessary legal and institutional framework to govern CDM projects locally. This situation has been responsible for the seeming neglect of African countries in the booming CDM market and their reputation as unattractive CDM countries. Aside from the fact that virtually all African countries lack legislation specifically designed to manage CDM projects, some existing national laws act as barriers and disincentives to foreign investments and technology transfer to Africa in this sector. The following discussion canvasses these legal barriers and analyses their impact on the implementation of the CDM, specifically with reference to Nigeria as a case study.

The Legal Barriers

Absence of CDM laws

One of the most significant barriers to CDM implementation in African countries is the absence of a specific legislative or regulatory framework to bring the mechanism into effect and to govern its implementation. The CDM has simply not found its way into the statute books. Without such recognition, the implementation of CDM projects remains legally precarious. The lack of domestic legal reform has meant that African implementation of the Kyoto Protocol including its CDM provisions has tended to rest on less stable policy-based initiatives. Ratification of an international agreement alone may not provide sufficient basis for domestic action.

Section 7 of Ghana's Constitution, for example, stipulates that there must be national legislation in place to give effect to international treaty obligations before they can be binding domestically. Similarly, section 12 of the 1999 Constitution of the Federal Republic of Nigeria provides that 'no treaty between the federation and any other country shall have the force of law except to the extent to which such a treaty has been enacted into law by the National Assembly'. Nigerian courts have consistently maintained that ratification by Nigeria of an international treaty per se does not automatically make it domestic law. In 2000, the Supreme Court of Nigeria in the *locus classicus* case of *General Sani Abacha and 3 others v. Chief Gani Fawehinmi* held that an international treaty can only be said to have come into effect in Nigeria if the provisions of such treaty have been enacted into law by the Nigerian National Assembly.[8] According to the Supreme Court 'when we have an international treaty of this nature, it only becomes binding when enacted into law by our National Assembly . . . it is such law that breathes life into it in Nigeria'.[9]

It cannot be overemphasized that in the absence of a CDM legal framework, investors will be wary of investing in African countries where there may be no legal protection for their investments. Of course, a country does not have to have a dedicated CDM law; the relevant legal provisions could be grafted onto a miscellany of existing legislation. In India, for example, even though there is no domestic statute specifically directed to the CDM, its existing investment and environmental regulations: (a) define clearly the procedure for proposing CDM projects and for getting national approval for such projects; (b) define the nature of projects that can be proposed as CDM projects in the country, and establish the Indian DNA stating the scope of its authorities, objectives, organizational structure, functions, priorities and mode of operation; and (c) grant the DNA full power in relation to the control, registration and discussion of CDM projects in the country (Babu and Michaelowa, 2003). The regulations also provide for easy administrative linkages between the DNA and relevant governmental ministries and parastatals, such as the Energy Ministry, the Environment Ministry, the Immigration Ministry, the Indian National Bank and all the other agencies that will likely be involved during a CDM project cycle. This simplifies the CDM process in India and makes it relatively stress-free for prospective investors. It is therefore unsurprising that the Indian DNA has granted host country approval to more than 300 project proposals, with an emission reduction potential of over 297 million tonnes of CO_2. So far, the CDM Executive Board has also registered over 200 Indian projects.[10]

Conflicting environmental regulations

Another prominent disincentive to CDM investment in African countries is the parallel approval system for environmental issues laid down in the constitutions of most of these countries. CDM investors have met a brick wall in the form of division of powers arrangements, which permit federal and state governments to legislate separately on environmental issues. Emmanuel Kasimbazi's chapter in this book on CDM projects in the Ugandan forestry sector illustrates this situation. Consequently, while the federal government can lay down its own environmental policies, the states can also set out separate environmental policies and adopt separate approval processes for CDM implementation. Thus, CDM project implementation in these countries would generally require approvals from several layers of government. This duplication has made the approval process for CDM investments very cumbersome, compared to the system in some other developing countries such as India and China.

Problematic laws on technology transfer

Another identifiable legal barrier to achieving sustainable development in Africa through the CDM is the narrow scope of the laws relating to technology transfer. Generally, technology transfer includes the process of proliferating technology across the border of two entities. The package may consist of a physical transfer of equipment, technical knowledge, skills and expertise that underline the country's capacity to undertake contemporary *clean* and *end-of-pipe* activities.

Technology transfer to African countries through the CDM portfolio has been difficult because technology transfer laws in most African countries are either archaic or completely silent on the possibility of transferring clean technologies through the CDM. Such laws do not stretch to cover environmental matters, nor do they specifically refer to the CDM or to transfer of technologies under the CDM portfolio. This failure to accord the CDM any recognition means that there is no legal support for transferring technology to these countries through CDM projects. Developing countries must make efforts towards removing all legal barriers to the active process of transferring technology across their borders. Studies show that 67 per cent of developing countries are not benefiting from the CDM because of the existence of domestic laws which make technology transfer very difficult (Silayan, 2006, p. 10). The absence of legal backing for such technology transfer is undoubtedly a major disincentive to prospective investors. Agyemang-Bonsu (2002, p. 7) was making this point about Ghana when he said that the absence of regulations on technology transfer in this country had engendered a lack of awareness among both regulators and investors on energy efficient technologies that could be transferred to Ghana through the CDM. According to him, the barriers to technology transfer in Ghana also include inadequate local capacity (technical know-how) for the installation, operation and maintenance of some of the technologies; lack of a production base for energy efficient technologies; an uneconomic utility tariff; and the absence of a national policy on energy efficient technologies.

Similarly, in Nigeria the regulatory framework for the transfer of technologies from any country is embodied in the National Office for Technology Acquisition and Promotion (NOTAP) Act of 1979. The Act established NOTAP with the responsibility of ensuring the acquisition of foreign technologies. It also has a legal mandate to implement the acquisition, promotion and development of technology through an efficient absorption and adaptation of foreign technology. The narrowness of these legislative provisions has undoubtedly contributed to the limited transfer of cleaner technologies to Nigeria through the CDM. Apart from the fact that the Act does not recognize or mention the transfer of technology into

Nigeria under the CDM, it has also been criticized for not stating clearly the procedures for transferring technologies into Nigeria, and for its use of inadequate modes of screening for incoming technologies. According to Dayo (2001, p. 8):

> the existing protocols used by NOTAP to screen technology acquisition are largely inadequate to handle CDM technologies. Consequently they may impair the smooth transfer of technology into Nigeria if NOTAP's capacity in this regard is not strengthened.

The situations in Ghana and Nigeria are not isolated. Countries like Tanzania, Zimbabwe, Kenya and Mozambique also fall into this category. This must change if African countries aspire to achieve a transfer of cleaner and modern technologies through the CDM.

Gaps in contract laws

Another notable barrier to effective CDM implementation in Africa is the gap in contract laws which affect the preparation of Emission Reduction Purchase Agreements (ERPA). Emissions purchase, being a novel concept, has not been incorporated into the laws on contractual relations in most African countries. Previous studies confirm that it is difficult to use existing contract law to govern the transfer of interests in CERs, as it must be established that the law of the host country recognizes CERs as commodities that can be bought and sold (Sullivan, 2006). For example, the sale of goods in Nigeria is regulated by the *Sales of Goods Act* of 1893, a colonial-era statute of general application received into Nigeria from England. This Act only defines 'goods' as anything capable of being bought, including chattels, items and choses in action. The question however is: do CERs qualify as items capable of being bought in Nigeria? In the absence of a law clarifying this subject, the trading of CERs remains officially unrecognized. Without legal recognition for the exchange of CERs, CDM investors will remain wary of investing in countries where they are not assured of a transfer of the returns of their investments.

Flowing from this is the fact that African lawyers often lack expertise in the preparation of CER transfer agreements (Dayo, 2001, p. 19). The knowledge base on CDM issues among government and private lawyers is still very low. Few African lawyers are conversant with the details of the Kyoto Protocol and its mechanisms. This weakness has constituted a barrier to the smooth transfer of CDM technologies. It cannot be overemphasized that if CERs are to be traded in Africa under the CDM, this requires the expertise and skills of legal draftspersons capable of putting such agreements into writing. The role of emission reduction purchase agreements (ERPAs) in the CDM and the importance of lawyers in CDM project transactions at all levels, including advising on risk allocation ranging from traditional project risks such as project financing and under-performance to developing international legal framework and adequate enforcement of agreements, has continuously been emphasized (Mouchi, 2006, p. 2). It is also vital that CDM project agreements, whether they are build-own-and-transfer, build-own-and-operate or bilateral joint ventures, be drawn up and negotiated by African lawyers who are sufficiently knowledgeable in industrial processes, climate change issues including GHG emissions, international climate law and CDM issues, and technology transfer arrangements, to name some of the pertinent issues. Knowledge of such issues will facilitate the incorporation of key elements in the legal documents.

Institutional Barriers

The effective implementation of the CDM also requires the strengthening of existing government institutions, the creation of special CDM implementation institutions and the establishment of a direct synergy among government institutions, both new and old. In Africa, the absence of legally recognized CDM national authorities, coupled with the lack of coordination amongst existing government institutions and ministries, has posed serious institutional challenges to the successful implementation of the CDM. This section addresses these institutional lacunae and weaknesses.

Absence of a legally recognized DNA

The Designated National Authority is the CDM monitoring body which the CDM rules require to be formed by every developing country for the purpose of approving a prospective CDM investment. The DNA is designed to ensure that the host country participates in a CDM project voluntarily, and to confirm that a project contributes towards the country's sustainable development, in accordance with the standards set by that host country.

According to Manso (2003), the other functions of the DNA include: serving as the focal point between investors and the host country government; providing potential projects for investors; processing framework agreements with investors; ensuring that an environmental impact assessment is carried out before approving projects and considering the assessment reports carefully; providing legal advice for project investors; coordinating with other relevant official entities and authorities within the host country; drawing up standardized baselines; monitoring ongoing

CDM projects; granting export of emissions rights (CERs); and conducting public relations and providing information on CDM implementation in the host country through advertisements in the media and through a website. The DNA is also required to design and establish an evaluation procedure that adopts international eligibility criteria to assess the contribution of the prospective CDM projects to sustainable development in the host country (Manso, 2003, p. 3).

The approach taken by developing countries so far is to establish the DNA within an existing government department or ministry (particularly those that deal with direct foreign investment and trade, environment or energy), through a specific governmental minister, within the existing domestic UN Framework Convention on Climate Change (UNFCCC) focal point for the CDM, or as an independent and new office (UNEP, 2004, p. 24). Whichever approach is adopted, it is essential for the host country to lay down a legal framework that establishes the DNA, equipping it with the necessary powers and decision-making procedures to govern the CDM. Such a law should contain clear statements regarding the DNA's authority, objectives, organizational structure, functions, priorities and mode of operation. Further, it should provide guidelines or directions for conducting public consultation processes to underpin the policies developed by the DNA. The government of a developing country is also expected to decide its national CDM investment priorities and make them easily identifiable in its CDM laws. As Figueres (2002, p. 55) explains:

> [t]his legal instrument shall contain justifications, authorities, objectives and organizational structure, financing functions and procedures that will be the platform for the development and sustainability of the DNA. Furthermore, the DNA should have the authority to grant export of emission rights (CERs).

The DNA must therefore be enabled by legislation which clearly defines the extent of its powers and functions, including the authority to grant export of CERs. In this respect, Figures (2002, p. 56) observes that:

> it is critical to obtain official governmental sanction of the DNA and to enlist the support of key political figures such as the Ministers of Environment, Energy, Transportation, Natural Resources, Agriculture, Trade and Foreign Affairs. Official recognition may come in different forms, depending on the degree of knowledge and acceptance of the CDM and administrative and legal procedures of the country in question. Validation of the DNA may come *from the legislature*, a presidential or ministerial decree or other similar legal instrument. However, it is important to realize that the approval of CDM projects implies allowing the export of emission rights. Therefore the DNA must have the authority to grant this permit. (Emphasis added.)

Most African countries have yet to create their DNAs; of those which have been created, some do not enjoy legal recognition under national laws. This serves as a major disincentive to CDM investments in Africa. Without a functional and legally recognized DNA investors will be reluctant to commit to CDM projects, because returns for investments are not guaranteed, there is no functional contact body that will oversee the successful execution of the project investments, and national laws do not authorize or safeguard their investments.

Absence of a CDM master plan

Preparation of a 'master plan' for a national CDM strategy is another institutional reform that developing countries must undertake in order to boost their attractiveness for CDM project investments (Kituyi, 2006, p. 10). This master plan should contain a description of projects that can be proposed to the DNA under the CDM. It should also elaborate the sustainable development criteria for such projects and provide a clear explanation of the investment procedures in that country. Such a master plan would thereby promote participation in CDM projects and improve coordination among those industries involved in the investments.

This kind of master plan does not currently exist in most African countries. Tanzania, Mozambique, Rwanda, Congo, Ghana and Nigeria for example presently lack a policy document that summarizes their positions on CDM investments. Without such a master plan, CDM implementation in these countries remains largely uncoordinated while investors struggle to know how to take part in CDM investments there. Most African nations have failed to provide clear guidelines on the opportunities and projects that can be explored under the CDM portfolio. They have also not formulated a list of activities that are eligible for the CDM nor a list of business incentives available to CDM investors. Consequently, for prospective investors in Africa the entire CDM investment process is shrouded in uncertainty.

Lack of intergovernmental linkages and coordination

As discussed above, different government ministries and organizations have prominent roles to play in the effective implementation of CDM projects. For example, the finance ministries have a role in supervising currency importation and approval of payment instruments, while the ministry of internal affairs is needed for granting approvals for project execution. Also, regulatory bodies in charge of company registration will be needed to grant approvals for foreign agencies to carry out investment activities, while ministries responsible for energy and transportation must be involved in approving projects involving resources or undertakings within their portfolio jurisdiction. This multilayered set of institutional responsibilities underscores the importance of coordination among these ministries. With solid intergovernmental coordination, granting approvals for CDM projects will be less cumbersome and more straightforward. Such coordination will also simplify the process of sharing information among government ministries, thereby reducing bureaucratic delays.

Sadly, such intergovernmental linkages are very weak in Africa. The execution of CDM projects in many African countries will require the sanction of several government parastatals and government ministries, as is the case in Nigeria, Ghana and Uganda, among other countries. It becomes a laborious task for CDM investors to obtain approvals from each of these ministries because of the weak informational linkages and coordination existing among these agencies and ministries. These weak institutional ties can also cause these bodies to work at cross purposes, and they often require CDM investors to go through separate approval processes.

5. REMOVING BARRIERS TO CDM IMPLEMENTATION IN AFRICA

This chapter has analysed the legal and institutional barriers to CDM implementation in Africa. Thus, how can these barriers to CDM be removed or mitigated? The following recommended actions may be of assistance:

(i) *Domesticating the Kyoto Protocol and the UNFCCC*: It is necessary as a starting point for African states to domesticate and enforce their obligations under the Kyoto Protocol and the UNFCCC. Without domestication, the Kyoto Protocol remains unrecognized as a binding document in many African states, making it difficult for the executive arm of government to create a CDM plan or to carry out binding actions as far as the CDM is concerned. Law-making bodies in Africa should act without delay by domesticating both climate regimes so as to give them life as legally binding documents.

(ii) *Enacting CDM Laws*: African countries should develop CDM laws which comprehensively deal with all the necessary issues hindering the implementation of the CDM in their territories. Such laws should:

(i) Define clearly the procedure for proposing CDM projects and for obtaining national approval.
(ii) Define the nature of projects that can be proposed as CDM projects in that country.

(iii) Establish the DNA and state the scope of its authority, objectives, organizational structure, functions, priorities and mode of operation; the DNA in turn must develop enforceable environmental regulations on CDM projects.
(iv) Grant the DNA full power over the control, registration and implementation of CDM projects in the country.
(v) Provide clear guidelines for the process of consultation, discussion and partnership between the DNA and relevant governmental ministries and parastatals, such as the Energy Ministry, Environment Ministry, Immigration Ministry, National Bank and all other agencies that will likely be involved during a CDM project cycle.
(vi) Clearly recognize CERs as a form of property that can be held and traded within the country.
(vii) Provide a system under which CERs can be traded and elaborate how contracts involving the sale of CERs can be formed and couched, including what is to be sold under the contract and who has the legal entitlement to the CERs.
(viii) Contain provisions which ensure that CERs are properly transferred to the party who is purchasing them.
(ix) Contain clear provisions on determining how CERs are to be sold and transferred; the terms of payment and purchase for the rights; appropriate warranties and indemnities; managing a shortfall in the delivery of CERs; and resolution of disputes related to the sale and purchase of CERs.

(iii) *Amending outdated laws*: Law makers across Africa should also amend existing national laws that hinder CDM implementation. Many African states have rudimentary environmental laws enacted several decades before the Kyoto Protocol was negotiated and, as such, those laws do not contain provisions on climate change, not to mention the CDM. For example, the NOTAP Act which has continued to hinder the transfer of useful environmental technologies to Nigeria through the CDM should be amended to give recognition to technology transfer through the CDM and to make such transfer as straightforward as possible. Similarly, property laws should be amended to recognize CERs as a form of property that can be traded and owned. Concomitantly, contract laws should address how contracts involving the sale of CERs can be formed and couched, and who has the legal entitlement to the CER. Legal reform should also provide for appropriate warranties and indemnities, including how to manage a shortfall in the delivery of CERs and how to resolve disputes relating to the purchase of CERs. To train lawyers in such matters, African law schools should offer courses that will educate future and even current lawyers on these legal requirements and skills.

There is also a need to provide more trade incentives for CDM investors. For example, India amended its Companies Act to include a viable system of trade incentives which reduce the overall cost of carrying out CDM investments in the country. The Indian Companies Act contains different provisions that serve as huge incentives to foreign companies. For example, no government approval is required for foreign direct investments in virtually all the sectors/activities, except a small negative list notified by the government; the use of foreign brand names/trademarks for the sale of goods in India is permitted; and Indian companies are permitted to raise funds from international capital markets while free repatriation of capital investment and profits thereon is permitted provided the original investment was made in convertible foreign exchange (Gupta, 2004; Reed, 2002). Also, special investment and tax incentives are given for exports and sectors such as power, electronics, software and food processing.

In Africa, such attractive trade incentives for the CDM have not yet been put in place. This has to change if African states wish to compete favourably with other developing countries in the struggle to gain the patronage of CDM investors. There is a need to amend existing investment laws by providing more trade incentives which lessen the overall costs of carrying out CDM investments in the respective African states. By offering more trade incentives and more efficient investment regulation, CDM investors will view African countries in a new light: as cost effective locations for such investments.

(iv) *Institutional rearrangements:* Flowing from the above considerations is a need for institutional rearrangements that allow for a smooth implementation of the CDM at all levels of government. African states will have to establish and empower stand alone DNAs on CDM implementation. If possible, the DNA should be made a 'one-stop shop' for all permits and approvals that have to do with the CDM. In India for instance, the DNA serves such a unified function. The Indian DNA, known as the National CDM Authority,[11] offers all the registration, approval and clearance processes for CDM investors. This means that a CDM investor in India needs to deal only with the DNA office to get all approvals needed to execute a CDM project in India. The DNA thus offers a 'single window' clearance facility, and provides 'investor escort services' to simplify the approval process for new ventures. Apart from this, the Indian DNA promises to make decisions on CDM investments within 30 days of a complete application. In most African states, by contrast, CDM investors usually must obtain registration with different government authorities before any project can be executed, and obtain other operational approvals and permits. Each of these approval stages is often tedious and time consuming. Restructuring the DNA in each African state such that it becomes a one stop approval shop for any form of CDM investment will go a long way in simplifying the stress and time associated with getting CDM approvals. Also, further to remove the delays associated with bureaucratic processes in Africa, the DNA should be granted its autonomy as a separate, stand-alone agency, and not an appendage of any government ministry.

(v) *Developing a CDM master plan*: Every African state should devise a comprehensive CDM master plan. In India such a plan, which was drafted after consultations with ministries, agencies and government parastatals, gives a comprehensive summary of the area of priorities, sustainable development needs, procedures for investing and the incentives offered to CDM investors. This master plan is an easy reference document which simplifies all that prospective CDM investors need to know about the process and procedure for investing in India. Likewise, each DNA in Africa should, in consultation with other government agencies and other stakeholders including business and environmental organizations, devise a clear and comprehensive master plan that describes the process and procedure for undertaking CDM investments.

(vi) *Capacity-building*: There is a need for capacity-building and greater public awareness about the CDM in Africa. In Nigeria, there is almost no awareness of the CDM in government ministries, organizations, agencies and even among legal practitioners. Without broad understanding of the role and purpose of the CDM, there is little scope for private participation and public–private partnerships on CDM investments. This situation must of course change if Africa is to have a tangible presence in the CDM market. African DNAs should partner government ministries responsible for public information and education to build greater public awareness of the potential sustainable development gains of the CDM and how the private sector can get involved.

African states must learn from countries like India, China and other nations that have had substantial success in the CDM market, in carrying out the legal, institutional and policy reforms canvassed in this chapter. In a continent often beset with environmental problems, the growing environmental awakening across Africa may make it easier for environmental authorities and policy makers to convince their national governments of the sustainable development benefits of the CDM. Most African countries, like their counterparts in other areas of the South, have a very high potential to mitigate GHGs and vast scope for projects to generate large CERs. At the same time, African leaders must appreciate that realizing those opportunities requires an appropriate legal and institutional framework to implement CDM projects. This chapter is a wake-up call to African legislators and policy makers to take these steps with utmost priority, so that their people can start benefiting from the many sustainable development prospects waiting to be tapped through the CDM.

NOTES

* LL.M (Calgary), Alberta Law Foundation Scholar and Fellow, Canadian Institute of Resources Law.

1. See the UNFCCC Secretariat's list of CDM statistics, http://cdm.unfccc.int/Statistics/index.html (visited 2 April 2009).
2. Ibid.
3. CDMINDIA (2007), 'Why CDM?', www.cdmindia.com/about-background.htm?I1.x=34&I1.y=16 (visited 2 April 2009).
4. Ibid.
5. M. Bess (2005), 'CDM Failing to Deliver for Africa' (23 August), http://copnairobi.energyprojects.net/Documents/environmental_finance_julaug_050714.pdf (visited 2 April 2009).
6. See the World Bank's 'Investment Climate Assessment' portal, www.worldbank.org/investmentclimate (visited 2 April 2009).
7. P. Collier (2004), 'Natural Resources and Conflict in Africa', www.crimesofwar.org/africa-mag/index.html (visited 2 April 2009).
8. [2000] NWLR 228.
9. Ibid.
10. UNFCCC Secretariat (2006), 'Status of Ratifications', http://unfccc.int/files/essential_background/kyoto_protocol/status_of_ratification/application/pdf/kpstats.pdf (visited 2 April 2009).
11. See www.cdmindia.com and http://cdmindia.nic.in (visited 2 April 2009).

REFERENCES

Aaron, C. (2005), *Realizing the Development Dividends: Making the CDM Work for Developing Countries* (International Institute for Sustainable Development).

Agyemang-Bonsu, W.K. (2002), *Ghana's Technology Transfer Needs Assessment: Report on Scoping Phase* (Ghana Environmental Protection Agency).

Alao, C.A. (1999), 'The Problem of the Failed State in Africa', in M. Alagappa and T. Inoguchi (eds), *International Security Management and the United Nations* (United Nations University Press).

Aslam, M. (1998), *The Clean Development Mechanism: Unravelling the 'Mystery'* (Mimeo Press).

Babu, N.Y.D. and A. Michaelowa (2003), *Removing Barriers for Renewable Energy CDM Projects in India and Capacity-Building at the State Level*, HWWA Report 237 (Hamburg Institute of International Economics).

CDM Watch (2004), *Market Failure: Why the Clean Development Mechanism Won't Promote Clean Development* (CDM Watch).

Climate Justice Programme and Friends of the Earth Nigeria (2007), *Gas Flaring in Nigeria: A Human Rights, Environmental and Economic Monstrosity* (Climate Justice Programme and Friends of the Earth Nigeria).

Dayo, F.B. (2001), *Nigeria Clean Development Mechanism (CDM) Phase II Report* (Submitted to the United Nations Industrial Development Organization).

Dhakal, S. (2006), *CDM Market: Size, Barriers and Prospects* (Institute for Global Environmental Strategies).

Doelle, M. (2005), *From Hot Air to Action? Climate Change, Compliance and the Future of International Environmental Law* (Carswell).

Ellis, J., et al. (2004), *Taking Stock of Progress Under the Clean Development Mechanism* (OECD).

Figueres, C. (2002), *Establishing National Authorities for the CDM – A Guide to Developing Countries* (International Institute for Sustainable Development).

Friends of the Earth (2004), 'Gas Flaring in Nigeria', Media Briefing (October).

Gómez-Echeverri, L. (2006), 'Most Developing Countries are Neither Prepared to Address Nor Interested in Climate Change', in Gómez-Echeverri, L. (ed.), *Climate and Development* (UNDP Regional Bureau for Latin America and the Caribbean, and the Yale School of Forestry and Environmental Studies).

Grubb, M. (1999), *The Kyoto Protocol: A Guide and Assessment* (Earthscan).

Gupta, S. (2004), 'Implementing Kyoto-Type Flexibility Mechanisms for India; Problems and Prospects', in Figueres, C. (ed.), *Greenhouse Gas Emission Trading and Project-based Mechanisms* (OECD).

Gupta, R., S. Kazi and J. Cheatle (2005), 'Newest Biggest Deal', *Down to Earth* (November 15).

Hanh, D. and A. Michaelowa (2006), *From GHG Abatement Potentials to Viable CDM Projects – The Cases of Cambodia, Lao PDR and Vietnam*, HWWA Report 259 (Hamburg Institute of International Economics).

Hebst, J. (1994), 'Is Nigeria a Viable State?', *The Washington Quarterly* 19(2), 152.

International Institute for Sustainable Development (IISD) (1997), *Earth Negotiations Bulletin,* No. 68 (December 2).

Jung, M. (2006), 'Host Country Attractiveness for CDM Non-sink Projects', *Energy Policy* 34(15), 2173.

Kituyi, E. (2006), *Attracting Clean Development Mechanism Projects, Prerequisites for African Governments* (African Centre for Technology Studies).

Lecocq, F. and P. Ambrosi (2007), 'The Clean Development Mechanism: History, Status and Prospects', *Review of Environmental Economics and Policy* 1(1), 134.

Manso, P. (2003), *Establishing a National Authority for the Clean Development Mechanism (CDM): The Costa-Rican Experience* (OECD).

Michaelowa, A. (2003), 'CDM Host Country Institution Building, Mitigation and Adaptation Strategies for Global Change', *Mitigation and Adaptation Strategies for Global Change* 8, 201.

Michaelowa, A. and F. Jotzo (2005), 'Transaction Costs, Institutional Rigidities and the Size of the Clean Development Mechanism', *Energy Policy* 33, 511.

Mouchi, E. (2006), 'Legal Aspects of Emissions Reduction Purchase Agreements (ERPAS)' (Presentation at the 1st International Conference on the CDM, Riyadh, Saudi Arabia, 19–21 September 2006).

Odumosu, I. (2006), *Reforming Gas Flaring Laws in Nigeria – The Transferability of Alberta Regulatory Framework,* LLM Thesis (Faculty of Law).

Olhoff, A., et al. (2004), *CDM Sustainable Development Impacts* (UNEP Risoe Centre).

Olsen, K. (2007), 'The Clean Development Mechanism's Contribution to Sustainable Development – A Review of the Literature', *Climatic Change* 84(1), 59.
Point Carbon (2002), 'Is there a Business Case for Small Scale CDM Projects?', *Carbon Market Analyst* (1 December).
Reed, A.M. (2002), 'Corporate Governance Reforms in India', *Journal of Business Ethics* 37(3), 249.
Reno, W. (1998), *Warlord Politics and African States* (Lynne Rienner).
Reuters News Service (1997), 'Delegates Say Prospects Brighten for CO_2 Treaty', (10 November).
Silayan, A. (2006), *Equitable Distribution of CDM Projects Among Developing Countries*, HWWA Report 255 (Hamburg Institute of International Economics).
Sullivan, R. (2006), 'CERSPA: A New Template Agreement for the Sale and Purchase of Certified Emission Reductions (CERs)', *Environmental Liability* 2, 120.
Sylvester, J.H. (1994), 'Sub-Saharan Africa: Economic Stagnation, Political Disintegration and the Specter of Recolonization', *Loyola of Los Angeles Law Review* 27, 1299.
United Nations Conference on Trade and Development (UNCTAD) and the Earth Carbon Market Program Council (2006), *A Layperson's Guide to the CDM* (UNCTAD).
United Nations Environment Programme (UNEP) (2004), *CDM Information Book* (UNEP Risoe Centre).
Weiner, J. (1999), 'Global Environmental Regulation: Instrument Choice in Legal Context', *Yale Law Journal* 108, 677.
Werksman, J. (1998), 'Unwrapping the Kyoto Surprise', *Review of European Community and International Environmental Law* 7(2), 147.
World Bank, et al. (2004), *Clean Development Mechanism in China: Taking a Sustainable and Proactive Approach* (World Bank).
Zhang, Z. (2000), *Estimating the Size of Potential Market for the Kyoto Flexibility Mechanisms* (Proceedings of the IGES International Workshop on the Clean Development Mechanism, 26–27 January 2000, Hayama, Japan).

12. Policy and legal dimensions of CDM projects in the forestry sector: implications for climate change mitigation and adaptation in Uganda

Emmanuel B. Kasimbazi*

1. INTRODUCTION

It has become an important focus in the climate change debate to consider how carbon sequestration through the forestry sector can help to mitigate climate change. For a developing country such as Uganda, carbon sequestration also represents an opportunity to fund sustainable development through investment in afforestation and reforestation. Projects in the forestry sector and environmentally sound land-use practices have the potential to help mitigate climate change by acting as sinks for greenhouse gases (GHGs), particularly carbon dioxide.

Uganda, as a party to the Kyoto Protocol,[1] has undertaken to implement Clean Development Mechanism (CDM) projects in the forestry sector. These projects also bring economic and social benefits to the local communities where they are implemented and to the country at large, thereby promoting climate change mitigation and adaptation. Like many other countries in Africa, Uganda has suffered extensive deforestation in recent decades, resulting in not only the release of GHGs, but also collateral impacts on biodiversity, soil conservation and other environmental values.

The parties to United Nations Framework Convention on Climate Change (UNFCCC)[2] and the Kyoto Protocol recognize that state parties should enact effective environmental legislation to combat climate change.[3] The state parties also commit to taking climate change considerations into account to the extent feasible in their relevant social, economic and environmental policies and actions, and to employ appropriate methods to mitigate and adapt to climate change.[4] As a state party to the UNFCCC and the Kyoto Protocol,[5] Uganda, in compliance with this requirement, has developed a legal and policy framework for sustainable forestry resources management which has implications for the implementation of CDM projects.

The purpose of this chapter is to examine the policy and legal interventions for the implementation of CDM projects in Uganda's forestry sector. The chapter analyses the forestry projects that have been implemented, and how the policy and legal framework contribute to investments of CDM projects in order to mitigate and adapt to climate change in Uganda. The chapter is divided into five further sections. Section 2 provides an overview of climate change in Uganda. Section 3 examines the role of forests in mitigating and adapting to climate change, and identifies CDM projects in the forestry sector. Section 4 reviews the policy and legal framework for CDM projects in Uganda. Section 5 considers challenges for implementing CDM forestry projects in Uganda. The final part concludes with some brief recommendations to address the challenges identified.

2. OVERVIEW OF CLIMATE CHANGE IN UGANDA

Winston Churchill once famously described Uganda, then a British Protectorate, as the 'Pearl of Africa', a remark inspired by his impressions of the beauty and uniqueness of the country's lush environment. Uganda is situated in the mid-eastern part of Africa. It has an area of approximately 241,038 km^2, of which open water and swamps constitute 43,941 km^2 or 17 per cent (Ministry of Agriculture, Animal Industry and Fisheries (MAAIF), 2002, p. 15). Most parts of the land-locked country lie at an average altitude of 1,200 metres above sea level. Owing to its location straddling the equator, two rainy seasons are experienced annually, although they merge as one moves away from the equator (Waiswa, 2003). Mean annual rainfall varies from 750 to 2000 mm between different parts of the country, shaping the geographic distribution of social and economic activities (Orindi and Eriksen, 2005, p. 22). In the highlands and around mountains the elevated landmass exerts a local influence on climate, producing rainfall and temperature patterns that are distinct from those of the humid and warmer lowlands.

Climate in Uganda, particularly rainfall, has been increasingly erratic since the early 1990s. The incidence, duration and amount of rainfall have all exhibited abnormal departures from long-term means. While rainfall in some years was far short of long-term means, thereby causing droughts, in other years it was excessive and produced catastrophic floods. The heaviest rains in recent years were recorded in 1994 and were associated with the

El Niño phenomenon (Republic of Uganda, 2005, p. 3). The rains led to sharp rises in lake levels, widespread flooding, washing away of roads and bridges, and extensive soil erosion and landslides. These events caused the deaths of some 1,000 people, and a further 11,000 people were hospitalized and treated for cholera, and about 150,000 people were displaced from their homes. Damage to the infrastructure was estimated at some US$400 million (Consultancy Africa Intelligence, 2007).

Again, between July and September 2007, floods swept across the eastern region of Teso owing to excessive rainfall. The floods were caused by the La Niña weather pattern in the Pacific Ocean, a phenomenon during which oceans cool down faster due to extremes in temperatures, causing the rainy season to start earlier (Consultancy Africa Intelligence, 2007). The resultant effects have seen people having to deal with floods that are more extreme than usual. Climate scientists believe that global warming may increase the intensity and frequency of El Niño and La Niña cycles, resulting in more unusual extremes in weather patterns as Uganda has recently experienced (Republic of Uganda, 2005, p. 3).

In addition to the variability in precipitation, there have been confusing shifts in the seasons since the early 1990s, with heavy rains falling in the months expected to be dry, and persistent desiccating sunshine experienced in the months that are normally wet and cold (Republic of Uganda, 2005; Directorate of Water Development, 2006). Since Uganda's economy largely depends on agriculture, which is heavily dependent on reliable rainfall patterns, these erratic climatic swings have caused an increase in the frequency of food and water shortages in the country, with the worst hit area being the dry cattle corridor that stretches from the Uganda–Tanzania border to the Karamoja region. In the Karamoja and Teso regions, several deaths from starvation have been recorded in recent years (Republic of Uganda, 2005). Between 1991 and 2000, Uganda experienced seven droughts in a period of ten years compared with eight droughts recorded over the 80-year period between 1911 and 1990. Another effect of climate change is that that the snow cap on the Rwenzori Mountains in western Uganda is receding fast (Gwage, 2004).

In the energy sector, already hydroelectric power generation in Uganda has been significantly affected by changes in precipitation. The low power generation in Uganda is partly attributable to the low water volume of dams, caused by diminished rainfall. This decreased power generation has translated into massive load-shedding which has caused huge economic losses to parts of the national economy dependent on reliable electricity supply.

Some parts of Uganda are arid and semi-arid, commonly experiencing rainfall of below 500 mm (Gwage, 2004, p. 13). Such areas already face problems of water scarcity for domestic, agricultural and hydropower generation. Climate change affects, and will continue to affect, both the quantity and quality of water available in these places. Over-extraction of ground water resources, increased competition and conflicts over water may become common in parts of Uganda where per capita water storage is already meagre. In pastoral areas, severe water shortages resulting from drying up of rivers and reservoirs have contributed to death of livestock from hunger, thirst and disease, and have led to increased conflicts over grazing belts (Gwage, 2004, p. 13).[6]

Natural disasters cannot easily be prevented, but adequate adaptation measures can reduce the magnitude of the impact on the national economy. Whereas developing countries like Uganda can and must engage in climate change mitigation, their priority for now should be adaptation to the impacts of climate change.

3. THE ROLE OF FORESTRY RESOURCES MANAGEMENT IN MITIGATING AND ADAPTING TO CLIMATE CHANGE

Introduction

Forests play a role directly in mitigating climate change by absorbing and storing CO_2, and indirectly by offsetting other emissions of GHGs. Growing forests and plants have enormous capacities for carbon sequestration through photosynthesis (FAO, 2008, p. 14). Long established old-growth and mature forests can store significant amounts of carbon for long periods of time. Nonetheless, when disturbed by human activities, forests no longer play a mitigation role in global warming. Rather, they become part of the problem by becoming a considerable source of CO_2[7] (FAO, 2008, p. 18). Forests act as sources of CO_2 whenever the capacity of the ecosystem to uptake carbon is limited, the rates of photosynthesis can no longer match the increasing rates of carbon concentrations in the atmosphere, or due to anthropogenic or natural ecosystem degradation.

Forests serve as carbon sinks by absorbing carbon from the atmosphere and storing it in the wood, soil and other organic material. Forests also serve as carbon sinks through an increase in biomass, and through management and establishment of forests. Given the role of forests as major carbon sinks, their loss is predicted to increase levels of carbon emissions and continued loss of biodiversity (Magezi, 1998).

Uganda's Forestry Resources

Uganda has about 49,500 km^2 (4.9 million hectares (ha)) of forests and woodlands, covering approximately 24 per cent of its total land area. The vast majority of this area is woodland, while the remainder is tropical high forest and forestry plantations.[8] The western region of the country (Kibale, Kabarole and Bundibugyo districts) has more than 60 per cent of the country's closed forests, while the central region has a little more than 20 per cent (National Environment Management Authority (NEMA), 2007). About 70 per cent of the forest and woodland resources of Uganda lie outside protected areas, making their conservation precarious. These forest and woodland areas often consist of small, scattered patches, making it difficult to protect them through governmental measures.

The Permanent Forest Estate (PFE) covers about 1.9 million ha, including all forest reserve land and all forested areas in the National Parks and Wildlife Reserves.[9] The PFE represents about 9 per cent of the land area of Uganda. These areas are set aside permanently for the conservation of biodiversity, the protection of environmental services and the sustainable production of domestic and commercial forest produce. Half of the PFE comprises the gazetted central and local forest reserves, land that is held in trust and managed by the NFA and local authorities. The other half includes the forested areas of national parks and wild reserve land held in trust and managed by the Uganda Wildlife Authority (Republic of Uganda, Ministry of Water, Land and Environment, 2001). The 70 per cent of forested land in Uganda that grows on private or customary owned areas is not part of the PFE; there is no formal governmental policy that determines forestry management on these areas.

Forests and Climate Change Mitigation and Adaptation

Sustainable forestry management contributes significantly to the mitigation of harmful effects of GHGs. The Intergovernmental Panel on Climate Change (IPCC) has identified several methods for carbon mitigation in the forestry sector. These include: managing forests with high carbon uptake potential, expanding such forests through reforestation and afforestation, reducing deforestation, providing an enabling environment for investments and market access to sustainable forest-based products, as well as increasing the use of forest-based products such as bio-energy and durable wood products, and substituting these for less eco-efficient materials (IPCC, 2001).

Forest-related mitigation activities also provide functions other than climate change mitigation. They support other national development and poverty alleviation priorities because of the multiple benefits forests provide. Uganda's forests and woodlands are central to sustainable development. They play important roles in the social and economic development of the country (DFID, 2004). They also create employment for a large portion of the population. Approximately 100,000 people are involved in the informal sector in fuel wood and charcoal production.

The current annual turnover of business in forestry is about 356 billion Uganda Shillings, with a further estimated annual value of 112 billion Shillings as environmental services (DFID, 2004). It is expected that with the full operationalization of the activities of the Forest Reserve management institutions like the NFA and the District Forestry Service, the contribution of the forestry sector to the economy will triple.

In the energy sector, over 90 per cent of the country's energy demands are met from wood fuels. About 18 million tonnes of firewood and nearly 500,000 tonnes of charcoal are consumed annually (Moyini, et al., 2008). Large volumes of timber are also used for construction, furniture making and other manufacturing, estimated at 800,000 m^3 per year (DFID, 2004). The value of non-timber products derived from forest reserves and other forests, such as medicines, craft materials, and food, are also known to be significant.

Forests also provide environmental services, although these services and values are not easily quantifiable. They are recognized as integral to agricultural productivity, soil and water conservation, and nutrient recycling. Forests also act as reservoirs of the country's biodiversity, including its unique genetic resources and diverse ecosystems (Moyini, et al., 2008). Forest reserves are also crucial for the Ugandan tourism industry Tourism contributes to economic and social development and to resource conservation (Kayanja and Byarugaba, 2001). The Uganda Wildlife Authority reveals that ecotourism revenues are approximately 2.7 billion Uganda Shillings (Kasimbazi, Kateregga and Busingye, 2006).

Forests also have the potential to contribute to national adaptation strategies. Planting forests and sustainable forest management can aid in the protection of soil and land against the detrimental impacts of flooding. In addition, forests can be used to rehabilitate degraded land and maintain water quality by trapping sediments, taking up nutrients, and immobilizing toxic substances. Adaptation strategies that promote sustainable forest management and community-based forest management have the potential not only to protect land and people from some of the harmful effects of rising temperatures, but also to provide opportunities for greater, more sustainable rural development and poverty alleviation through income generation and employment opportunities (Robledo and Forner, 2005).

CDM Investment in Forestry Projects in Uganda

The CDM allows industrialized countries in Annex I to the Kyoto Protocol to implement projects that reduce GHG emissions or, subject to constraints, remove GHG by carbon sequestration or 'sinks', in the territory of non-Annex I Party developing countries such as Uganda. The resulting certified emission reductions (CERs) can then be used by the sponsoring states to help them meet their emission reduction targets. CDM projects must fulfill the following conditions: be approved by all parties involved, lead to sustainable development in the host countries, and result in 'real, measurable and long-term benefits in terms of climate change mitigation'.[10] Crucially, the reductions in GHG emissions must also be 'additional' to any activity that would have occurred without the project. Presently, the eligibility of CDM projects for land use change and forestry activities is limited to afforestation and reforestation.[11]

In order to participate in CDM projects, there are certain eligibility criteria that countries must meet. All parties must meet three basic requirements: permit voluntary participation in the CDM projects, establish a National CDM Authority, and ratify the Kyoto Protocol.[12] Uganda has fulfilled these requirements, and has started to host some CDM projects in the forestry sector, as outlined below.

- *Forest Rehabilitation in Mount Elgon and Reforestation of Rainforest in Kibale National Park Project.*[13] This project is a joint venture between the Uganda Wildlife Authority as the implementing agency and the FACE Foundation of Holland as the investor. The FACE Foundation signed an agreement for 99 years (1994–2093) with the government of Uganda to replant the deforested areas of Mount Elgon and Kibale National Parks in order to sequester carbon, manage water resources and recreate a habitat for diverse wildlife. In return, the government allows FACE to sell the carbon offsets generated. The emissions saved from 27,000 ha of regenerated forest in Kibale and Mount Elgon National Parks amount to 1.5 million tonnes of CO_2, worth US$ 45 million in the prevailing international carbon market.
- *The Nile Basin Reforestation Proj*ect.[14] This project, started in 2006, is implemented by the Uganda National Forestry Authority (NFA) in the Rwoho Central Forest Reserve in South Western Uganda and covers about 2,000 ha. It involves the World Bank BioCarbon Fund as the investor and the NFA as the implementing agency. The project is designed to plant pine and mixed native species. The timber benefits are shared with the local community and the carbon credits are given to the World Bank.

- *The International Small Group (TIST).*[15] This project, begun in 1999, involves the World Bank Biocarbon Fund as the implementing agency, and United States Agency for International Development (USAID) and Dow Chemical Company as the investors. It covers the districts of Bushenyi, Kabale and Kanungu. Under the project about 150,000 trees had been planted as of early 2009.
- *Trees for Global Benefits–Plan Vivo Project.*[16] This project is set to last for nine years (2003–2012). It involves the UK Department for International Development (DFID), USAID, START[17] and Tetra Pak UK as investors, and Ecotrust Uganda, the Edinburgh Centre for Carbon Management (ECCM) and the International Centre for Research in Agroforestry (ICRAF) as implementing agencies. It covers the Bushenyi, Hoima and Masindi districts of Uganda. The project is designed to achieve carbon sequestration through small scale tree planting on 5,000 ha. The timber and other benefits go to the farmers. In 2003 Tetra Pak bought 14,000 tonnes of CO_2 from the project.
- *Bukaleba Forestry Project.*[18] This is a project provided by Tree Farms, a Norwegian forestry company that operates in African countries, and operates in Uganda under the subsidiary name of Busoga Forestry Company. It was established in 1995 with grant aid from the Norwegian Agency for Development Cooperation (Norad) to offset emissions from new gas-fired power stations to be built in Norway. The company obtained a 50 year lease on 5,160 ha in the Bukalcba Forest Reserve on Lake Victoria from the NFA at an extremely low-cost. The project commits to planting between 80,000 and 100,000 ha with eucalyptus and fast-growing pines. The firm anticipates that it will be able to sell 500 tonnes of CO_2 credits per ha, or 2.13 million tonnes of CO_2 to protect the forests in Uganda.
- *Bakojja New Wood County Forest Plantation project.*[19] This project is implemented by the Bakojja New Wood County Forest Plantation Company. It involves plantation of mixed species, and is located in the Mubende District. The total investment is approximately USS4 million, and 4,160 tonnes of CO_2 credits are expected to be generated.

Experience from the above reviewed CDM forestry projects indicates that the projects have the potential to provide opportunities for afforestation and reforestation in Uganda. However, they still face challenges in relation to the technical and financial capacity to implement the projects. This is in addition to limited understanding in Uganda on the design and implementation of CDM projects.

4. POLICY AND LEGAL DIMENSIONS OF CDM FORESTRY PROJECTS AND THEIR IMPLICATIONS FOR ADDRESSING CLIMATE CHANGE

In order for the CDM to be successfully implemented in Uganda, there must be a supportive policy and legal framework. Since participation in CDM projects is voluntary,[20] it is up to individual countries to put in place such frameworks depending on whether they consider such CDM projects to be beneficial to them. On the other hand, arguably designing an effective policy and legal framework for the implementation of the CDM is mandatory once the country undertakes to host CDM projects. This section of the chapter reviews the policy and legal framework pertaining to the implementation of the CDM in the Ugandan forestry sector, focusing on its adequacy for CDM project implementation.

Specific Policy Frameworks

Although governmental policies in Uganda cannot per se be relied on as the basis for any legal action, they can be referred to by regulators and courts in order to clarify issues and proposed directions (Obitre-Gama, 2000). Policies guide government agencies in the implementation of environmental laws and in ensuring the sustainable use of natural resources. Uganda has a variety of policies that are relevant to the implementation of CDM projects in the forestry sector, as this section explains further.

Uganda's Vision 2025

The development aspirations for Uganda were expressed in the Vision 2025 report. Vision 2025 was launched in 1998 as a strategic framework for national development and is one of the strategic framework documents relevant in the implementation of the CDM. The Vision promotes strategies to limit GHG emissions, which include: improving energy efficiency so as to reduce demand and hence the amount of CO_2 generated during energy production; using cleaner energy sources and technologies to reduce emissions of CO_2 and pollutants that cause acid rain and other environmental problems; improving forest management, expanding forest areas and encouraging tree planting to increase the size of carbon sinks in the country; and adopting agricultural practices which reduce emissions of methane and nitrous oxide (Ministry of Finance Planning and Economic Development, 1998). These strategies indicate Uganda's commitment to recognize the role of the forestry sector in the implementation of CDM projects.

Poverty Eradication Action Programme (PEAP)

The PEAP is a strategic document to address poverty in Uganda. It provides the overall framework within which the government's planning and programming of development interventions take place. It guides the identification of priorities, allocation of resources as well as the assessment of progress and impacts of government development programmes, among others. As a mechanism that aims at meeting sustainable development goals, the CDM fits within the PEAP framework. The PEAP recognizes that economic development needs should take account of predicted climate change. It also recognizes the need to take actions such as reforestation and the preservation of wetlands in Uganda, which help to mitigate climate change (Ministry of Planning and Economic Development, 2004, p. 84). The government pledges to strengthen its data collection capacity to ensure the adequacy and timeliness of data needed to generate weather and climate information, with particular focus on reaching the rural poor; it has also promised to carry out an in-depth assessment of user needs, including those of the rural poor, and to investigate, establish and develop the appropriate institutions to take advantage of opportunities under the CDM.

The Uganda Forestry Policy 2001

The main objective of this policy is to establish an integrated forestry sector that achieves sustainable increases in the economic, social and environmental benefits from forests and trees for the people of Uganda, especially the poor and vulnerable (Ministry of Water, Lands and Environment, 2001, p. 15). The policy provides for the protection of the Permanent Forest Estate (PFE) under government trusteeship, as well as the development and sustainable management of natural forests on private land. It also provides for a wide cross section of stakeholder participation in the management of the forests. The policy further encourages partnerships with local forest communities to develop sustainable management of forests. This partnership provides an opportunity for foreign companies to work with the local communities to implement CDM projects through reforestation and afforestation.

Policy Statement No. 3 of the Forestry Policy is relevant to CDM implementation in so far as it provides for commercial forest plantations. The Statement notes that the private sector will play the major role in developing and managing commercial forest plantations. These will either be large-scale industrial plantations on government or private land, or small-scale plantations on farms. The role of government in this respect is to support and regulate this development. It is required to put in place a regulatory framework, which will create a positive investment climate to encourage private sector investment in commercial forest plantations. The government is required, amongst other tasks, to set out priority areas for the development of carbon storage plantations in various parts of Uganda. The Uganda Forestry Policy 2001 also indicates that the government will pursue the following principal strategies, amongst others:

- Strengthen legal agreements between government and private investors for use of the Forest Reserves, and review forest and tree tenure rules, to encourage development of commercial forest plantations.
- Encourage small to medium-scale commercial plantation development, to foster local economic benefits, especially for the poor, women and the youth.
- Strengthen private sector organisations to improve communications, access to market information and technical collaboration in the plantation industry.
- Progressively divest the management of existing commercial plantations on Forest Reserves to the private sector.
- Make the administrative procedures and allocation of permits more transparent, and minimize interference by government in market processes.
- Develop standards of best practice for commercial plantations and disseminate these to the private sector.
- Ensure that social and environmental impact assessments are observed when developing management plans and legal agreements. [Ministry of Water, Lands and Environment, 2001, p. 17].

The 2001 Forestry Policy sets out key elements of support for the implementation of its objectives. Many of these are relevant to CDM projects. For example, the Policy includes sector planning, whereby a National Forest Plan (NFP) will be drafted to provide a strategic framework for the development of the forest sector. Second, the Policy envisages sector investment, whereby the government will develop a favourable regulatory and policy framework for private and public investment in the sector. Section 3 of the Forestry Policy provides Forestry Policy Statements that provide strategies for the implementation of the policy. Policy Statement No. 3 on the role of encouraging profitable and productive forest plantations emphasizes the need for new financial incentives, the removal of market distributions, review of the Investment Code, review of legislation on land leases, the creation of a Forestry Fund, and the development of more transparent and accountable systems of administration and regulation. It is anticipated that this new framework, coupled with active investment promotion, will encourage a range of private investments in commercial forestry. The government will also pursue further sources of funding for sector development including carbon credits and international funds to support biodiversity conservation. Third, the Policy provides for international and regional co-operation. Under this component, the government is required to participate fully in the development and implementation of international obligations and cross-border co-operation agreements. One of the international agreements ear-marked by the Forestry Policy is the UNFCCC (Ministry of Water, Lands and Environment, 2001, p. 28).

A review of the Forestry Policy reveals that it provides a fairly adequate framework for the pursuit of large-scale afforestation and reforestation programmes relevant to the CDM. The Forestry Policy emphasizes the government's commitment to 'promote innovative approaches to community participation in forest management on both government and private forest land' (Ministry of Water, Lands and Environment, 2001, p. 25). The Policy puts a strong emphasis on public involvement, especially by forest adjacent communities, including that they should benefit from forest management including CDM foresty projects.

There are, however, some challenges to using the Policy to implement CDM projects. The first challenge relates to PFE or natural forests. Although sharing benefits from the forestry resources with local communities is provided for in the Policy, there are no guidelines governing such arrangements. The communities are often left with low value items (mushrooms, water ponds and medicinal species).[21] The high value products (reserved timber species and revenue from forestry services such as ecotourism) are acquired by the Responsible Body (the NFA). This may discourage communities from appreciating the protection of, and investment in, natural forests. Another challenge is that there are as yet no guidelines for investment in private forests. As a result, very few private forests have been registered at the local level.

The National Environment Management Policy (NEMP) 1994

The overall goal of the NEMP is 'to establish sustainable social and economic development, which maintains or enhances environmental quality and resource productivity on a long-term basis that meets the needs of the present generation without compromising the ability of future generations to meet their own needs' (Ministry of Water, Lands and Environment 1994, p. 3).[22] The Policy's guiding principles are relevant to the implementation of CDM projects. The Policy recognizes that Uganda's forests provide a wide range of environmental services and values such as the amelioration of climatic extremes and stabilization of soils, which are critical to agricultural development. It also recognizes that private forestry should be encouraged by appropriate incentives, extension services, marketing assistance and increased security of land and tree tenure. To this end, the policy may provide a basis for investment in private forestry.

The Policy also contains relevant strategies for the implementation of the CDM. For instance, it calls for improvement of local capacity to manage protected and gazetted forest reserves by encouraging community participation in forest planning and management. The development of local capacity helps to ensure the viability of CDM projects initiated in the forestry sector. The Policy also calls for economic incentives and the legal framework and technology necessary to encourage and facilitate rural communities, wood-fuel using industries and institutions, and the private sector to be self-sufficient in forest product requirements. The economic incentives under the Policy could include financial support from CDM supporting countries (Minang, et al., 2007, pp. 204–18).

Legal Framework for the Implementation of the CDM in the Ugandan Forestry Sector

The Constitution of the Republic of Uganda, 1995

The 1995 Constitution is the supreme law in Uganda, and has several provisions concerning the environment that are relevant to the conservation and management promotion of forestry resources and CDM projects in this sector. Under Article 27 of the Constitution's accompanying 'National Objectives and Directive Principles of State Policy', the state is expected to enact laws to preserve and protect the environment from abuse, pollution and degradation. It should also promote measures intended to manage the environment for sustainable development and to promote environmental awareness.[23] Conversely, Article 39 of the Constitution itself provides 'every Ugandan' with 'a right to a clean and healthy environment'. The Constitution also imposes a specific duty on the state under Article 13 to protect important resources, including land, water, wetlands, oil, minerals, fauna and flora on behalf of the people of Uganda. Further, under Article 237(2)(b), the national government (or where appropriate, local government) is required to hold in trust for the people and to protect natural forest reserves and any land to be reserved for ecological and tourism purposes for the common good of all citizens.

The trust obligation imposed on the state should facilitate the implementation of the CDM in two ways. First, CDM projects can more easily be implemented by the government than would be the case if the forestry resources were subject to private ownership, since this would involve a lengthy process of land acquisition for the implementation of projects.[24] Second, the trust obligation bars the government from leasing out or otherwise alienating the forests referred to. This constraint should help to ensure perpetuation of CDM projects in natural forests. Yet, the enforcement of the public trust doctrine faces many challenges in Uganda because the government has been degazetting some forest reserves in the name of promoting private investment and industrialization as opposed to preservation of natural forests. The following examples illustrate this position.

In 2000, the government sought to degazette 3,500 of the 6,500 ha protected forest estate of Bugala Island in the Kalangala district for the development of an oil palm estate by BIDCO Uganda. This was protested by civil society organizations and the NFA.[25] Eventually, an environmental impact assessment on the proposed degazettment was conducted and approved by the NEMA with conditions, one of which was that the forest reserves (particularly the Strict Nature Reserve) be excluded from the proposed development plans. Subsequently, local authorities donated the forest land and the project acquired land from the absentee landlords around the forest reserve. The BIDCO project is proceeding and is acquiring more land around the forest.

In June 2001, the government attempted to degazette the Butamira Forest Reserve for the benefit of commercial sugar cane growing by Kakira Sugar Works (KSW). Several ecological, socio-economic, legal and procedural concerns were raised by opponents of the plan. The area's Member of Parliament at the time petitioned Parliament[26] on behalf of local farmers to uphold a decision made in 2000 to evict KSW. However, the government disregarded this and gave a permit to KSW to proceed with the development. This prompted the Advocates' Coalition for Development and the Environment (ACODE) to institute litigation in the case of *ACODE v. Attorney General.*[27] ACODE contended that the government's issuance to KSW of a 50 year sugar cane growing permit in respect of Butamira Forest Reserve was unlawful. The Court considered the provisions of Article 237(2)(b) of the Constitution and section 44 of the *Land Act*. It held that they should be read together, as should Article 237(2)(a) and Section 42 of the *Land Act*. It was thus held by the Court that:

> Butamira Forest Reserve is land which the Government of Uganda holds in trust for the people of Uganda to be protected for the Common Good of the citizens. The Government has no authority to lease out or otherwise alienate it. However, the Government or a local government may grant concessions or licenses or permits in respect of land held under trust with authority from parliament and with consent from the local community in the area or district where the reserved land is situated. In the instant case there was evidence that the permit was granted to Kakira Sugar Works amidst protests from local communities which raised up a pressure group of over 1500 members who depended on the reserve for their livelihood through agro-forestry, and as a source of water, fuel and other forms of sustenance. There was therefore breach of the Public Trust doctrine.

Astoundingly, despite this judicial decision and the ongoing insistence of civil society groups, Parliament still proceeded to approve the permit.

Parliament's brazen actions reflect the reality that the rule of law can be precarious in a country accustomed to one-party rule. As a result, forest reserve land supposedly held in trust by the government for the people of Uganda was converted into a sugar cane plantation owned by a private entity (Tumushabe and Bainomugisha, 2004).

In another controversial example that had a more positive outcome, in 2006 the government proposed to allocate 7,100 of the 32,000 ha of the Mabira Central Forest Reserve to the Sugar Corporation of Uganda (SCOU), owned by the Mehta group, for investment. This attracted criticism from many civil society organizations, and a petition was instituted challenging the government's decision. Mugyenyi, an officer of ACODE, said in a press conference: '[t]he current plan to degazette part of Mabira is the epitome of abuse of Public Trust property vested in the Government to sustainably manage the natural resources for present and future generations' (Tenywa, 2007).

Environmental agencies such as NFA and NEMA protested the government move to degazette Mabira. NEMA strongly opposed the Mabira give-away on economic, environmental, social and ecological grounds.[28] This critcism was followed by public demonstrations on 17 April 2007 against the conveyance of the forest. The organizers of the demonstration had obtained permission from the police to lead a procession to Parliament to deliver a petition and thereafter to hold a protest rally. While the police sought to thwart the demonstration, resulting in violence and several deaths, the protest was ultimately successful in pressuring the government to withdraw the decision to degazette Mabira.[29]

These examples indicate that there remain political challenges to the implementation of CDM projects in natural forests because the will to protect them can be lacking where the government places a higher value on private investment.[30]

The National Environment Act (NEA),[31] 2000

The NEA is the principal legislation governing the environment in Uganda. Its purpose is to provide for sustainable management of the environment, and it establishes the National Environment Management Authority (NEMA) as a coordinating, monitoring and supervisory body. Under section 6(1) of the NEA, NEMA is the main institution responsible for the implementation of the Environmental Impact Assessment (EIA) regime. Projects subject to the EIA process include forestry-related activities such as reforestation and afforestation.[32]

While the NEA generally provides a comprehensive framework for environmental management, climate change issues are not given any special consideration. Nevertheless, some of the Act's provisions are relevant to the implementation of CDM forestry projects. Section 45(2) of the Act provides that 'NEMA shall, in consultation with the lead agency which is the NFA, issue guidelines and prescribe measures for the management of all forests and that all forests shall be managed in accordance with the principle of sustainable development'. These guidelines must take into account forests in protected areas, including forest reserves, national parks and game reserves; and forests on lands subject to interests held by private persons. Under section 45(2), NEA further empowers NEMA, in consultation with NFA, expressly to prohibit human activities in any forest area by declaring it a specially protected forest. Further, section 39 of NEA mandates the District Environment Committees to identify areas for afforestation and reforestation. This policy, however, is restricted to hilly and mountainous areas, and therefore does not adequately address deliberate large-scale afforestation and reforestation initiatives as envisaged by Article 3(3) of the Kyoto Protocol.

Notwithstanding this, the establishment of NEMA along with other institutions under the Act and the obligations imposed on it in relation to forest management ensure that the implementation of CDM projects can be integrated into the overall environmental legal framework of Uganda. This is advantageous to the extent that such integration guarantees facilitation of the implementation process by the government. Sustainable management of the environment including the forestry sector, which is the underlying objective of the NEA, has far reaching implications for climate change mitigation.

However, environmental bodies created to monitor and protect natural resources have not effectively executed their mandates. NEMA, the principal agency responsible for the management of the environment, is required to coordinate, monitor and supervise all activities in the field of the environment. Specifically, by section 19 of the Act, NEMA must ensure observance of proper safeguards in the planning and execution of all development projects, including those already in existence 'that have or are likely to have significant impact on the environment'. The execution of this mandate could require NEMA to halt projects including those in relation to forest resources. Section 67 of the NEA even empowers NEMA to issue environmental restoration orders; however, NEMA has not yet exercised any of these powers in respect of forestry protection.

Further, NFA is required to liaise with NEMA in the protection of Uganda's forest resources and the evaluation of any EIA. Such requirenments are sometimes not followed; in the Butamira case, an EIA was not carried out and the degazettement was undertaken. It can therefore be argued either that these authorities have not effectively executed their mandate in relation to forestry management or that real authority for determining environmental decisions lies elsewhere in the Ugandan government. EIA is designed as a tool for environmental protection, and its fulfillment influences progress in the application of the public trust doctrine. Arguably, if a full EIA were carried out, the environmental impacts that would ensue from the degazettement would have been foreseen and the process thus halted.

The NEMA also must undertake its responsibilities with regard to relevant regional environmental accords that the Ugandan government has entered into. One of these is the Protocol on Environment and Natural Resources Management of 2006.[33] Its objectives are to promote sustainable development and sustainable utilization of the Partner States' environment and natural resources. Article 23 of the Protocol, which is intended to combat desertification and mitigating effects of drought, specifically calls upon the parties to carry out afforestation, reforestation, tree planting and conservation programmes to prevent desertification or mitigate the effects of drought. This provision is relevant to CDM forestry projects because it encourages parties to combat climate change through forestry related activities such as afforestation and forestation.

The National Forestry and Tree Planting Act, 2003[34]

This is the main Act dealing with forestry resource management in Uganda. Its objective is 'to promote the conservation, sustainable management and development of forests for the benefit of the people of Uganda'. It provides for the establishment of the National Forestry Authority under section 52, and for the District Forestry Office under section 48, to manage Central Forest Reserves (CFR) and Local Forest Reserves (LFR) respectively.

Sections 22 and 23 of the Act contain features that are crucial to the implementation of CDM projects. Section 22 specifically provides for the establishment of private forest plantations and declares that all forest produce on a private forest plantation belongs to the owner of the plantation and that the owner is free to use the forest produce in any manner he or she may determine. Section 23 further provides that any person may enter into a contractual or other arrangement with the owner of an interest in a private forest for the right to harvest, purchase, sell all or any part of the forest produce in the private forest. These provisions can be used for private investment in CDM forestry projects.

The Act further provides, under section 46, for the government to extend technical services to persons involved in the development of private forests and forestry activities in general. Such assistance includes the provision of information, training and advice on the management of forests; the establishment and maintenance of nurseries and other facilities necessary for seeds and plants; the provision of materials and/or financial assistance; and assistance with cooperation and liaising with other lead agencies in the management of forests and forest produce. Section 48 further requires each district local government to establish a forestry office that is responsible for management of forest resources in the district. This office is important for providing technical advice on forestry management at the local level, including in relation to CDM investments. Section 49 of the Act requires the government to prepare a National Forest Plan, which shall be the framework for the implementation of the forestry policy and other programmes by the government and other stakeholders in the forest sector.

As noted above, the legal framework for the management of private forests as provided for under sections 21 to 27 of the Act is conducive to private sector investment in CDM forestry projects. These provisions require registration of private forests with the District Land Board. Unfortunately, however, there are no guidelines providing for the step-by-step procedure, nor are incentives for registering private forests provided. This has undoubtedly affected investment in private forests (Byarugaba, 2008).

The Land Act[35]

The *Land Act* governs the tenure, ownership and management of land. Under the Act, land use must comply with the various laws listed in section 43, including the *Wildlife Act*, the *Water Act* and the *National Environment Act*. Section 44 of the *Land Act* reiterates the constitutional provision in Article 237(2) creating a trust over environmentally sensitive areas such as forests. Section 44(4) prohibits the national or local governments from leasing out or otherwise alienating any natural resource referred to in this section. The central government may, pursuant to section 44(5), grant concessions, licenses or permits in respect of such land subject to any law. The permit would authorize use of the land in a manner stipulated by the relevant laws, but would not confer ownership of the land. These rules governing management of forest land in Uganda are important for the implementation of CDM projects because these provisions restrict the conversion of natural forests to other land uses.

The Local Government Act[36]

This Act consolidates and streamlines the existing law on local governments in line with the Constitution, to give effect to the decentralization and devolution of functions, powers and services which have been a long-standing priority of the Ugandan Government under President Museveni (Richardson, 1993). Schedule 2 to the *Local Government Act* provides that it is the responsibility of local governments to protect and preserve natural resources within their district from abuse, pollution and degradation, and to manage these resources for sustainable development. The District Council is the highest local political authority and Schedule 2 to the Act prescribes its functions. These include land surveying, land administration, physical planning, forests and wetlands, environment and sanitation, and protection of streams and lakeshores. The Act specifically assigns the management of forest resources to local authorities including sub-county councils. It is clear from these provisions that local governments should have a pivotal role to play in CDM projects, particularly in the design and implementation of projects which are to be located within their jurisdiction.

5. CHALLENGES FOR IMPLEMENTING CDM PROJECTS IN THE UGANDAN FORESTRY SECTOR

There exist a number of challenges in the implementation of the CDM in the Ugandan forestry sector. The first challenge relates to the impact of tree plantations. While CDM projects improve local incomes and natural resource management, there are concerns that conversion of land into tree plantations can also harm local ecosystems. Local people lose access to land they were using for cultivation as a result of commercial tree planting by CDM private investors. The carbon offsetting targets of certain projects may also be too optimistic. Farmers are evicted from forest reserves in order to create land for commercial planting of trees, potentially causing the farmers to move to new land which they clear to farm, thereby causing CO_2 emissions elsewhere.

Second, sequestered carbon is a relatively new ‘commodity’ introduced by the Kyoto Protocol and it has unique characteristics. Complex legal issues arise from defining the property rights it engenders and from drafting contracts for the sale of carbon sequestration rights. It is not easy to approve, monitor and verify that carbon credits have been earned without clear and transparent rules for permanence and sustainability, given the limited technical knowledge available in Uganda (Brown, et al., 2004). Further, concerns arise because participating in CDM projects requires the signing of international contracts. They require expert negotiation by people experienced in international contracts, commercial law and CDM legal issues, which most Ugandan lawyers are not familiar with.

Third, the implementation of CDM projects at the present stage has yet to deliver on the goals of poverty reduction and forest conservation because of a number of issues relating to conversion of arable land to plantation forestry. Most farmers do not have enough land to plant crops while at the same time planting trees for CDM purposes. The policy frameworks and management protocols to address such challenges have yet to be adequately developed in Uganda. For now, communities around protected forests do not see themselves as forest owners because of the management gap between NFA and themselves.

Fourth, Uganda lacks a strong national institutional framework to implement the Kyoto Protocol mechanisms. There are failures at different institutional and policy levels for environmental management in implementing the Kyoto mechanisms. Policies such as the National Environmental Policy and the National Forestry Policy lack financial and institutional capacity. While it is now largely accepted that the climate is a crucial environmental issue that must be addressed, and implementation of the Kyoto Protocol is expected to be done through a hierarchy of enforcement processes spanning the Ministry of Water and Environment, NEMA, local governments, and the community level. Yet, the enforcement capacity available at some of these levels of governance does not appear to be commensurate with the widespread nature of the problem. In particular, administration at the district and local environment committee levels lacks the financial and human resources necessary to implement and monitor all of the current and proposed CDM projects.

Fifth, there is inadequate private sector funding for CDM projects. As of early 2009, not more than two projects have been implemented solely by private actors in Uganda. Uganda mainly relies on donor funding to implement CDM projects. This adversely affects Uganda's ability strategically to position itself to attract CDM investment in pioneer projects. It is necessary for private Ugandans to form associations to enable them to engage in CDM projects. Private sector investment is also hindered by the limited awareness these stakeholders have about the potential economic benefits of the CDM.

Sixth, Uganda has limited capacity to assess its GHG emissions baseline. Under Article 12(5)(c) of the Kyoto Protocol, emissions reductions may be certified only if they are *additional* to any reductions that would have occurred anyway (the additionally concept). Uganda has not ascertained its emissions baseline against which surplus reductions can be measured. The Marrakech Accord defines the baseline for a CDM project activity as 'the scenario that reasonably represents the anthropogenic emissions by sources of greenhouse gases that would occur in the absence of the proposed project activity'.[37] The methodology used to approve the baseline scenario is complex, and with limited technical capacity it would be harder for most people in Uganda to determine the anthropogenic emissions that would have occurred if the trees had not been planted in a certain locality, thus making it difficult to appreciate the role of planting trees as carbon sinks.

6. CONCLUSION AND RECOMMENDATIONS

The CDM should be an excellent investment opportunity for a developing country like Uganda because the country is endowed with appropriate resources for carbon sinks such as forests and wetlands. Investment by means of CDM forestry projects is a way to ensure long-term sustainable development that is socially equitable and economically beneficial. It provides a more efficient and fair mechanism for international development, and for transfer of technology from the global North to the South. According to the UN Food and Agricultural Organization (FAO), Uganda is currently harvesting forests at 17 per cent above the sustainable level, but CDM investment may turn this round and improve carbon sequestration (Walkera, et al., 2008, p. 14).

However, the investment in carbon-sink projects is constrained by high costs for both the investors and host countries (Cacho, Marshall and Milne, 2005). This problem may be particularly severe for projects involving small landholders in Uganda. Of particular concern are the transaction costs incurred in developing projects, such as costs in relation to measuring, certifying, and selling the carbon sequestration services generated by such projects. Investment in CDM projects is further constrained by the lack of a comprehensive climate law and policy framework in Uganda. The legal and policy provisions are scattered across various instruments, making implementation complicated. This problem is compounded by the lack of transparency and poor implementation of these existing legal and policy provisions. To address this problem, Uganda needs to develop a specific climate law and policy framework which would encompass the forestry sector. The comprehensive legal and policy instruments needed to aim specifically at developing climate change mitigation measures are necessary for optimal implementation of CDM investments projects.

The limited financial, technical and human capacity of most government institutions must also be enhanced. The government must step up capacity-building efforts in all of the state entities involved in CDM implementation. This should go hand in hand with raising awareness about the role of CDM projects in mitigating climate change and how the projects can be implemented. The CDM is a highly technical mechanism which can only be understood and appreciated by those with specialized knowledge and skills. It is recommended that the Department of Meteorology, NEMA,

NFA and the recently created National Climate Change Secretariat, be tasked with these goals.

In addition, a CDM Fund to provide financial assistance is necessary for the mechanism to operate effectively. In view of its very limited financial capabilities, Uganda must lobby developed countries to contribute towards this CDM Fund through financial bodies established under the Kyoto Protocol, such as the GEF, the Special Climate Change Fund and the Least Developed Countries Fund. Presently, the international levy on CDM projects is earmarked for the Kyoto Protocol's Adaptation Fund, and such money is not available to enhance developing countries' capacity to address climate mitigation through CDM projects and other initiatives.

Capacity in the area of climate research and data collection must also be enhanced. It is clear that Uganda still has limited capacity to monitor weather and climate change, and yet such information is required to support adaptation activities as well as long-term monitoring of climate change. A good network of climate change monitoring is critical for the provision of accurate and timely information for adaptation activities, including early warning systems. It is therefore important that the national meteorological services be strengthened. This should be done through improving climate observation mechanisms (for example through adopting advanced technology in climate change observations); improving the communication systems for efficient climate data collection and exchange; improving the data processing and archiving system to ensure the availability of quality data for climate monitoring and operations research; and improving the capacity for climate data management systems through the training of personnel in data monitoring and processing.

Ultimately, CDM forestry projects cannot assist in mitigating or adapting to climate change unless there is a long-term action plan. Long-term mitigation/adaptation plans must take into account the vulnerability of various sectors of the economy. These plans should be consistent with, and should reinforce, existing principles and sectoral plans. The following goals should guide the development of such mitigation/adaptation action plans: strengthening of the early warning information capacity, especially for food security and short-term climate prediction; incorporating climate change and variability information and projections into long-term development plans; carrying out an inventory of existing practices and methods used to adapt to extreme climate events; and creating participatory and consultative approaches. The effectiveness of the above approaches depends on several considerations, including the underlying nature of climate risks, the technical knowledge about climate change, the development context of decision-making and time-lags to realize the benefits of implementing the specific approach.

This chapter has thus highlighted how the realities of CDM forestry projects in Uganda – and possibly in many other developing countries – can be far removed from the international rhetoric sometimes associated with the CDM as a vehicle for successful North–South collaboration in mitigating climate change. It is unrealistic to expect that the CDM can leverage significant improvements unless there are more comprehensive reforms to environmental law regimes in host countries. Such reforms, in turn, tend to hinge on addressing the broader challenges of institutional capacity-building, promotion of the rule of law, and addressing poverty and social injustice. Such challenges must therefore be recognized by the international community in the negotiations of the Copenhagen Protocol if it is to provide a tangible advance on the Kyoto regime.

NOTES

* Senior Lecturer, Faculty of Law, Makerere University, Uganda.
1. Kyoto Protocol to the United Nations Framework Convention on Climate Change (1997). (1998) ILM 37, 22.
2. United Nations Framework Convention on Climate Change (1992) ILM 31, 822.
3. Ibid., the Preamble.
4. Ibid., Art. 4 (1)(f).
5. Uganda signed the UNFCCC in 1992 and it ratified it on 8 September 1993; it acceded to the Kyoto Protocol on 25 March 2002.
6. There have been increased conflicts due to drought over grazing belts especially among the pastoralists on the boarders of Kenya and Democratic Republic of Congo.
7. According to the FAO, deforestation in the tropics is responsible for as much as one fifth of all greenhouse gas emissions.
8. A woodland is an area covered in trees, differentiated from a forest. A forest is a largely closed canopy and the branches and foliage of trees interlock overhead to provide extensive and nearly continuous shade. A woodland, on the other hand, allows sunlight to penetrate between the trees, limiting shade. Woodlands may support an understory of shrubs and herbaceous plants (often including grasses). Woodlands may form a transition to shrublands under drier conditions. Forestry plantations are trees grown by individuals, communities and state forestry authorities. See Millennium Environmental Assessment (2005), *Current State and Trends Assessment: Volume 1* (Island Press), 590–91.
9. Permanent Forest Estates or natural forests are governed by the *National Forestry and Tree Planting Act* 2003; those in wildlife reserves are governed by the *Uganda Wildlife Act*, chapter 200 of the Laws of Uganda.
10. These are requirements provided under Art. 12 of the Kyoto Protocol.
11. This was decision -7/CP.7 of the Report of the Conference of the Parties serving as the meeting of the Parties to the Kyoto Protocol on its first session, held at Montreal from 28 November to 10 December 2005.
12. Ibid., Decision 3/CMP.1.
13. The FACE Foundation ((Netherlands) is a non-profit organization established by the Dutch Electricity Generating Board. It thus works on behalf of the Dutch energy group. See www.stichtingface.nl.

14. Details about the project can be found at http://cdm.unfccc.int/Projects/Validation/DB/FQGS8PBL08KI02MK9PEC5TY2K3G1MS/view.html. See also the World Bank Carbon Finance website at http://wbcarbonfinance.org/Router.cfm?Page=Projport&ProjID=9644.
15. Details about the project can be found at www.tist.org.
16. For details about the project see www.erm.com/extranet/UK/resource.nsf/Files/VoluntaryStandards_PlanVivo_Ecotrust_PNantongo/$FILE/VoluntaryStandards_PlanVivo_Ecotrust_PNantongo.pdf.
17. SysTem for Analysis, Research and Training, 'sponsored by the Earth System Science Partnership (ESSP) comprising the International Geosphere-Biosphere Programme (IGBP), the World Climate Research Programme (WCRP), and the International Human Dimensions Programme on global environmental change (IHDP), and DIVERSITAS'. See www.start.org/index.html.
18. For details about the project see http://64.233.169.132/search?q=cache:ZzSK8jQxtYoJ:www.wdm.org.uk/resources/reports/climate/submissiontotheEAConoffsetting29012007.pdf + Bukaleba + Forestry + Project +in+Uganda+with+tree+farms+norway&hl=en&ct=clnk&cd=7&gl=us.
19. See www.icex.es/ProtocoloKIOTO/Uganda/Uganda%20Cartera%20de%20Proyectos%20mayo%202006%20.pdf.
20. See participation requirements set out in Decision 15/CP.7 of the Report of the Conference of the Parties on its 17th Session held in Marrakesh from 29 October to 10 November 2001 on the Principles, Nature and Scope of the Mechanisms pursuant to Arts 6, 12 and 17 of the Kyoto Protocol.
21. Empowering Civil Society for Participatory Forest Management in East Africa, www.care.org/careswork/projects/UGA088.asp.
22. This statement is closely related to that expressed by the Brundtland Commission, formally the World Commission on Environment and Development (WCED), created in 1983 to address growing concern 'about the accelerating deterioration of the *human environment* and *natural resources* and the consequences of that deterioration for *economic* and *social* development'.
23. Constitution of the Republic of Uganda of 1995, Art. 245.
24. This position is fortified by s. 44(4) of the *Land Act*, chapter 227 of Laws of Uganda 2000.
25. For example the NFA Executive Director, Olav Bjella, resigned as a result of his opposition to the degazettement of the forest: see http://64.233.169.104/search?q=cache:Hy9uDF1s78J:earthhopenetwork.net/Uganda_Forest_Faces_Bulldozers.htm+NFA+officials+resign+over+mabira&hl=en&ct=clnk&cd=3.
26. Dr Frank Nabwiso, MP Kagoma County Jinja District.
27. Miscellaneous Cause No. 0100 of 2004.
28. *Muyita Daily Monitor* (29 March 2007).
29. *Muyita Daily Monitor* (18 April 2007).
30. For instance President Yoweri Museveni commented that critics who opposed the forest give away were 'people who don't see where the future of Africa lies'. Further while responding to a question asked him about the Mabira give away he said, 'We know what we are doing and what we are doing is in the interest of the people'.
31. *National Environment Act*, chapter 153 of the Laws of Uganda 2000.
32. This requirement is provided under the third Schedule of NEA and the Environmental Impact Assessment Regulations, 1998.
33. Available at www.eac.int.
34. *National Forestry and Tree Planting Act* No. 8 of 2003.
35. *Land Act*, chapter 227 of the Laws of Uganda, 2000.
36. *Local Government Act* Chapter 243 of the Laws of Uganda 2000.
37. The baseline (or 'baseline scenario') for a CDM project activity is defined in 3/CMP.1, Annex, para. 44 of the Marrakesh Accords adopted at the COP in its seventh session, held at Marrakesh, Morocco from 29 October to 10 November 2001.

REFERENCES

African Development Bank, et al. (2003), *Poverty and Climate Change: Reducing the Vulnerability of the Poor through Adaptation* (African Development Bank).

Brown, K., et al. (2004), 'How do CDM Projects Contribute to Sustainable Development?' Technical Report 16 (Tyndall Centre for Climate Change Research, University of East Anglia).

Byarugaba, S.R. (2008), *Policy and Legal Provisions on Community Forests and Private Forests Community Forests* (report for Food and Agricultural Organization).

Cacho, O., G.R. Marshall and M. Milne (2005), *Transaction and Abatement Costs of Carbon-sink Projects in Developing Countries* (Cambridge University Press).

Consultancy Africa Intelligence (2007), *Climate Change and Increasing Floods in Africa: Implications for Africa's Development* (Consultancy Africa Intelligence).

Directorate of Water Development (2006), 'Water, a Shared Responsibility', *The Uganda National Water Development Report,* Prepared for the 2nd UN World Water Development Report (Directorate of Water Development and the World Water Assessment Programme).

Dutschke, M. (2002), 'Sustainable Forestry Investment Under The Clean Development Mechanism: The Malaysian Case', Discussion Paper No. 198 (Hamburgisches Welt-Wirtschafts-Archiv (Hamburg Institute of International Economics)).

Gwage, P. (2004), *Briefing Document on the UNFCCC and its Kyoto Protocol*, Department of Meteorology (Ministry of Water, Lands and Environment).

Huq, S., et al., (2003), *Mainstreaming Adaptation to Climate Change in Least Developed Countries (LDCs)* (International Institute for Environment and Development (IIED)).

Intergovernmental Panel on Climate Change (2001), *Climate Change: Impacts, Adaptation and Vulnerability. Contribution of Working Group II to the Third Assessment Report of IPCC* (Cambridge University Press).

Kasimbazi, E., E. Kateregga and G. Busingye (2006), *Synergy, Cost Benefit and Socio-Impact Assessment of National Environmental Policies and Laws, Rio MEAs and related MEAs and their efficiency in poverty alleviation in Uganda* (Consultancy report, submitted to the National Environment Management Authority and United Nations Environment Programme).

Kayanja, F.I.B. and D. Byarugaba (2001), 'Disappearing Forests of Uganda: The Way Forward', *Current Science* 81, 936.

Magezi, S. (1998), *Potential for Clean Development Mechanism Projects in Uganda* (Ministry of Water Lands and Environment).

Millennium Environmental Assessment (2005), *Current State and Trends Assessment: Volume 1* (Island Press).

Minang, P.A., M.K. McCall and H. Bressers (2007), 'Community Capacity for Implementing Clean Development Mechanism Projects Within Community Forests in Cameroon', *Environmental Management* 39(5), 615.

Ministry of Agriculture, Animal Industry and Fisheries (MAAIF) (2002), *The Second National Report to the Conference of Parties on the Implementation of the United Nations Convention to Combat Desertification (UNCCD) in Uganda* (MAAIF).

Ministry of Agriculture, Animal Industry and Fisheries and the Ministry of

Finance, Planning and Economic Development (2000), *Plan for Modernization of Agriculture: Government Strategy and Operational Framework* (MAAIF).
Ministry of Water, Lands and Environment (MWLE) (1994), *The National Environment Management Policy for Uganda* (MWLE).
Ministry of Water, Lands and Environment (MWLE) (2001), *The Uganda Forestry Policy* (MWLE).
Moyini, Y., et al. (2008), 'Economic Valuation of the Proposed Degazettment of Part of Mabira Central Forest Reserve', Nature Uganda.
National Environment Management Authority (NEMA) (2000), *State of Environment Report 2000/2001* (NEMA).
National Environment Management Authority (NEMA) (2006), *State of Environment Report 2006/2007* (NEMA).
Obitre-Gama, J. (2000), 'The Application of International Law into National Law, Policy and Practice', Paper presented at the WHO International Conference on Global Tobacco Control Law (New Delhi, India, 7–9 January 2000).
Orindi, V.A. and S. Eriksen (2005), 'Mainstreaming Adaptation to Climate Change in the Development Process in Uganda', Ecopolicy Series No. 15 (African Centre for Technology Studies).
Republic of Uganda (2005), *The Uganda National Water Development Report.*
Richardson, B. (1993), 'Environmental Management in Uganda: The Importance of Property Law and Local Government in Wetlands Conservation', *Journal of African Law* 127(2), 109.
Robledo, C. and C. Forner (2005), *Adaptation of Forest Ecosystems and the Forest Sector to Climate Change* (UN Food and Agricultural Organization).
Tenywa, G. (2007), 'Uganda: NRM MPs Oppose Mabira Giveaway', *Earthwire Africa* (4 April).
Tumushabe, G. and A. Bainomugisha (2004), *Constitutional Reform and Environmental Legislative. Representation in Uganda: A Case Study of Butamira Forest*, ACODE Policy Research Series, No. 10 (Advocates' Coalition for Development and Environment).
UK Department for International Development (DFID) (2004), *Climate Change and Poverty. Making Development Resilient to Climate Change* (DFID).
UN Food and Agricultural Organization (FAO) (2008), *Forests and Energy: Key Issues* (FAO).
Waiswa, M.M. (2003), 'Strategic Choices for Enhancing Capacity of Rural Communities to Adapt to Climate Variability: A Case for Uganda', Contribution to the NOAA Office of Global Programs Workshop, *Insights and Tools for Adaptation: Learning from Climate Variability* (Washington, DC, 18–19 November 2003).
Walkera, S., et al. (2008), *Carbon Market Opportunities for the Forestry Sector of Africa* (Winrock International).

13. Brazilian policy on biodiesels: a sound means of mitigating climate change?

Solange Teles da Silva* and Carolina Dutra**

1. THE MAJOR ISSUES

Climate change, resulting from the accumulation of greenhouse gases (GHG) in the atmosphere from anthropogenic activities, is the major twenty-first century challenge for humankind. Changes in precipitation, leading to floods, droughts and heat waves, as well as sea level rises and other ecological impacts, will have a profound impact on human life and the entire biosphere. Under the international regime of the United Nations Framework Convention on Climate Change (UNFCCC) and its Kyoto Protocol, the climate system is presented as a shared resource the stability of which is a 'common but differentiated responsibility' of all nations to ensure. Yet, while developing countries do not have quantified GHG emission limitations or reduction obligations under the current international regime, their governments, including that of Brazil, have introduced some measures that could help address climate change. One such measure is the development of biofuels, which are touted by some people as a clean, renewable energy source.

Already, biofuels have emerged as significant sources of energy in some countries. In 2003, renewable energy provided 13.3 per cent of the world's total primary energy supply, with biofuels accounting for nearly 80 per cent of all renewable energy (Food and Agricultural Organization, 2007, p. 98). For several decades, Brazil has been the world's leading producer and consumer of biofuels. Yet, even here, fossil fuels such as oil, coal and natural gas still account for about half of Brazil's energy matrix (Brazil, Ministério de Minas e Energia, 2007). The economic sector that consumes most fossil fuels is transportation, accounting for almost 50 per cent of Brazil's oil consumption (Bajay and Badanham, 2005). The adulteration of such fuels with 'bio' components may both promote environmental benefits and reduce national dependence on fossil fuel imports.

Many governmental authorities and scientists consider biodiesel and other types of biofuels a renewable energy that can help reduce emissions of GHGs (Farrell, et al., 2006). Thus, projects to expand their use could be important candidates for Clean Development Mechanism (CDM) projects under the Kyoto Protocol. Apart from their climate benefits, biofuels may help reduce local air pollution from motor vehicles, particularly in the heavily polluted cities of São Paulo and Rio de Janeiro (Goldemberg, 2003). Using biodiesel in commercial and mass transportation can improve air quality because, besides its lower sulfur oxides content, it can reduce emissions of carbon monoxide, hydrocarbons and particulates (CENBIO, 2003).

Besides these environmental benefits, the biofuels industry may provide an important economic stimulus to the agricultural sector through the creation of new markets, employment opportunities and enhanced rural incomes (National Biodiesel Board, 2006). A further anticipated economic benefit of a local biofuels industry is that it can promote national energy security by enabling a country to reduce its reliance on foreign oil imports of increasingly uncertain supply and price. The United States' Energy Policy Act[1] of 2005 and the Energy Independence and Security Act[2] of 2007 are premised on this concern, and include provisions to stimulate the country's biofuels industry. In the European Union, authorities are also legislating provisions to increase the uptake of biofuels, including requiring fuel producers or importers to meet minimum biofuel content standards in their fuels (Schlegel and Kaphengst, 2007). Therefore, biofuels promise to allow states to meet the policy challenges of global warming while concomitantly achieving economic gains that do not seriously jeopardize the economic status quo.

Yet, the enthusiasm which the biofuels sector has received in recent years from some governments and industry groups has exceeded the necessary careful scrutiny of all the potential social and environmental consequences of these alternative fuels. Eagerly embraced and promoted by Brazil and other states despite evidence that biofuels may simply lead to swapping one set of environmental problems for another, the need for regulatory standards in this sector should be apparent. The criticisms of biofuels are clustered around two sets of concerns: one involves the science of energy generation and the other the social and environmental context of production. The first of these concerns challenges whether biofuels actually offer a net energy output when one considers their entire production and consumption (Pimentel, et al., 2002). Some scientists believe that biofuels require more energy, including that of fossil fuels, than they produce (Kaltschmitt, Reinhard and Stelzer, 1997). The second concern, discussed more fully later in this chapter, relates to deforestation and other environmental impacts that may accompany biofuels production. Therefore, properly to evaluate biofuels' contribution to addressing climate change, it is important to examine their production and use in a comprehensive life-cycle context (i.e., from well-to-tank or well-to-wheel, as some may describe the cycle).

Biofuels are comprised of two principal families – ethanol and biodiesel. Ethanol is commonly produced from corn and sugar cane, while biodiesel is made from oilseeds, such as soybeans and palms, as well as from waste oils and fats of organic origins. Because particular crops or other raw materials are associated with each type of biofuel, they therefore may involve different production methods with divergent environmental impacts. For example, production of biodiesel from vegetable oils requires tracts of land to cultivate the seeds, while production of biodiesel from waste oils and fats simply utilizes (with modifications) existing waste management systems. The primary environmental policy challenges presented by biofuels are thus to find the source materials that best reduce GHG emissions while minimizing any collateral environmental impacts, such as on biodiversity in areas dedicated for production of raw materials.

While the Brazilian government has long emphasized the need to retain family-based agricultural systems, Brazilian biodiesel production from vegetable oils has come mostly from large scale plantations. The industrial scale of some of these mammoth farms, such as those producing soybeans, has been associated with increased deforestation. This problem has been particularly evident in the northern and central regions in Brazil, respectively in the biodiversity-rich Amazon and Cerrado biomes. Approximately 75 per cent of Brazil's carbon dioxide (CO_2) emissions in recent decades have been tied to forestry and land use changes, and of that total almost 90 per cent correspond to conversion of forests to other uses, particularly agriculture and cattle ranching (Brazil, Ministry of External Relations, et al., 2007).

Thus, any discussion of biofuels in Brazil must recognize this complex context, which requires multi-faceted responses that draw upon economic and energy policies, as well as natural resources management. This chapter examines the regulatory and policy framework in Brazil concerning biofuels, with particular reference to biodiesel. It evaluates whether it can truly be considered a 'green fuel' contributing to climate change mitigation. In order to analyse the environmental integrity of biodiesel, the chapter firstly reviews the different forms of biodiesel produced, including the attendant risks and impacts on deforestation, largely with reference to the soybean production areas in northern and central Brazil. Secondly, the chapter analyses various aspects of biodiesel production in Brazil, highlighting the importance of good environmental management in this sector. Finally, some Brazilian activities pursuant to the UNFCCC and Kyoto Protocol are examined, including bioenergy projects advanced through the CDM. The central message of this chapter is that an environmentally sustainable biofuels industry that contributes positively to climate policy cannot be based on an unregulated market. Rather, Brazil like other states must closely regulate the industry, and do so on a life-cycle basis that ensures that all environmental impacts are factored into the entire biofuels production process.

2. BRAZILIAN BIODIESEL: A 'GREEN FUEL'?

Historical and Economic Context of Biodiesels Development

Biofuels have been promoted as a technologically viable and clean energy alternative to aid the transition from the oil-based economy to an ecologically sustainable, post-fossil fuel economy. Yet, whether they are ethanol or biodiesel, biofuels do not offer a miraculous solution. The following analysis of the Brazilian biodiesel industry examines the entire life-cycle of its production, highlighting the various costs and problems along the way, and the difficulty of regulating them.

Biodiesel has a history as almost as long as conventional gasoline. During the 1900 World Exhibition, in Paris, inventor Rudolph Diesel unveiled an engine powered by a mix of petroleum and natural oils derived from peanuts, vegetables and even fish (Grosser, 1978). Diesel fuel appeared only with the advent of the direct injection engine in the 1950s. Gradually, any use of organic oils as a staple transportation fuel was abandoned in favor of petroleum-based fuels due to their commercial and technical advantages and the absence of any environmental considerations until recently.

However, it became evident that the resulting dependence on such fossil fuels had drawbacks when the international oil crisis erupted in 1973, as Middle East countries reduced supply and raised crude oil prices by more than 300 per cent (Yergin, 1991). The crisis affected nearly all countries including Brazil, whose foreign debt consequently ballooned by more than 40 per cent to pay for the costlier oil imports. This upheaval was a milestone in international energy history, leading many to appreciate that energy is a 'master resource' for the economy, and indeed the maintenance of human civilization generally (Homer-Dixon, 2007, p. 80). In 1979, a second great oil shock rocked international markets, when oil prices again spiked sharply. As a result of these oil crises, governments and businesses worldwide initiated programs to save energy and find alternative sources of power (Gazzoni, 2008).

Since this period, Brazil has adopted a range of policy measures to stimulate a national biofuels industry in the quest for energy security and stronger economic development. Unable to rely on market forces alone, the Brazilian government has essentially underwritten parts of the industry, especially ethanol production, by providing highly subsidized financing through a guaranteed market, price support for producers and subsidies for consumers (Geller et al., 2004). However, the health of its biofuels sector has also fluctuated somewhat, depending on the scarcity and price of oil, which remains a strong competitor (Plummer, 2006). Brazil has grown and exported sugarcane, from which ethanol is commonly produced, since the seventeenth century. Brazil was also once the holder of a significant patent for biodiesel production, based on research at the Federal University of Ceara in the early 1970s.[3] Yet, it was not until the international oil crises of the 1970s that Brazilian authorities began to develop strategically a national biofuels sector.

In 1975 the government launched the National Alcohol Program (Proálcool) as a response to the international oil crisis. Its main elements were: a guarantee that the state-owned oil company, Petrobrás, would buy a certain volume of ethanol; fiscal incentives such as tax concessions and low interest loans to agricultural businesses that produced ethanol; and subsidization of the retail price of ethanol so that it remained competitive with gasoline and diesel oil. The effects of the successful program were such that '[b]y 1989, almost all cars manufactured in Brazil ran on hydrated alcohol, with ethanol production peaking at 12.3 billion liters in 1986–87' (Oritz, 2008, p. 10). However, Brazil's ethanol boom collapsed in the late 1980s, as the relative economic advantages of conventional gasoline increased in an era of increasingly low oil prices.

A variety of other energy policy initiatives were undertaken in response to the turmoil in international oil markets (Nass, Periera and Ellis, 2007). One example is the 1975 Plan for Production of Vegetable Oils Production for Energy Purposes (also known as the Pró-Óleo Program). It was designed to increase production of vegetable oils for blending with conventional diesel fuel (Conselho Empresarial Brasileiro para o Desenvolvimento Sustentável (CEBDS), 2007, p.8). In 1980, a new research program was created, involving Petrobrás and the Ministry of Aeronautics, which led to a new patent of biodiesel called Probiodiesel (Pousa, Santos and Suarez, 2007). By 1983, the National Program for Energy from Vegetable Oils (known as 'OVEG') was launched by the government. Although the scheme had the participation of relevant industry groups, including those in the automobile and fuel sectors, the costs of biodiesel remained higher than that of conventional fuels and therefore the scheme struggled to achieve commercial viability (Pousa, Santos and Suarez, 2007).

While Brazil's technological expertise in biofuels production has grown markedly, the country also faces serious social and environmental problems resulting from the increasing preference to produce biofuels on large-scale, monoculture plantations. With further expansion of the sector planned, its environmental sustainability must surely be questioned. Although the Brazilian government has become a strong advocate of biofuel technologies, it has made comparatively little progress in addressing these collateral problems. A debate is needed – and has been called for by many civil societal groups in Brazil – to assess the effectiveness of current biofuel policies for tackling climate change, promoting social inclusion, enhancing rural development and contributing to a more efficient and sustainable energy policy (Oritz, 2008). Brazilian biodiesel itself is certainly not a perfect solution to climate change, since it is partly composed of fossil fuel-based diesel.[4] It also contains high levels of sulfur, a chemical compound less evident in the diesel composition found in other countries (Almeida, Bomtempo and Silva, 2007). Some of the principal advantages and drawbacks of biodiesel used in Brazil are summarized in Table 13.1:

Some other drawbacks of biofuels production relating to the agricultural economy have been highlighted. One of these, explains Ottinger (2008, p. 11) is 'the increasing concentration of sugarcane lands in the ownership of a few large landowners in Brazil and many other developing countries, combined with the takeover of land for biofuel cultivation by large international agribusinesses'. These large agribusinesses, explains Ottinger (2008, p. 11), 'remov[e] small farmers from their lands, throwing them into poverty; are mechanizing the harvesting of sugarcane, thus throwing local labor out of their jobs and eliminating the prospect of local economic growth from biofuels cultivation; paying substandard wages; and siphoning off most of the profits from biofuel production and processing'. The development of biofuels may also take land out of food production, leading to food shortages and higher commodity prices (Chakrabaty, 2008). As Brown explains (2009, p. 28), '[t]he line between the food and energy economies is becoming blurred as the two begin to merge. As a result, the world price of grains is now moving up toward its oil price equivalent. If the food value of a commodity is less than its fuel value, the market will move it into the energy economy'. Thus, commercial biofuels production may not be a socially sustainable model for local communities. The economic benefits to developing countries are also hindered by the protectionist trade measures of some developed countries, particularly the United States and the European Union. They create tariff barriers on biofuel imports and provide domestic subsidies to protect their national biofuel producers, thereby impeding the expansion of a viable biofuels industry in developing countries (Coelho, 2005).

The critical question that must therefore be asked is whether Brazil's policies on biodiesels are being reformed to address these potential and emerging problems. The following sections explore this issue.

Table 13.1 Main features of feedstocks for production of biodiesel

	Advantages	Disadvantages
Energy Supply and Security	Reduced use of conventional oils. Diversification and decentralization of energy supplies.	Higher production and distribution costs than of petrol.
Food Security	Maintenance and improvement of rural income and/or agricultural jobs.	Inflationary pressure on prices of food and other agricultural products.
Industry	Increase in the octane (ethanol) or cetane content (biodiesel) of fuel, thus improving its lubricity. Ease of application in some sectors (transportation and manufacturing)	Cannot be mixed with conventional fuels beyond certain thresholds without problems such as increased volatility and decreased fuel performance.
Environment	Reduction of GHG emissions. Reduction of HC, particulates (ethanol and biodiesel), SO_2, benzene, butane, isobutene, toluene, xylene (ethanol), aromatics (biodiesel). Recovery of waste by-products or residues from forestry and municipal activities.	Increased deforestation. Pollution from agricultural inputs, including fertilizers and pesticides. Increased emissions of volatile organic compounds.

Source: Adapted after Dorin and Gitz (2007).

Guidelines of the National Program for Production and Use of Biodiesel

In recent years, the Brazilian government has begun to take a more strategic approach to policy-making on biofuels with a view to taking into account the industry's broader economic, social and environmental ramifications. Thus, in July 2003 an Inter-Ministerial Working Group was created with the mandate to review biodiesel production and its use in the country. Its December 2003 report created the basis for the National

Program for Production and Use of Biodiesel (PNPB) and creation of the Biodiesel Inter-Ministerial Executive Commission to coordinate action (Segall and Artz, 2007). Further, in 2004, the Brazilian National Congress enacted Provisional Measure No. 214 of 2004, subsequently enshrined in Federal Law No. 11.097 of 2005, to promote biodiesel as a key renewable source in Brazil's energy matrix. Further, in late 2005 the National Plan of Agroenergy for 2006–2011 was unveiled, setting ambitious targets to harness the agricultural sector to make an even larger contribution to the supply of renewable energy in Brazil.

Federal Law No. 11.097, which amended Law No. 9.478 of 1997 (Regulation of the Brazilian Petroleum Industry), now provides that one of the core legal principles of the national energy policy is 'increasing the contribution of biofuels in the national energy matrix, on the basis of economic, social and environmental terms' (Article 1(12)). In mandating incorporation of biodiesel into the Brazilian energy matrix and establishing biodiesel use targets, the Law defines a biofuel as 'a fuel produced from biomass for use in internal combustion engines or, in accordance with regulations, for other types of energy generation, that could partially or totally substitute fossil fuels' (Article 6(24)). The Federal Law empowered the restructured National Petroleum, Natural Gas and Biofuels Agency (Agência Nacional do Petróleo (ANP)) to regulate and supervise the sale of biofuels, including establishing the permissible technical characteristics of biofuels and administrative penalties (fines) for infractions of the ANP's standards. The ANP, however, only has jurisdiction to regulate the actual production and commercialization of biodiesel, but not to supervise the broader agricultural and environmental policy context of its production.

Brazilian biofuels policy remains deeply wedded to serving economic and social goals, pushing environmental or climate change considerations into a much lower priority. The officially proclaimed goals of the PNPB as stated in government policy guidelines include to: increase jobs, particularly in the small, family-based farming sector; create a competitive market, ensuring quality and supply, from use of different raw materials; increase exports of biofuels; and stimulate regional economic development, while respecting environmental values in Brazil (Brazil, 2003; Brazil, 2004). Under the PNPB, Brazil has also established the benchmark goal of adding 5 per cent biofuel content to all diesel by January 2013. Zapata and Nieuwenhuis (2009, p. 10) explain that the PNPB 'is essentially different from Proalcool . . . The mandate has been designed with the explicit aim of job creation and wealth distribution to the most deprived areas of the country and thus fits in with the Lula government's broader social policies'. The aim, therefore, has been not to replicate the problematic model of sugarcane production on large monoculture plantations. The new policy reflects awareness of the limitations of Proalcool, which did not lead to sufficient job creation in rural areas and did not materially assist community-based, small and medium-size feedstock producers (Zapata and Nieuwenhuis, 2008, p. 10).

In theory, the environmental dimensions of biofuels production in Brazil may be addressed through other policy and regulatory frameworks. According to Federal Law No. 6.983 of 1981, which established the National Council on the Environment, the Council may establish rules, criteria and standards relating to environmental quality control.[5] So far, it has developed several standards applicable to biodiesel production. First, in relation to environmental licensing, it has issued Resolution 237/97 which specifies general technical parameters of environmental licensing, and Resolution 284/2001 which determines which types of agricultural activities are subject to environmental licensing. The Resolution may extend to small, family-based faming operations. Secondly, in relation to air quality, Resolution 315/2002 establishes a new phase of the National Program of Vehicular Emissions and addresses the need for further improvements in automotive and combustible (also diesel) technologies to improve air quality.

Other regulatory standards pertaining to biofuels address the social and economic dimensions of fuel production. For example, Federal Decree No. 5.297 of 2004 provides for issuance of a Fuel Social Certificate to 'producers or importers' who purchase raw materials from farmers affiliated to the National Program for Family-based Agriculture, in exchange for economic benefits such as tax exemptions (Silva and Dutra, 2007). Further criteria for entitlement to these economic benefits were elaborated in Federal Law No. 11.116 of 2005, including with regard to the region planted with crops, feedstock species adopted and the method of production. This Federal Law provided additional incentives for biofuel producers in the form of access to concessional financing from the National Bank of Development (Brazil, 2003; Brazil, 2004). Between May 2005 and February 2008, 24 of 55 biodiesel producers in Brazil earned this Fuel Social Certificate, while two producers lost their certificates because of regulatory infractions.

3. BIODIESEL PRODUCTION IN BRAZIL AND ITS ENVIRONMENTAL DIMENSIONS

Feedstock Production for Biodiesels

Biodiesel is made from the addition of oil extracted from natural and renewable sources to mineral diesel. According to the National Plan of

Agroenergy for 2006–2011, Brazil is well placed to produce a wide range of vegetable and residual oils for energy purposes. Yet, as a country of continental dimensions, with diverse environmental conditions, public policy in Brazil needs to take into account its variable regional potentialities for biofuels production. Depending on the region, Brazil is able to produce biodiesel from traditional crops such as soybean, peanut, sunflower, castor bean and palm, as well as from new crop alternatives such as jatropha, nabo forrageiro, pequi, buriti and macaúba (Santos, 2007).

To date, soybean has been the main source material for biodiesel in Brazil. It has many advantages for such production, including climatic conditions, land availability and local market demand for both the fuel and the food crop. Other plants such as pequi, buriti and macaúba are not yet economically viable to be produced on a large scale to meet the goals of the PNPB. In addition, the Brazilian Corporation for Research in Agriculture has not yet completed its research on the botanical and agronomic cycles of these species in order to enable them to be fully harnessed as biofuel crops (Brazil, 2005). Many Brazilian farmers are experimenting with jatropha, although its potential environmental impacts and economic potential have yet to be fully assessed (Fairless, 2007). Biodiesel may also be produced from waste oils and fats of organic origin, which has economic and environmental benefits; but production of biofuels from these sources in Brazil is presently relatively small.

This wide array of potential feedstocks for production into biodiesel is advantageous for several reasons. Not being dependent on any single crop, which would be more prone to climatic shifts, disease and other variables, Brazil is able to ensure that production is not jeopardized and that market demand can thus be securely met. Furthermore, as the feedstocks have different properties, the technical barriers to utilizing them as biodiesel are more likely to be overcome. On the other hand, most research on the feedstocks available in Brazil has yet to take account of the numerous environmental effects and interactions associated with their production. These include local impacts associated with increased water consumption and soil erosion, as well as larger issues surrounding food security and loss of biodiversity (Santos, 2007). Therefore, commentators suggest that new types of feedstock should be cultivated on vacant or under-utilized lands not currently used for food production (Almeida, et al., 2007).

Information on the use of feedstocks and regional patterns of biodiesel production obtained from the PNPB, ANP and other government bodies, as well as from biodiesel producers, are summarized in Tables 13.2 and 13.3. The data reflect the situation as of February 2008.

Table 13.2 shows that soybean is grown primarily for biodiesel production, with some two-thirds of the crop harvest allocated for this purpose.

Table 13.2 Portion of crop harvest used for biodiesels

Raw Materials	Utilization (%)
Soybean	67,27
Residual Waste Oil	40,0
Sunflower	30,90
Castor Bean	25,45
Jatropha	23,63
Cotton	21,81
Palm	9,09
Nabo Forrageiro	9,09
Colza	3,63
Babassu	1,81
Microalgae	1,81

Its production occurs in many parts of Brazil, contrary to the advice of the National Plan of Agroenergy Guidelines, which recommend growing other crops such as castor bean and palm in some regions presently dominated by soybean. This suggests that there is still some way to go to achieve the goal of the PNPB to diversify the range of feedstocks grown for biofuels (Almeida, Bomtempo and Silva, 2007). Some further diversification away from the traditionally heavy reliance on soybean is likely as research into alternative source materials progresses. For instance, some Brazilian researchers such as at the University of São Paulo are exploring how to use vegetable oil from fried food.

Table 13.3 reveals that most biodiesel producers are concentrated in the mid-west region of Brazil, particularly in Mato Grosso State, which is situated in the officially demarcated greater 'Legal Amazon' area.[6] Altogether this area, covering 61 per cent of Brazilian territory, contains 21 biodiesel producers. This represents nearly half of the number of such producers in the country. While the PNPB recognizes the contribution that family-based agriculture can make to biodiesel production, most of the feedstocks are grown on large farms, which enjoy economies of scale and other efficiency advantages.

Environmental Controls in Biodiesel Production

One of the most severe environmental impacts associated with biodiesel production is the loss of forests to make way for feedstock plantations (Food and Agricultural Organization, 2007, p. 79). In Brazil, this impact is particularly evident in soybean production, although less closely correlated with some other feedstocks, such as castor bean and jatropha.

Table 13.3 Number of producers of biodiesel by geographic regions and states

Geographic Regions	Number of producers	States	Number of producers
Central West	21	Goiás	3
		Mato Grosso	18
North	6	Pará	2
		Rondônia	2
		Tocantins	2
Northeast	7	Bahia	2
		Ceará	2
		Maranhão	1
		Piauí	2
Southeast	14	Minas Gerais	4
		Rio de Janeiro	1
		São Paulo	9
South	7	Paraná	3
		Rio Grande do Sul	4
Total	55		

Deforestation associated with biofuels production has been observed in other countries. In Malaysia, the expansion of palm oil production, principally to meet demand from the European Union, has been identified as the main cause of the country's significant deforestation between 1985 and 2000 (Brown, 2008; Monbiot, 2006). Biofuels are also a significant cause of deforestation in Indonesia. Vast tracts of forest in Indonesia and other Southeast countries have been cleared to grow oil palm. Greenpeace's 'Cooking the Climate' report (2008) has concluded that forest clearance in Indonesia for palm plantations has made the country the third largest producer of GHG emissions in the world, behind the United States and China.

On this basis, Brazil ranks fourth (Greenpeace, 2008, p. 19). Some of the deforestation in Brazil is related to the biofuels industry, particularly in the Amazon (Medina, 2006; Oliveira, 2007). According to the National Institute for Space Research (INPE), 11,968 km^2 of forest in the Legal Amazonia were cleared from August 2007 to July 2008, an increase of 3.8 per cent in the area cleared over the previous assessment period (INPE, 2008). Research in recent years has affirmed a clear link between soybean production and deforestation in this region (Fearnside, 2002b; Morton, et al., 2006). However, while biofuels production is a source of much environmental concern, it appears that other factors principally drive deforestation. For the period 2000–2005, it is estimated that commercial agricultural including biofuel feedstock accounted for only 1 per cent of deforestation in the Amazon, whereas cattle ranches were responsible for 60 per cent of the forest loss (Butler, 2008). Yet, it has also been observed that soybean farms have insidious indirect effects 'by consuming cleared land, savanna, and transitional forests, thereby pushing ranchers and slash-and-burn farmers ever deeper into the forest frontier. Soybean farming also provides a key economic and political impetus for new highways and infrastructure projects, which accelerate deforestation by other actors' (Butler, 2008). And any offsetting contribution that soybean production for biodiesel makes to mitigating climate change is negligible. According to La Rovere (2006), Brazil's production of soybeans for biodiesel should avoid some 1.3 million tonnes of CO_2 per year in 2008 and about 3.9 million tonnes in 2011. Crucially, however, this projection does not count for the impact of deforestation in soybean production.

In theory, the regulation of the environmental impacts of biodiesel production may be undertaken through an environmental licensing scheme implemented under the Brazilian National Environment Policy.[7] Environmental controls may also be imposed pursuant to energy sector legislation. Under Federal Law No. 11.097/05, the ANP has jurisdiction to authorize and control activities related to the production, importation, exportation, storage, distribution, sale and marketing of biodiesel, overseeing them directly or through agreements with other federal, state or municipal authorities. According to Decree No. 5.297 of 2004, supplemented by ANP Resolution 41/04, any company, cooperative or consortium of companies in Brazil may produce biodiesel only with the permission of the ANP. Presently, activities in the biofuels industry that are controlled pursuant to these provisions include: crop field burning; soil protection; herbicides and insecticides storage and usage; liquid waste application for fertilizer; forest preservation; surface and ground water quality; feedstock storage; water usage; and transportation (Martines-Filho, Burnquist, and Vian, 2006, p. 95).

In theory, Brazil's forestry and nature conservation legislation also can help control the environmental impacts associated with biofuels production. The National Forest Code, Law No. 4.771 of 1965, and the National System of Protected Areas, Law No. 9.985 of 1981, provide for the designation of forest legal reserves and other protected areas (Crawford and Pignataro, 2007). As of 2007, some 90 million hectares comprising about 11 per cent of Brazil are set aside as protected areas (Capobianco, 2008). Further, the Forest Code can be used to require private land owners to set aside areas for permanent protection. Together, these provisions can be used to protect forested lands against conversion into large scale plantations for the biofuels industry. However, the agribusiness sector is pushing for substantial changes to the Forest Code to minimize environmental preservation standards for rural areas.

However, because Brazilian environmental law is often relatively weak and ineffectual, many environmental controls over the biofuels industry are therefore at best nominal. Even though the Brazilian Federal Constitution (e.g., Chapter VI and Title VIII) contains several provisions that suggest the environment and sustainable development are of national importance, and the country has an extensive body of environmental laws including the comprehensive National Environment Policy Law No. 6938 of 1981 and the Environmental Crimes Law No. 9.605 of 1998, implementation of environmental standards is often poor. On paper, Brazil is also well stocked with environmental agencies and administrative capacity. At the federal level, the principal authorities are the Brazilian Environmental Council (CONAMA, the lead normative and decision-making agency), the Ministry of the Environment (responsible for the coordination and supervision of environmental policy), and the Brazilian Institute for the Environment and Renewable Natural Resources (IBAMA) and the Chico Mendes Institute for the Conservation of Biodiversity (both of them, the federal executive authority). Other environmental bodies have been established at the state level.

The lack of effective environmental regulation in Brazil is attributable to a variety of factors, including weak, segregated and under-resourced environmental agencies, and the political power of business elites hostile to any interference in their economic activities (Fernandes, 1992; Romano, 1999; Drummond and Barros-Platiau, 2005). In these respects, the precarious state of Brazilian environmental law is little different from that of many other developing countries which have struggled to build regulatory capacity to ensure sustainable development (Richardson, 2000; Roberts and Thanos, 2003). Consequently, sometimes new methods to work collaboratively with the private sector including farming organizations are being pursued in Brazil to improve the design and implementation of environmental standards (Brannstrom, 2005). Further, Brazilian environmental authorities are also increasingly working with Indigenous peoples in the Amazon to improve implementation of environmental goals (Fearnside, 2002a).

Nonetheless, better implementation of environmental standards must begin with improved environmental standards tailored specifically to the biofuels industry. Clearly, an unregulated market for biodiesel production will not enable this sector to provide an environmentally acceptable contribution to mitigating climate change. Comprehensive and enforceable regulatory standards should be enacted to control the cultivation and processing of biofuels (Ottinger, 2008). Voluntary instruments, such as best practice guidelines and eco-labeling systems, could support action by state authorities to encourage producers to act responsibly. In designing further regulations for a sustainable biofuels industry in Brazil, detailed studies about the environmental impacts of large scale production of feedstocks, particularly soybean, jatropha and castor bean, are also needed, so that regulations can target the most acute problems. At the same time, environmental regulation of the biofuels sector must address the pressing collateral issues of eradicating poverty and social marginalization, and creating sustainable rural economies, if it is to achieve legitimate and secure popular support (Silva, 2004). Poor rural people need jobs and economic opportunities, and will resist regulatory interventions that give them no other viable means of livelihood.

4. THE CLIMATE CHANGE CONVENTION AND BIOFUELS

Do the UNFCCC and the Kyoto Protocol provide a framework to facilitate environmentally sound biofuel production in Brazil? While the international climate law agreements do not require Parties to promote biofuels, they contain provisions which may facilitate the sector. Under the UNFCCC, developing countries such as Brazil presently have commitments only to compile and periodically update their national inventories of anthropogenic emissions by sources and removals by sinks of GHGs not controlled by the Montreal Protocol (Article 4), as well as to produce a general description of steps taken or envisaged to implement the Convention (Article 12). The Brazilian government has initiated several measures to meet these commitments, despite its historical intransigence and opposition in international climate law negotiations to acceptance of any meaningful obligations for itself (Johnson, 2001).

In 2004, Brazil set up a multi-institutional team, under the auspices of the Ministry of Science and Technology, with responsibility to prepare the Brazilian National Communication for the UNFCCC. Its first National Communication, submitted in 2005, details the national circumstances and special arrangements in Brazil that affect how it can respond to climate change, provides a national GHG inventory for the period 1990 to 1994, and explains steps taken or envisioned to help achieve the goals of the UNFCCC. The National Communication refers to enhancement of biodiesel and ethanol production as one of these steps, and it also mentions the collateral benefits that arise from using local raw materials, providing jobs and harnessing local expertise and technology (Brazil, Ministry of Science and Technology, 2004, pp. 173–75).

In November 2007, the Brazilian Inter-Ministerial Committee for Climate Change (Comitê Interministerial sobre Mudança do Clima) (CIMC) was created to formulate the National Plan on Climate Change, pursuant to Federal Decree No. 6.263 of November 2007. In its preliminary version, this National Plan, under public consultation in 2008, identifies biodiesels as a valuable part of Brazil's alternative energy plans (CIMC, 2008). The Plan highlights Brazil's considerable experience in developing liquid biofuels, including ethanol, through the 'Pro-Álcool' Program and the National Program of Production and Use of Biodiesel (CIMC, 2008).

While it is crucial for developing countries such as Brazil increasingly to take their own steps to reduce GHG emission domestically, the UNFCCC's Kyoto Protocol envisions that the most concrete means by which such nations can contribute to global efforts to mitigate climate change is through the Clean Development Mechanism (CDM). The function of the CDM, which is to help Annex I Parties achieve compliance with their quantified GHG emission limitation and reduction commitments by financing off-setting projects in developing countries, is explained more fully in other chapters in this book. However, it should be noted here that the Kyoto Protocol does not provide any incentives for reduced deforestation in non-Annex I countries, since 'avoided deforestation' is not an eligible project activity under the CDM. Apart from the fact that Brazil and other major developing countries may be subject to some GHG emission limitations under the post-Kyoto instrument, such an instrument may also include provisions to allow Annex I countries to acquire emission rights through projects that reduce deforestation rates in the developing countries. During the Kyoto negotiations and at subsequent UNFCCC Conferences of the Parties, the Brazilian delegation has opposed the inclusion of avoided deforestation under the CDM. Its official stance is that conserving forests would not help significantly to mitigate climate change and the impracticalities of verifying the quality of CDM projects in this sector (Persson and Azar, 2004, p. 4). However, critics including Brazilian environmental organizations attribute Brazil's opposition to the 'government's fear of internationalization of the Amazon' (ibid.). In other words, Brazilian authorities fear that expanding the ambit of the CDM to cover forests would allow greater foreign influence in the development and management of the Amazon. The potential for CDM projects in this sector in Brazil is considered vast (Teixiera, Murray and Carvalho, 2007).

Nonetheless, Brazil has welcomed other types of CDM projects, particularly in the energy sector, and was one of the first countries to begin receiving them. Biofuel projects, however, have hardly been included in the CDM project portfolio because the necessary biofuel baseline and monitoring methodologies have not yet been approved by the CDM Executive Board, which is a necessary requirement for project validation (Bakker, 2006). The one exception is the development of a methodology for 'Production of Waste Cooking Oil-based Biodiesel for Use as Fuel'. This methodology is based on three CDM experimental projects, of which one involves Brazil, namely the Biolux Benji Biodiesel Beijing Project, proposed by Biolux Benji Energy and Recycling Company; Agrenco Biodiesel Project in Alta Araguaia, proposed by Agrenco do Brasil S/A; and the Palm Methyl Ester Biodiesel Fuel production and use for transportation in Thailand, proposed by a Japanese firm (Bakker, 2006, p. 13; UNFCCC, 2008).

The production and use of biofuels may eventually be more comprehensively eligible for CDM projects, and thereby it could provide a significant economic stimulus to the Brazilian biofuels market through the sale of Certified Emission Reductions from projects. However, such projects would need to satisfy the CDM 'additionality' criterion, namely of avoiding GHG emissions that otherwise would have occurred in the absence of the project in question. If the full life cycle biofuel production is taken into account, this condition may not be met. Indeed, such projects may actually intensify, not mitigate, climate change.

This is also problematic because the CDM not only is a mechanism to address global warming, but is intended as a means to promote sustainable development in host countries. Biofuel projects conducted under the CDM should not be approved 'if standards are not adopted to provide against substitution of fuel for food crops, endangerment of clean water supplies, deterioration of the land and inequitable distribution of the profits from biofuel production' (Ottinger, 2008, p. 17). Developing countries need to be made more aware of the wider relationship between the biofuels sector and sustainable development, and cease viewing biofuels projects as simply the provision of a renewable energy source. CDM biofuels projects must function within a set of higher scale policies and regulations that take account of collateral impacts on the rural economy, social justice and biodiversity protection (Jürgens et al., 2004).

5. CONCLUSIONS

Ostensibly, the burgeoning ethanol and biodiesel industries in Brazil have placed the country in a comfortable position to make a meaningful contribution to the global challenge of mitigating climate change (Zapata and

Nieuwenhuis, 2008). Yet, it is crucial to treat biofuels as not just an energy policy issue, but rather to evaluate the sector in a comprehensive manner that takes account of the production and use of biofuels over their entire life-cycle. This chapter has attempted to lift the veil on some aspects of this life-cycle in Brazil.

Although some official policies and regulations appear serious about addressing the environmental dimensions of biodiesel and other biofuels, their practical implementation and enforcement are another matter (Zapata and Nieuwenhuis, 2008). Even if we concede that the replacement of fossil fuels by biofuels can help lower GHG emissions – which is a problematic argument, of course – a fuller picture of the life-cycle of biofuels production that takes into account all of the other environmental and social impacts suggests that the sector can make only a modest contribution to sustainable development under the most optimal conditions (Sachs, 2007).

So, at least on the Brazilian experience, the potential contribution of biodiesel as a 'green fuel' is doubtful. Soybean production in Brazil is associated with deforestation in the Amazon. If this impact and other environmental harms are not resolved soon, the biofuels industry will eventually be viewed as one of the most environmentally misguided initiatives ever to be promoted. The emergence of second-generation feedstocks such as jatropha may be less environmentally problematic, but further research into their impact is needed. Biodiesel production based on discarded oils, fats and other organic waste materials is likely to be the most environmentally sound approach. More investment in technological development coupled with stronger public policy and enforcement is essential to achieve sustainable development in this sector.

NOTES

* Professor of Sustainable Development at Mackenzie Presbyterian University, and Professor of Environmental Law at State University of Amazonas, Brazil.

** Lawyer, Master of Laws in International Law and Environmental Law at Catholic University of Santos, Brazil.

1. Pub. L. 109-58.
2. Pub L. 110-140.
3. British Council (2008), 'O que é biodiesel?', www.deolhonoclima.com.br/CMSMediaCenter/Arquivos/Documentos/faq.pdf (visited 10 March 2009).
4. Each liter of mineral diesel in Brazil emits about 2.7 kg of CO_2 (La Rovere, 2006).
5. See www.mma.gov.br/conama (visited 15 March 2009).
6. The 'Legal Amazon' area has been created to promote the development of this region by Federal Law No. 1.806, January 1953, modified by Federal Law No. 5.173, October 1966 and Federal Complementary Law No. 31 October 1977.
7. Brazilian law generally provides that the construction, installation, expansion and operation of facilities that may pollute or use natural resources must be licensed by the appropriate environmental agency (Federal Law No. 6.938/81).

REFERENCES

Almeida, E.F., J.V. Bomtempo and C.M.S. Silva (2007), *The Performance of Brazilian Biofuels: an Economic, Environmental and Social Analysis*, Discussion Paper No. 5. (Organisation for Economic Co-operation and Development, and International Transport Forum).

Bajay, S.V. and L.F. Badanham (2005), *Energia no Brasil: os próximos dez anos* (Ministério das Minas e Energia (MME)).

Bakker, S.J.A. (2006), *CDM and Biofuels. Can the CDM Assist Biofuel Production and Deployment?* (Energy Research Centre of the Netherlands).

Brannstrom, C. (2005), 'Environmental Policy Reform on North-eastern Brazil's Agricultural Frontier', *Geoforum* 36(2), 257.

Brazil (2003), *Biodiesel como fonte alternativa de energia* (Relatório Final do Grupo de Trabalho Interministerial).

Brazil (2004), *O Biodiesel e a inclusão social* (Câmara dos Deputados).

Brazil, Agência Nacional do Petróleo (ANP), Gás Natural e Biocombustíveis (2008), 'Importações e Exportações – Dados Estatísticos', Superintendência de Planejamento e Pesquisa.

Brazil, Casa Civil da Presidência da República (CIMC) (2008), *Plano Nacional de Mudança do Clima – PNMUC* (CIMC).

Brazil, Ministry of Science and Technology, General Coordination on Global Climate Change (2004), *Brazil's Initial National Communication to the United Nations Framework Convention on Climate Change* (Ministry of Science and Technology).

Brazil, Ministério da Agricultura, Pecuária e Abastecimento (MAPA) (2005), *Plano Nacional de Agroenergia 2006–2011* (Ministério da Agricultura, Pecuária e Abastecimento).

Brazil, Ministério de Minas e Energia (MME) (2007), *Balanço Energético Nacional*, www.mme.gov.br/site/menu/select_main_menu_item.do?channelId=1432&pageId=15304.

Brazil, Ministry of External Relations (MRE), et al. (2007), *Brazil's Contribution to Prevent Climate Change*, White Paper (MRE).

Brown, L. (2008), *Plan B 3.0 – Mobilizing to Save Civilization* (W.W. Norton & Company).

Bucci, M. (2006), 'O conceito de política pública em direito', in M. Bucci (ed.), *Políticas Públicas: reflexões sobre o conceito jurídico* (Saraiva).

Butler, R. (2008), 'Deforestation in the Amazon', www.mongabay.com/brazil.html (visited 1 April 2009).

Capobianco, J.P. (2008), 'Monitoring and Responding to Biodiversity Loss: The Brazilian Experience', in *OECD World Forum – Measuring and Fostering the Progress of Societies* (OECD).

Centro Nacional de Referência em Biomassa (CENBIO) (2003), *Estudo de pré-viabilidade ambiental comparativo entre o diesel tipo D e o éster etílico de óleo de soja* (CENBIO).

Chakrabaty, A. (2008), 'Secret Report: Biofuel Caused Food Crisis. Internal World Bank Study Delivers Blow to Plant Energy Drive', *Guardian*, 4 July.

Clean Development Mechanism (CDM) Executive Board (2008), 'Production of Waste Cooking Oil-based Biodiesel for Use as Fuel', AM0047 (CDM Executive Board).

Coelho, S.T. (2005), 'Biofuels – Advantages and Trade Barriers', in United Nations Conference on Trade and Development (UNCTAD), *Proceedings*.
Conselho Empresarial Brasileiro para o Desenvolvimento Sustentável (CEBDS) (2007), *Biofuels and Climate Change – Interfaces and Potentialities* (CEBDS).
Crawford, C. and G. Pignataro (2007), 'The Insistent (and Unrelenting) Challenges of Protecting Biodiversity in Brazil: Finding "The Law That Sticks"', *University of Miami Inter-American Law Review* 39, 1.
Dorin, B. and V. Gitz (2007), 'Ecobilans des biocarburants: une revue des controverses et des enjeux agronomiques mondiaux', in *Natures, Sciences and Sociétés*, 5 December.
Drummond, J. and A.F. Barros-Platiau (2005), 'Brazilian Environmental Laws and Policies, 1934–2002: A Critical Overview', *Law and Policy* 28(1), 83.
Fairless, D. (2007), 'Biofuel: The Little Shrub That Could – Maybe', *Nature* 449, 652.
Farrell, A., et al. (2006), 'Ethanol Can Contribute to Energy and Environmental Goals', *Science* 27, 506.
Fearnside, P. (2002a), 'Conservation Policy in Brazilian Amazonia: Understanding the Dilemmas', *World Development* 31(5), 757.
Fearnside, P. (2002b), 'Soybean Cultivation as a Threat to the Environment in Brazil', *Environmental Conservation* 28(1), 23.
Fernandes, E. (1992), 'Law, Politics and Environmental Protection in Brazil', *Journal of Environmental Law* 4(1), 41.
Focault, M. (2001), *A ordem do discurso* (10th edn, Loyola).
Food and Agricultural Organization (FAO) (2007), *State of the World's Forests 2007* (FAO).
Food and Agricultural Organization (FAO) (2008), *Biofuels: Prospects, Risks and Opportunities* (FAO).
Gazzoni, D.L. (2008), 'History of Biodiesel', in *Revista Biodiesel BR*, 2 October 2008, www.biodieselbr.net/biodiesel/history/history.
Geller, H., R. Schaeffer, A. Szklo and M. Tolmasequin (2004), 'Policies for Advancing Energy Efficiency and Renewable Energy Use in Brazil', *Energy Policy* 32(12), 1437.
Goldemberg, J. (2003), 'Ethanol Learning Curve – the Brazilian Experience', *Biomass and Bioenergy* 26, 301.
Greenpeace (2008), *How the Palm Oil Industry is Cooking the Climate* (Greenpeace).
Grosser, M. (1978), *Diesel: The Man and the Engine* (Atheneum).
Homer-Dixon, T. (2007), *The Upside of Down* (Vintage Canada).
INPE, Instituto Nacional de Pesquisas Espaciais (2008), 'Projeto PRODES – Monitoramento da Floresta Amazônica por Satélite, Ano 2006–2007', www.obt.inpe.br/prodes.
Johnson, K. (2001), 'Brazil and the Politics of the Climate Change Negotiations', *Journal of Environment and Development* 10(2), 178.
Jürgens, I., G. Best and L. Lipper (2004), 'Bioenergy Projects for Climate Change Mitigation: Eligibility, Additionality and Baselines', in *2nd World Conference on Biomass for Energy, Industry and Climate Protection* (Rome, 10–14 May 2004).
Kaltschmitt, M., G.A. Reinhard and T. Stelzer (1997), 'Life Cycle Analysis of Biofuels Under Different Environmental Aspects', *Biomass and Bioenergy* 12(2), 121.

Klein, R., E.L. Schipper and S. Dessai (2003), 'Integrating Mitigation and Adaptation into Climate and Development Policy: Three Research Questions', Working Paper No. 40 (Tyndall Centre for Climate Change Research).
Konchinski, V. (2008), 'Ministério Público quer que Petrobras venda diesel com menos enxofre', *Agência Brasil*, www.agenciabrasil.gov.br/noticias/2008/01/29/materia.2008-01-29.7027249679/view.
La Rovere, E. (2006), 'Aligning Climate Change and Sustainable Development Objectives in Brazil', in *Workshop Developing Visions for a Low-Carbon Society Through Sustainable Development* (Tokyo, 14–16 June).
Marques, G. (2008), 'Biodiesel não é óleo vegetal', *Revista BiodieselBR*, www.biodieselbr.com/noticias/biodiesel/biodiesel-nao-oleo-vegetal-03-04-07.htm, 11 May 2008.
Martines-Filho, J., H.L. Burnquist and C.E. Vian (2006), 'Bioenergy and the Rise of Sugarcane-Based Ethanol in Brazil', *Choices* 21(2), 91.
Medina, H. (2006) 'Soja avança sobre o mercado do biodiesel', in *Folha de São Paulo*, 19 Nov. 2008, http://www.biodieselbr.com/noticias/biodiesel/soja-avan-ca-sobre-mercado-biodiesel-19-11-06.htm (visited 4 June 2009).
Midiateca, S.B. (2007), 'Ciclo Temático Civilização da Biomassa', Instituto de Estudos Avançados (IEA) da Universidade de São Paulo (USP), www.iea.usp.br/iea/online/midiateca/biomassa/index.html, 10 September 2008.
Monbiot, G. (2006), 'Biocarburants: Attention, Danger!', *Courrier International* 797, 9–15 February.
Morton, D.C., et al. (2006), 'Cropland Expansion Changes Deforestation Dynamics in the Southern Brazilian Amazon', in *Proceedings of the Natural Academy of Sciences of the United States of America* (Natural Academy of Sciences of the United States of America), 103(39), 14637.
Nass, L., P.A.A. Periera and D. Ellis (2007), 'Biofuels in Brazil: An Overview', *Crop Science* (7 November), 2228.
National Biodiesel Board (2006), *Biodiesel's Contributions to the U.S. Economy* (National Biodiesel Board).
OECD, Organisation for Economic Co-operation and Development, FAO, Food and Agriculture Organization of United Nations (2007), 'Agricultural Outlook 2007–2016', www.oecd.org/dataoecd/6/10/38893266.pdf, 29 September 2007.
Oliveira, K. (2007), 'Produção de soja para biodiesel deve diminuir nos próximos anos, segundo entidade', *Agência Brasil* (26 September).
Oritz, L. (2008), 'Sugarcane Ethanol Production', in *Biofuels Fueling Destruction in Latin America* (Friends of the Earth International).
Ottinger, R.L. (2008), 'Climate Policy Energy Solutions for Developing Countries – Be Careful What You Wish For', paper presented at the conference IUCN Academy of Environmental Law, *Climate Law in Developing Countries Post-2012: North and South Perspectives* (University of Ottawa, 26–28 September).
Persson, M. and C. Azar (2004), *Brazil Beyond Kyoto: Prospects and Problems in Handling Tropical Deforestation in a Second Commitment Period* (Report to the Swedish Environmental Protection Agency).
Pimentel, D., et al. (2002), 'Renewable Energy: Current and Potential Issues', *BioScience* 52(12), 1111.
Plummer, R. (2006), 'The Rise, Fall and Rise of Brazil's Biofuel', *BBC News Online*, 24 January, http://news.bbc.co.uk/2/hi/business/4581955.stm.

Pousa, G., A. Santos and P. Suarez (2007), 'History and Policy of Biodiesel in Brazil', *Energy Policy* 25(11), 5393.
Richardson, B.J. (2000), 'Environmental Law in Postcolonial Societies: Straddling the Local – Global Institutional Spectrum', *Colorado Journal of International Environmental Law and Policy* 11(1), 1.
Roberts, J.T. and N.D. Thanos (2003), *Trouble in Paradise: Globalization and Environmental Crises in Latin America* (Routledge).
Romano, C.S. (1999), 'Brazilian Government Policies toward the Amazon Rain Forest: From a Development Ideology to an Environmental Consciousness', *Colorado Journal of International Environmental Law and Policy*, 10, 65.
Sachs, I. (2007), 'A revolução energética do século XXI', *Revista de Estudos Avançados da Universidade de São Paulo (USP)*, 21(59). São Paulo, www.scielo.br/scielo.php?pid=S010340142007000100003&script=sci_arttext, 29 September 2007.
Santos, M.A. (2007), 'Inserção do biodiesel na matriz energética brasileira: aspectos técnicos e ambientais relacionados ao seu uso em motores de combustão', Dissertação de Mestrado do Programa de Interunidades de Pós-Graduação em Energia da Universidade de São Paulo (Universidade de São Paulo).
Schlegel, S. and T. Kaphengst (2007), 'European Union Bioenergy Policy', *Journal of Agricultural and Food Industrial Organisation* 5(2), 1193.
Segall, S.D. and W.E. Artz (2007), 'The Brazilian Experience with Biofuels', *Liquid Technology* 19(1), 12.
Silva, S.T. (2004), 'Princípio de precaução: uma nova postura em face dos riscos e incertezas científicas', in M.D. Varella and A.F. Platiau (eds), *Princípio da Precaução* (Del Rey).
Silva, S.T. and C. Dutra (2007), 'Analisis legal del Programa Brasileño de Producción y Uso de Biodiesel y del Desarrollo Sostenible', *Revista Ecología Política – Agrocombustibles*, No. 34, 103.
Silva, S.T. and C. Dutra (2008), 'Direito e Energia: Biodiesel a partir de Oleaginosas, Desafios e Perspectivas', in *IV Encontro Nacional da Associação Nacional de Pós-Graduação e Pesquisa em Ambiente e Sociedade* (ANPPAS), Brasília, 4–6 June.
Teixiera, M.A., M.L. Murray and M.G. Carvalho (2007), 'Assessment of Land Use and Land Use Change and Forestry (LULUCF) as CDM projects in Brazil', *Ecological Economics* 60(1), 2600.
Teles da Silva, S. (2008). 'La pollution atmosphérique dans la Région Amazonienne: l'expérience de la législation brésilienne', *Revue Juridique de l'Environnement* 2, 191.
United Nations Framework Convention on Climate Change (UNFCCC) (2008), 'Draft Revision to the Approved Baseline and Monitoring Methodology AM0047 – Production of biodiesel on waste oils and/or waste fats from biogenic origin or from oil seeds cultivated in dedicated plantations for use as fuel' (UNFCCC).
Viola, E., A. Platiau and H. Leis (2008), 'Governança e Segurança Climática na América do Sul', in R. Neubauer and G.W. Silveria, *Uma Nova Agenda Econômica e Social para a América Latina*, Instituto Fernando Henrique Cardoso (IFCH), Corporacion de Estudios para Latinoamerica (CIEPLAN), Banco Interamericano de Desenvolvimento (BID), Programa das Nações Unidas para o Desenvolvimento (PNUD).

World Wide Fund for Nature (2008), 'Cerrado', www.worldwildlife.org/wildworld/profiles/terrestrial/nt/nt0704_full.html (visited 1 September 2008).

Yergin, D. (1991), *The Prize: The Epic Quest for Oil, Money and Power* (Simon and Schuster).

Zapata, C. and P. Nieuwenhuis (2009), 'Driving on Liquid Sunshine – the Brazilian Biofuel Experience: A Policy Driven Analysis', *Business Strategy and the Environment*, forthcoming.

PART 4

Climate policy bridging the North and South

14. Improving citizen responsibility in the North and its consequences for the South: voluntary carbon offsets and government involvement

Marjan Peeters*

1. INTRODUCTION

Traditionally, environmental law has dealt mainly with industrial behaviour. Governments have established impressive packages of regulation governing industrial operations and product characteristics, with the aim of ensuring environmental protection. In the field of climate change policy however citizen behaviour is increasingly seen as one of the main sources of greenhouse gas (GHG) emissions that should be targeted by policies.[1] Personal transportation, consumption of food and consumer products, and heating and cooling of homes cause a substantial amount of GHG emissions, in particular in developed countries where incomes and living standards are high. Consequently, governments are developing regulatory approaches to reduce GHG emissions caused by citizen behaviour.[2] At the same time, nongovernmental actors are developing an environmental market initiative in which consumers and others purport to offset their GHG emissions by purchasing so-called 'carbon credits'. This market is growing remarkably quickly, yet in relative obscurity. Offsets are being offered for a wide range of consumer activities including air travel,[3] credit card purchases[4] and entertainment.[5] It is even possible to purchase carbon credits as a gift.[6]

The compensatory emissions reductions represented by these carbon credits are achieved predominantly in developing countries. Because it is normally much cheaper to achieve GHG emissions reductions in developing rather than developed countries, it is primarily there that the much wealthier citizens of developed countries compensate for their carbon emitting behaviour. This raises the question whether these voluntary offset mechanisms might lead to perverse outcomes. The key concern is whether citizen responsibility in the global North, in the form of voluntary offsetting of carbon emissions, will indeed benefit the climate, and at the very least not be disadvantageous to citizens in the global South. For example, execution of offsets might not take sufficient account of the concerns of local citizens, or might even lead to unacceptable adverse impacts on them. The main question for this chapter therefore is whether there is a need for government intervention to ensure that private carbon offset mechanisms indeed make a useful contribution to climate change policies and ultimately to the pursuit of sustainable development.

Perverse outcomes may also affect the consumers of carbon credits: those who are buying carbon offsets should of course not be misled. The agreed offsets for which consumers pay may not actually be achieved or, if achieved, may not be an effective solution to greenhouse gas reduction. Negative experiences with carbon offsets have already received media attention, and constitute a disincentive for individuals and commercial organisations to buy such carbon credits.

The chapter investigates these questions in relation to current practices in Western Europe, with particular focus on the United Kingdom (UK) and the Netherlands. This exploration starts from the premise that Northern consumers are not sufficiently equipped to ensure the integrity of the carbon offset offer, especially when the offsets will be achieved in another part of the world. Secondly, some or most suppliers of carbon offsets have commercial interests, which means that they will compete for sales of carbon credits. This leads to the proposition that there should be guarantees against a race to the bottom regarding the credibility and integrity of the carbon offers.[7] Credibility refers to whether the agreed offsets will in fact be achieved. Integrity refers to whether their achievement will lead to unacceptable negative effects contravening the principle of sustainable development.

Consumers must have confidence in the credibility and integrity of carbon offers, or they may become reluctant to take responsibility to offset their emissions. This, in turn, may lead to less investment in climate change mitigation in developing countries. This chapter does not, however, explore the determinants or extent of demand for voluntary carbon offsets among Northern consumers, especially in the situation where carbon prices increase due to provisions ensuring the integrity of carbon offers.[8] Neither does it explore the related question whether the decision to purchase carbon offsets would lead to a decreased allocation of personal budget expenditures to general funds concerned with developing countries (for example, donations to international development agencies). Such questions illustrate the need for additional research into consumer behaviour with regard to voluntary carbon offsets.

This study takes mainly a legal perspective, by discussing the question whether and, if so, how governments should intervene in order to enhance the credibility and integrity of the private carbon market, not only to ensure consumer confidence but also with a close eye on the concerns of the developing countries and the citizens living there. The voluntary market may have the beneficial effect of channelling funds to developing countries for innovations and low carbon technologies. We will touch upon the tension between, on the one hand, such reasons for stimulating voluntary investments into developing countries and, on the other hand, ensuring the robustness of carbon credits, which would lead to more expensive offsets due to the need for safeguards such as verification, validation and certification. The 'extra' money represented by these higher prices would not go directly to the developing world. Before being able to shed light on this dilemma, the chapter first describes the carbon offset market, and governments' current approaches to it. The focus will be on the strikingly divergent policy developments in the United Kingdom and the Netherlands.

The findings of this chapter are also relevant to a discussion of voluntary industrial investments into carbon offset projects, for example to comply with Corporate Social Responsibility policies (see Richardson, 2008, pp. 500–01, 507). Individual citizens and companies are in fact buying credits on the same markets. Companies can be expected to have better capacity to check and control the credibility and integrity of offsets. This proposition and the usefulness of voluntary carbon offsets alongside regulatory approaches to target industrial carbon emitting behaviour deserve attention in another paper. The same is true of policies to make governments themselves carbon neutral, including the option to offset governmental carbon emissions.

Finally, it is of utmost importance that governments of developed countries implement policies to reduce their own citizens' emissions. Voluntary carbon offsets are only complementary to reducing emissions at home by, for example, lowering the energy use of buildings and reducing emissions from personal transportation. If however such carbon reducing measures are very costly, investment in developing countries could be a better option for the time being, provided this investment would benefit both the climate and the citizens of developing countries. In the long run one might expect that such voluntary investments will be only a transitional phenomenon until regulators have put in place meaningful policies targeting citizen behaviour, and, ultimately, industrialised societies have achieved a low carbon economy. In the meantime, given the large transitions required to get to that point, it is worthwhile to investigate how citizens, notably citizens in countries with minimal climate change mitigation requirements such as the United States (US), but also citizens in European countries who want to take responsibility and wealthy citizens in developing countries which lack meaningful carbon regulations, could offset their emissions.[9] This chapter focuses on European citizens and investigates the convergence or divergence between the voluntary offset market and European greenhouse gas regulation. The final section explores the significance of the European greenhouse gas ETS for the voluntary offset market.

2. THE CARBON OFFSET MARKET

The Case for Government Involvement

The economic and legal literature on emissions trading generally starts from the proposition that the market will be established by a regulatory framework. The literature examines how governments can establish emissions trading markets, and, subsequently, how private actors can act within such a market. The private carbon offset market is however a clear example of bottom-up development of a voluntary credit-based emissions trading market. Such a private initiative in fact occurred in the context of the Clean Air Act in the US.[10] Polluting firms, looking for investment options and consequently room for emissions, explored the idea of financing emission reduction measures at older facilities, even those managed by other operators. This gave rise to the possibility of buying and selling pollution credits. Hahn and Hester (1989) conclude that legal certainty about the value and lifetime of the credits is needed for the successful development of such a market. This certainty can be provided by a public law framework facilitating the private market. This is what has happened subsequently, as governmental policies and regulations have established legal frameworks for emissions trading.

The voluntary carbon offset market can be seen as a spin-off of the international and domestic legal frameworks for carbon emissions reduction accompanied by emissions trading. In line with the regulatory frameworks for emissions trading, the market is now offering carbon credits, to be used for voluntary emission offsets. Yet the market, rather than government, is determining the basic conditions such as the content of the carbon credit, assurance of the promised emission reductions, and calculation of the carbon emissions to be offset. Being unregulated and uncoordinated, the market has already produced an impressive range of carbon offers with different conditions.

Divergent opinions have been expressed in Europe about the proper role of government in relation to the emergence of the carbon offsets offered to citizens. On the one hand, the idea of a government-initiated Carbon

Offset Code has been put forward in the UK (Defra, 2008). The Code aims to provide a legal framework for ensuring the credibility and integrity of carbon offsets. On the other hand, the Netherlands government has no present intention to set up such a regime for ensuring the credibility and integrity of voluntary offsets.

Voluntary carbon offsets have also developed quickly in other parts of the world. The Australian government, for its part, has issued official guidance on carbon claims (ACCC, 2008).[11] In the US a variety of products and services offer carbon neutral options, ranging from carpet, wood and yogurt to rental cars (Kerr, 2007, p. 618). The offset market is not targeted exclusively at citizens: industries are increasingly interested in offsetting part of their emissions. Many observers of the American voluntary offset market are however sceptical whether this wide array of carbon offset offers is environmentally sound and trustworthy. Consequently there are also calls in the US for common offset standards that would enhance transparency and boost consumer and business confidence in the market (ibid., p. 619).

These developments raise at least two important questions about private offset markets. First, are the suppliers of offsets able to establish a framework that would garner sufficient confidence for the public to enter the private market in substantial numbers? Such a framework should at least consist of (a) transparent information regarding the content of the offsets, and (b) reliable verification provisions to check whether the agreed offsets will be achieved in a sound way.

Second, does the private market indeed contribute to combating the climate problem, and, in a wider sense, to sustainable development? Climate change is a fundamental concern of governments, and governments should intervene if:

- offsets are not executed as agreed, and consumers of offsets are not equipped to inspect and enforce the agreements themselves;
- offsets, even if executed as agreed, would not contribute to solving climate change because of problems such as 'leakage' (which would arise for example where payments to conserve forest in one place cause deforestation to take place somewhere else);
- offsets would be contrary to sustainable development or human rights, such as where local residents are forcibly displaced without due process or adequate compensation, to make way for forestation.[12]

At the moment there is no international legal framework to control the credibility and integrity of the private carbon offset market, aside from the fact that the Kyoto Protocol provides for credit-based emissions reduction trading in the form of Joint Implementation (JI) and the Clean Development Mechanism (CDM).[13] Citizens can choose to purchase credits that will be achieved through such Joint Implementation and Clean Development Mechanism projects. The market is also however offering offsets that are not connected to the Kyoto framework.

The Executive Secretary to the United Nations Framework Convention on Climate Change (UNFCCC)[14] has proposed that voluntary carbon offsets should be certified through the Convention.[15] This could lead to a transparent and well controlled private carbon offset market. In the absence of such a supranational framework, individual governments can decide whether they wish to issue policies or regulations to steer the voluntary carbon offset market. The market however is already supranational, as most of the offsets will be achieved abroad. This supranational character makes it difficult for national governments to regulate the credibility and integrity of offsets. On the other hand, supranational harmonisation might discourage new developments outside the Convention regime and make the system more expensive, driving down demand for carbon offsets. To encourage citizens (and firms) to get involved in the carbon market, it may be important to give them choice of the kind of offsetting they wish to purchase.[16] Any movement toward an international regime for regulating offsets should leave room for such choice. The bottom-up, uncoordinated character of the voluntary offset market at least has the virtue of encouraging a wide variety of choice.

Another possible advantage of the bottom-up and uncoordinated development of the voluntary carbon market is that project developers may have more flexibility to implement projects which might not otherwise be viable, because for example they are too small or disaggregated (Kollmuss, et al., 2008, p. vi). Too heavy a control regime could to this extent mean less investment in the developing world.

Finally, now that the United Nations is committed to moving toward carbon neutrality, it will be interesting to see how this organisation will offset its carbon emissions.[17] The behaviour of governments and intergovernmental organisations as buyers of carbon offsets could send important signals to citizens about which offers are preferable and why.[18]

Current Examples

Despite the idea of an international 'UNFCCC' provision for ensuring the integrity of carbon offsets, international private approaches are already contributing to the development of common criteria for ensuring the integrity of carbon offsets.[19] Two examples are the Gold Standard and the Voluntary Carbon Standard (VCS).

The Gold Standard[20]

The Gold Standard was developed under the leadership of the World Wildlife Fund (WWF) to certify high quality CDM credits. It is now operated by a Swiss non-profit foundation. It has been endorsed by 60 non-governmental organisations and charitable organisations worldwide, including Greenpeace International.[21] It is essentially a set of tools used by project proponents to ensure that projects meet specified criteria related to renewable energy, additionality (going beyond 'business as usual') and sustainable development. Independent verifiers check that the criteria have been met and certify the projects as meeting the Gold Standard. Although originally designed for CDM and JI projects, the Gold Standard has been adapted for the voluntary carbon market. A Gold Standard for voluntary offsets was launched in May 2006. It is marketed as 'the world's only quality standard for generating verified emission reductions (VERs) for the voluntary carbon market'.[22] In March 2008, the Gold Standard launched a registry which allows the creation, tracking and trading of VERs, to ensure the transparency, quality and security of carbon commodities in the voluntary carbon market.[23]

Crucially, the Gold Standard's requirements are more stringent than those of the CDM. Carbon sinks and sequestration projects for example are not eligible for the Gold Standard because the Gold Standard presumes that climate change should be tackled through an energy transition. To qualify for the Gold Standard, therefore, offsets must support renewable energy or energy efficiency. Moreover, carbon accounting methodologies for forestation are less solid than those for energy. Finally, sinks and sequestration projects are often relatively easy to achieve, and thus generate 'cheap' credits crowding out the more expensive credits from renewable energy and energy efficiency projects.[24]

The Gold Standard for voluntary offsets differs from the Gold Standard for CDM projects in some respects, including simplified guidelines for very small projects, broader eligibility of host countries, a broader range of acceptable methodologies for establishing project baselines, no need for formal host country approval, and the issue of credits directly by the Gold Standard instead of a third party.[25]

The Voluntary Carbon Standard[26]

While the Gold Standard is an initiative of leading civil society organisations, the Voluntary Carbon Standard (VCS) is driven mainly by the global business community. It is an initiative of The Climate Group, an international non-profit membership organisation the mission of which is to get business and government leaders talking to each other about climate change.[27] The Climate Group seeks to build the business case for action on climate change by focusing on 'game changing' initiatives that will make the biggest difference in the least amount of time.[28] It sees the VCS as such an initiative.[29] The founding partners of the VCS are The Climate Group, the International Emissions Trading Association (IETA) and the World Business Council for Sustainable Development. The VCS was released in March 2006, just two months before the Gold Standard for voluntary offsets. A second edition was published in 2007, and revised in 2008.[30]

The purpose of the VCS is to provide a rigorous, trustworthy global standard, and validation and verification programme, for voluntary greenhouse gas offsets. To meet the standard, offsets 'must be real (have happened), additional (beyond business-as-usual activities), measurable, permanent (not temporarily displace emissions), independently verified and unique (not used more than once to offset emissions)'.[31] Unlike the Gold Standard, the VCS is not restricted to renewable energy and energy efficiency projects, but extends to 15 sectors including manufacturing, construction, transport, agriculture, and forestry and land use. Reduced emissions from deforestation and forest degradation (REDD) can qualify for VCS credits. Also unlike the Gold Standard, the VCS does not require that projects contribute to sustainable development. On the other hand, as with the Gold Standard, conformity to the VCS is verified by an independent third party and the VCS itself issues the resulting credits (called Voluntary Carbon Units (VCUs)) to qualifying projects. All VCUs are issued, held and cancelled in VCS registries. As of late 2008, VCS expected to launch the global VCS project database and registry system in early 2009. Until then, no VCUs can be issued.[32]

The growth of voluntary offset standards

The Gold Standard and the Voluntary Carbon Standard are just two of several prominent voluntary carbon offset initiatives, and the number is growing rapidly.[33] The market for carbon offsets is young, and the many initiatives may puzzle consumers. At present the vast majority of voluntary offsets are not certified under any third-party standard (Kollmuss, et al., 2008). A report prepared for the WWF noted that the lack of a standard body to approve projects exacerbates the problem of conflict of interest, particularly where verifiers are selected and paid by the project developer and compete with each other for repeat business. None of the voluntary standards surveyed for the report have specific procedures in place to review the approved verifiers or to impose sanctions against those who under-perform (ibid., p. viii).

In general governments do not get involved in these controversies, but media coverage in various countries has unveiled negative experiences with unregulated offsets. This has encouraged increasing attention

Table 14.1 Sample Gold Standard Carbon Offset Offers for Brussels–Ottawa round trip, September 2008

Offset vendor	Calculated emissions (t CO_2)	Total Price for Gold Standard offsets (Euros)	Price/tonne (Euros)
Vendor 1	2.5	62.76[35]	25.10
Vendor 2	3.9	91.00	23.33
Vendor 3	3.4	45.80	13.47

towards building credibility. Dutch suppliers of carbon offsets for example have developed a joint project to enhance the credibility of offsets.[34] In practice, each offset provider is free to choose the integrity mechanism it wishes. This is not necessarily a disadvantage. Predictably, different initiatives are already competing to be the leading standard. What will motivate the buyers of carbon offsets and what will ultimately determine the 'winning' standard are open questions. Will it be the one that offers the cheapest credits or the one that contributes most to sustainable development? In this sense the freedom to choose among different standards is not a disadvantage. On the other hand, the differences among standards can puzzle consumers and possibly discourage them from buying.

It is easy to get a taste of the confusion facing consumers of voluntary carbon offsets. If one compares just a few sellers of carbon offsets for air travel, one finds substantial differences in both the calculation of carbon emissions for the same journey and the price of offsets, even when all the offsets purport to meet the Gold Standard (see Table 14.1).

In sum, we can identify the key concerns about voluntary carbon offsets as follows:

- the transparency and credibility of the offer (will the offset indeed be realised, will it not be double counted, and is the amount of greenhouse gas emissions to be offset calculated accurately?); and
- the integrity of the offer (will the offset, if realised, lead to a carbon emission reduction without resulting in perverse outcomes at the place where it is realised – that is, without contravening the principle of sustainable development?).

Government intervention may be warranted to ensure the credibility and integrity of carbon offsets. The next section examines how and to what extent the governments of the Netherlands and the UK have intervened in the voluntary carbon market.

3. THE NETHERLANDS: NO GOVERNMENT INVOLVEMENT

The Dutch government has not yet established a regulatory programme or anything like a voluntary code to ensure the credibility or integrity of offset offers. Meanwhile numerous Dutch suppliers have entered the market, while recent media coverage has warned of the risks of buying such credits. The media coverage emphasised the risk both that consumers might be misled and that offsets might have negative effects in developing countries. Following this negative coverage, the Dutch parliament submitted questions to the environment Minister about the government's responsibility to ensure the credibility and integrity of the voluntary offset market. The government was thus forced to explain its position toward the private carbon market. The Minister emphasised that suppliers have the responsibility to ensure the quality of their products.[36] She stated that the government considers the uncertainties regarding realisation of agreed greenhouse gas reductions acceptable, taking into account the overall positive effects of voluntary carbon offsets.[37] She expressed her support for the existence of several types of carbon offers, with differing degrees of certainty regarding the carbon offset.

The discussion of the credibility and integrity of the carbon offset offers has been extended to the Dutch government's own use of offsets. It appears that the use of offsets varies among several ministries. The Minister for the Environment has however announced plans to coordinate and harmonise governmental offsetting of carbon emissions, in particular with regard to air travel by politicians and civil servants. The government has promised to present minimum conditions for government use of carbon offsets by February 2009.[38]

In the meantime consultations and discussions between the Minister concerned with environmental policy and a group of carbon offset providers resulted in an agreement that the carbon offset offers will be assessed by a private expert.[39] One such assessment confirmed the credibility of most offers, though a small portion had some problems.[40] The problematic offers involved projects in developing countries (Jamaica and Nicaragua). The offset provider announced that it will adhere in future only to broadly recognised standards such as the CDM, the Gold Standard and VCS. Interestingly, the supplier of the problematic offsets has expressed its wish for stricter rules to ensure the quality of offsets, for instance through the establishment of a quality mark.[41] It is also worth noting that the assessment did not cover 35 per cent of the carbon offset portfolio, which specifically contains only forestry projects.

The Minister has acknowledged that the CDM is the most reliable offset standard and that most carbon offers on the voluntary market have lower standards, which means that they ensure less certainty and credibility.[42] At the same time the Minister noted that supporting non-CDM offsets could be advantageous too, as the CDM process is quite elaborate and therefore expensive. The voluntary market could stimulate small-scale projects for which the CDM process would be too cumbersome or expensive, though they might nevertheless be worthwhile.

According to the Minister, one important aspect should be improved: the transparency of the carbon market. The carbon offset suppliers should give full disclosure of the content of their products, including whether the offset is permanent and what procedures are in place to ensure its credibility. Such transparency would enable purchasers to evaluate non-Kyoto carbon offsets with CDM offsets.[43] The Minister also stated that the government lacks the power to enforce such transparency and, moreover, has no intent to establish such competence.[44]

One may ask whether, even with full disclosure, consumers are capable of using this information to choose the best offset supplier, and, moreover, whether they would be willing to spend the time required for such a comparative investigation. The Netherlands has strong consumer organisations, which play an important role in controlling the integrity of products and services in general. Perhaps one of these organisations or a governmental authority is needed to fulfill this task (van Boom and Loos, 2007). Regarding consumer law in general, there is a general view that '[a] proper mix of private and public collective enforcement will yield the best results' (ibid., p. 8, citing van den Bergh, 2007). This could indeed be a guiding principle when developing a policy for ensuring the credibility and integrity of the voluntary carbon market. It has been argued that Dutch private organisations play an effective role in the case of bona fide market participants; that is however less effective in the case of mala fide participants (Ogus, Faure and Philipsen, 2006).

It is curious that the formal discussion between the Minister and the Parliament has not yet considered the extent to which private law, in particular consumer law, already provides consumers and consumer organisations with the legal means to address non-compliance. In recent years EU law has strengthened the position of consumers and collective consumer concerns to help balance the position of 'weak' consumers against powerful suppliers of products and services. Following EU legislation, specific consumer law provisions have been established throughout the EU, with some differences among Member States. In the Netherlands, a Consumer Authority has been established with specific enforcement tasks and competences, including a 'naming and shaming' competence. Research is needed to determine whether these developments in consumer law would help consumers, consumer organisations and the newly established Consumer Authority to enforce the credibility and integrity of carbon offers.

In the meantime additional consumer law provisions have been proposed to correct unfair business-to-consumer commercial practices. This new development, which is obligatory to implement EC law, ensures that businesses which do not comply with their initial offers can be more easily sanctioned.[45] Under this initiative, the Dutch Consumer Authority will receive some new competences and the maximum fine will be €450,000 or 10 per cent of turnover, whichever is greater. The extent to which these developments in consumer legislation can help ensure the credibility and integrity of the voluntary carbon offset market and establish effective control of offset providers has yet to be seen. Such analysis is needed before the government decides to establish new competences.

The Dutch government has signalled its desire to trust in the responsibility of carbon offers and voluntary certification mechanisms, and thus far no governmental initiatives or policies have been announced. This means that only existing consumer law and other existing provisions can be used to regulate the market. The Minister has emphasised mostly the responsibility of offset providers to develop procedures to ensure the credibility and integrity of the market. In the Minister's view, avoidance of possible conflicts with human rights should equally be ensured through such (private) certification provisions, and specific government investigations of the human rights impacts of carbon offset offers are not indicated.[46]

4. THE UK QUALITY ASSURANCE SCHEME FOR CARBON OFFSETTING

The UK government favours a policy to ensure the credibility and integrity of carbon offsets. In contrast to the Netherlands, the government has expressed a positive attitude toward the necessity of governmental involvement in the offsets market. Its preferred policy instrument is a Code of Best Practice for Carbon Offset Providers aimed at ensuring consumer confidence in the emerging voluntary offset market and continued growth of that market. On 19 February 2008 the Secretary of State of the Department for Environment, Food and Rural Affairs (Defra) presented a draft Code (Defra, 2008).[47] This followed an earlier report by the Environmental Audit Committee of the House of Commons, which emphasised the need for government leadership:

> There are many divergent and often loud opinions about the role of the voluntary offset market. Both individuals and businesses are very likely to be

> confused by the mixed messages available. They need clear guidance about the extent to which offsetting can help meet their responsibilities to reduce carbon emissions. We recommend strongly that the Government grasps the opportunity to show leadership here. It must set out its own view on the role that the voluntary offset market can play in reducing emissions and why offsetting is a positive thing. The view should be unambiguous, well-publicised and prominent in all Government communications concerning offsetting and climate change. (House of Commons Environmental Audit Committee, 2008, p. 50.)

The committee even suggested that businesses should be required to give consumers a choice to purchase carbon offsets when buying products and services. The committee supported particularly the idea of requiring all sellers of air tickets in the UK to include in the price offered the cost of an offset, and to sell that offset along with the ticket unless the customer requests otherwise (ibid., p. 54).

In presenting the draft Code of Best Practice for Carbon Offset Providers, the government considered that carbon offsetting could help to raise awareness of and reduce the carbon impact of consumer activities. The Code is intended to increase consumer confidence in the integrity and value for money of the offset products available to them. Ultimately it aims to provide confidence and clarity to consumers. Importantly, the government acknowledges that voluntary offsetting is not a cure for climate change, as the most effective way to combat climate change is to reduce emissions. In fact, the government is already taking into account that offsetting is only an interim measure, until other regulatory measures have been adopted for instance concerning air travel.[48]

The Code sets standards for:

- Robust and verifiable emission reduction credits;
- Accurate calculation of emissions to be offset, using statistics and factors published for this purpose by the Government (ensuring that providers follow a uniform formula to calculate how much carbon is emitted by a particular activity);
- Clear information for consumers regarding the mechanism and/or projects supported;
- Transparent pricing, and
- Timescales for cancelling credits.[49]

Voluntary offset products that meet these requirements may be accredited under the Code. Accredited products are awarded a quality mark so that individual or business consumers of offsets can easily recognise that they comply with the Code. Development of the Code was initiated and funded by the government. An independent Accreditation Body was established after a competitive tender to assist the government in finalising the Code, develop criteria to demonstrate compliance with the Code, and accredit offset products (Defra, 2007b). Offset providers that wish to have their offsets assessed and accredited to receive the quality mark must pay a fee. The government intends the accreditation programme to be self-financing (ibid.).

Accreditation under the Code is voluntary. Offset providers who sell accredited offsets are not prevented from selling offsets that do not meet the Code, but they may use the quality mark only for accredited products. The Code will be reviewed annually to ensure continual improvement.

The Code initially applies only to certified emission reductions generated via the CDM. This allows the government to adhere to the procedure established by the Kyoto Protocol, instead of developing a complex procedure of its own. The government however has challenged business to develop a standard for Voluntary Emissions Reduction credits (VERs) which could be included in the Code in the future, subject to verification of its robustness. In an open letter urging industry to develop a standard for VERs, the Secretary of State for Environment, Food and Rural Affairs emphasised the value of non-Kyoto projects as a means of developing new and innovative offset projects.[50]

To stimulate industry to develop a standard for VERs, the Government inter alia announced that it would sponsor a workshop to help industry come together and start discussion. The Government thus left the responsibility to the market though it stimulated such private standards development by announcing that projects and procedures will be assessed to consider whether credits approved under the industry standard can be used as offsets meeting the Code.[51] In this vein it is important not to overlook the opinion of the Environmental Audit Committee, which believed that a focus only on CDM credits would leave unregulated those portions of the market where there is the greatest innovation and greatest environmental or sustainable development benefit. This approach would also leave unregulated and unconstrained 'the activities of so called carbon cowboys peddling flimsy VERs' (House of Commons Environmental Audit Committee, 2008, p. 53).

The Government has invited members of the offsetting industry and other stakeholders to set up an expert panel to discuss the operation of the Code. It is intended that there should be no involvement from Government or the Accreditation Body, but that the Government and the Accreditation Body should liaise with the panel when developing and revising the Code. This panel may also consider input to the development of a standard for VERs (Defra, 2008, p. 7).

The draft Code also proposed an enforcement regime. If an accredited offset no longer meets the requirements against which it was accredited but the offset provider continues to use the quality mark the offset provider will be in breach of the Code. Similarly if accreditation is gained through false or misleading statements accreditation will be suspended. The requirements suppliers must meet regarding accredited offsets will be part of a licence agreement. Where offset providers fail to comply with the Code, they will be informed and offered the chance to correct the error within 10 working days. In the event that no corrective action is taken, the right to use the quality mark will be withdrawn by the Accreditation Body. In case of a breach of the licence agreement the offset provider may be taken to court. The Accreditation Body will perform bimonthly Internet reviews to check for breaches of the Code including incorrect use of the quality mark (Defra, 2008, p. 17). Compliance will also be checked through targeted or random audits. Offset providers will be given four weeks' notice that they will be audited. It is expected that the audit exercise will be completed within 10 weeks of notice being given. To pass the audit, offset providers must be able to provide information to prove that they meet the requirements of the Code by providing data on offsets sold and how they met the other requirements of the Code (Defra, 2008, p. 18).

If consumers or other stakeholders have complaints, for example that offset providers are not using the quality mark correctly, they can provide details to the Accreditation Body. The Accreditation Body will then assess the issue and take any required action. Organisations deemed to be in breach of the Code which persist in using the quality mark and any organisation using the quality mark without having applied for accreditation will be named on both the accreditation body's and Defra's websites (Defra, 2008, p. 17).

Offset providers may complain to the Accreditation Body about the application or audit process. The complaint will be considered independently of the application and audit process. A response will be sent within a further 15 working days. If the complaint is not resolved satisfactorily it may be presented to the government (Defra, 2008, p. 18).

The Draft Code was finalised in late 2008 as the Quality Assurance Scheme for Carbon Offsetting (QAS).[52] Around the same time, responsibility for the Code and for carbon offsetting policy was transferred to a new Department of Energy and Climate Change (DECC). The first carbon offsets to be approved under the new Quality Assurance Scheme for Carbon Offsetting (QAS) were accredited on 5 February 2009.[53]

5. DISCUSSION

Fragmented Approaches to Integrity and Credibility

The voluntary carbon offset market allows citizens (and businesses and governments) to offset their carbon emissions. The worldwide voluntary carbon offset market is fragmented, with a variety of quality standards and procedures. The purpose of these divergent mechanisms is basically to assure customers of the quality of the offsets they are purchasing (Ribón and Scott, 2007). Most of the offsetting measures will be conducted in developing countries, since offsets are much cheaper there. To the extent that these offsets will be realised outside the Clean Development Mechanism of the Kyoto regime, one can wonder what guarantees exist that such offsets are reliable and effective, and, moreover, are not contrary to the interests of local residents. Indeed the credibility and integrity of carbon offset offers should not be overlooked, and procedures should be in place to ensure these values. Of course the CDM as such is not without criticism, but it is at least intended to ensure real greenhouse gas reductions and promotion of sustainable development. Based on experiences with the CDM, there is of course always the opportunity to improve this mechanism. Voluntary carbon offsets that fall outside this scheme are not (yet) regulated under laws specifically addressing the integrity and credibility of the offset offers. Of course, consumer law, including advertising standards, is applicable and can be used to regulate carbon offset offers, for instance to redress misleading information.

In practice, remarkable differences exist among governmental policies regarding the credibility and integrity of the voluntary offset market. The EU has thus far not provided a specific approach towards regulation of voluntary offsets, and the EU Member States are still competent to issue regulations regarding those offsets. Within the UK we see the development of a Code of Best Practice for Carbon Offset Providers, while in the Netherlands the government mostly relies on offset providers to take responsibility themselves. The latter approach facilitates the development of a variety of offset mechanisms as alternatives to the Kyoto credits. But the problem is whether such alternative mechanisms are credible and sustainable, and whether consumer law is strong enough to control offset suppliers. The former approach is likely to be more expensive for government (not least because of choosing the CDM as the minimum standard), but has a better chance of being credible and sustainable. Moreover the UK government still encourages the development of alternative voluntary offset mechanisms, as long as basic criteria are met. In conclusion we see that the UK government is much more emphatic that private carbon offsets need to comply with criteria which ensure the credibility and integrity of the scheme, and, moreover, has taken an initiative to develop a mechanism for voluntary offsets.

In the meantime we have equally seen that the Executive Secretary of the UNFCCC has proposed the intriguing idea that the UNFCCC could provide a framework for awarding accreditation of voluntary offsets.[54] Such a supranational approach seems very appropriate. The advantage of an international mechanism would be that developing countries, being parties to the UNFCCC, would be able to express their opinion and their vote regarding the content of such a mechanism. Protection against misrepresentation and corruption also merit careful attention in any international mechanism. And, in case of a possible international harmonisation of the voluntary carbon offset market, the advantages of developing a portfolio of offset offers should not be overlooked, to make room for new and innovative, but by their nature more risky, approaches. Such new developments need full transparency and timely assessment to decide whether they should be supported.

Government as Influential Buyer in the Market

Besides providing a regulatory framework for accrediting offsets, government could play another important role in the voluntary offset market. Governments themselves are very large carbon emitters. As buyers of carbon offsets, they could have much influence on the development of the carbon market and could send important signals to consumers about which providers to choose.

The UK Government has developed a Government Carbon Offsetting Fund (GCOF) to meet its commitment to offset emissions arising from official and ministerial air travel from April 2006.[55] All central government official and ministerial air travel is to be offset by purchasing credits 'that meet strict international standards in terms of the certification and monitoring of emission reductions'.[56] All the credits come from CDM projects located in developing countries.[57] The British Government is also looking at the role carbon offsetting might play in the commitment for the Government office estate to go carbon neutral by 2012.[58] Provided that the carbon offset projects indeed meet the standards, such projects may contribute to the wellbeing of the local people. However, effects of carbon offset projects on local people, environment and development in the developing world should be assessed and reported on, as we still need to gain further experience with the effects of world-wide carbon trading on developing countries.

In the Netherlands the coordination of government carbon offsetting has just begun. Following some discussion with the Second Chamber of the parliament, the Minister for the environment started to coordinate the offsetting of air travel and other carbon emissions. Also here, the question can be raised whether a multi-track approach would be beneficial, as adhering to a credible and effective but costly and bureaucratic approach may mean that the development of the market will be hindered, which after all could not be beneficial for the developing world.

Interference with Regulatory Measures

As the ultimate goal of climate policy is to decrease the carbon footprint of the global North, the voluntary market must be seen as an interim measure. The near future of the voluntary emissions trading market will depend on how other regulatory regimes evolve. One can think of initiatives that would support the market: the European Parliament recently demonstrated its interest in disclosure of carbon information that would enable citizens to offset their carbon emissions, as it called for airline reservation systems to provide environmental information about flights such as carbon dioxide emissions per passenger.[59] Other regulatory measures could impede the willingness of citizens to offset their emissions: it can for instance be imagined that the decision to include air transport within the EU ETS means that citizens will decide not to pay twice for carbon: once for the price increase of the ticket as a consequence of the EU ETS, and, second, for the voluntary offset of the real emissions caused. And an air travel tax as imposed in the Netherlands in 2007 means that citizens are being discouraged from offsetting their emissions, as we can imagine again that they are not willing to pay twice: first for the obligatory tax, second for the voluntary offset. The difference between the tax and the offset is that the money in the first case will go to the national government (with no guarantee that it will be allocated to climate policy goals), while in the second case it will go to climate investment in a developing country.

Indeed the usefulness of the voluntary market will decrease as soon as other policy tools or other effects (like, for instance, rising energy prices) stimulate wealthy citizens to reduce their carbon footprint. Within the EU, some policy approaches (the effectiveness of which is an open question) already target emissions from buildings and cars (for an overview see Peeters, 2007). At the national level, governmental initiatives are increasingly being considered, such as the French minister for the Environment's proposal to widen the 'bonus-malus' tax and subsidy scheme for encouraging low carbon emission cars and products such as televisions, computers, tyres and light bulbs. This proposal however does not appear to have garnered wide support because of severe budgetary disadvantages.[60]

Moreover the content of the offset measure compared to national regulations matters too: it seems quite odd that offsetting would lead to cases in which a citizen is paying for replacement of fluorescent bulbs[61] without any obligation to have replaced them in the North. It could also be that citizens are paying for a wind farm, to be established in a developing country, without being ready to accept such wind farms in their own neighbourhoods. These policy questions should not be overlooked by governments when developing the voluntary offset market.

The EU Greenhouse Gas ETS as an Opportunity for Citizens

Another interesting thought is that EU citizens would be able to buy greenhouse gas allowances provided by the EU greenhouse gas ETS. The investment would then go to an industry that has reduced its emissions, provided that the problem of free allocation of allowances, leading to over-allocation, is avoided. Such free allocation occurred in the first period of the scheme in 2005–2007. The second period (2008–2012) also employs predominantly free allocation of allowances, although over-allocation might have been prevented this time. EU citizens may nevertheless feel reluctant to buy rights that were given for free to industries, even if this would lead to the reduction of greenhouse gases. In the meantime, the European Commission has proposed that auctioning of greenhouse gas allowances should be the default approach starting from 2013 onwards.[62] This could imply that citizens could buy EU greenhouse gas allowances directly from the issuing authority (provided that the regulatory framework would allow this), which they could then retire. In this way citizens would be able to contribute to the reduction of industrial emissions within the EU ETS scheme, as there would be fewer allowances available for such industrial emissions. They could use the allowances to offset their own carbon emissions.[63] Of course the price of the allowances compared to the prices of the voluntary offset mechanisms will influence their decisions.

6. CONCLUSION

As long as the voluntary carbon market is developing, the credibility and integrity of the scheme pose for governments the questions whether and how they should get involved. The strength of consumer law in this area and the additional role of governmental policies, including international policies, are however yet to be determined. The voluntary market has important benefits, since it enables citizens who want to take responsibility for their carbon emissions to do so. Moreover, if properly managed and controlled, this market can lead to beneficial investments in the developing world. The idea of including compulsory choice options for Northern consumers seems furthermore an interesting thought which could stimulate the market. Even the option of compulsory offsetting could be considered by Northern governments, for instance for some high-carbon services or products. But this is insufficient. The fact that citizens cannot easily check the integrity of the offset offer and the behaviour of the offset providers, including the execution of the offsets, makes it necessary for government to take responsibility. The government should at least check whether the agreed offset offers are complied with, in particular if consumer law appears not to be effective enough on its own. An even more appealing role for government is to implement policies to ensure that offsets bought by Northern consumers do not lead to detrimental effects for local people or nature in the developing world.

NOTES

* Marjan Peeters is a Professor at the Department of Public Law at Maastricht University, The Netherlands.

1. According to the Fourth Assessment Report of the Intergovernmental Panel on Climate Change (2007), there is high agreement that changes in lifestyle and behaviour patterns can contribute to climate change mitigation across all sectors. Intergovernmental Panel on Climate Change, 2007, p. 59.
2. See, for example, Directive 2002/91/EC of the European Parliament and of the Council of 16 December 2002 on the energy performance of buildings [2003] OJL001/65; Directive 1999/94/EC of the European Parliament and of the Council of 13 December 1999 relating to the availability of consumer information on fuel economy and CO_2 emissions in respect of the marketing of new passenger cars [2000] OJL012/16, as amended by Commission Directive 2003/73/EC of 24 July 2003 amending Annex III to Directive 1999/94/EC of the European Parliament and of the Council [2003] OJL186/34.
3. See the section of this chapter on 'The Growth of Voluntary Offset Standards', below.
4. See, for example, Visa GreenCard, www.visagreencard.nl (Netherlands) (visited 5 January 2009).
5. The Montreal International Jazz Festival 2008 for example used carbon offsets to achieve certification as a carbon neutral event: Planetair, 'Montreal Jazz Festival certified carbon neutral' (25 June 2008), http://planetair.ca/modules/smartsection/item.php?itemid=28&keywords=montreal (visited 23 December 2008).
6. See, for example, CarbonNeutral Company, www.carbonneutral.com (visited 23 December 2008).
7. See, for example, Schneider (2007) for a discussion of the negative effects of competition between Designated Operational Entities in the context of the Clean Development Mechanism (CDM).
8. There is evidence that consumer interest is very low: House of Commons Environmental Audit Committee, 2007, p. 13.
9. As I have argued elsewhere (Peeters, 2008), enhancing citizen responsibility is also critical to ensuring democratic accountability of carbon policies. Greenhouse gas emissions trading, the most popular market mechanism for reducing carbon emissions, is in fact

a very abstract tool with no public participation at the local level. Moreover, citizens usually have no information about the carbon footprint of consumer products. Carbon labelling may address this shortcoming, although it has its own difficulties. It would better enable citizens, if they wish to do so, to offset the carbon load of purchased products by investing voluntarily in carbon offsets. Such investment might involve 'double' compensation for the carbon emissions related to the product, since the cap and trade emissions trading scheme already ensures a reduction of emissions to some extent. The compensation by the citizen would thus indeed be an extra investment. I address the issue of double compensation in the final section of this chapter, in the case of air flights.

10. Clean Air Act, 42 USCs. 7401.
11. Discussion of Australian developments is however outside the scope of this chapter.
12. This phenomenon was explored in a Dutch television documentary in which it was suggested that local people in Uganda were illegally displaced to make way for a forestation project paid for by Dutch energy companies to offset their carbon emissions: Jos Van Dongen, Martin Maat and Kees Driehuis (Directors), 'Het CO2-alibi' ('The CO2 Alibi'), *Zembla* (Dutch current affairs television programme) (19 November 2007), Hilversum, The Netherlands: VARA/NPS (Dutch, with English subtitles), http://player.omroep.nl/?aflID=7320917 (visited 23 December 2008).
13. Kyoto Protocol to the United Nations Framework Convention on Climate Change (1997), (1998) ILM 37, 22, Arts, 6 and 12.
14. United Nations Framework Convention on Climate Change (UNFCCC) (1992), (1992) ILM 31, 848.
15. Y. De Boer, 'Climate Change: Mobilising Political Will', address presented to the OECD Forum 2008 on Climate Change, Growth, Stability (Paris, 3 June 2008), http://unfccc.int/files/press/news_room/statements/application/pdf/080603_speech_oecd.pdf (visited 13 January 2009).
16. Some suppliers offer a choice of different kind of offsets. See, for example, Planetair, http://planetair.ca/ (visited 23 December 2008).
17. United Nations General Assembly, Overview of United Nations activities in relation to climate change: Report of the Secretary-General, UN Doc. A/62/644 (10 January 2008), pp. 38–39. No deadline has been mentioned.
18. The United Nations General Assembly has its own criteria and methodology for carbon offset purchases: United Nations General Assembly, President of the 62nd Session, 'Reducing the Carbon Footprint of the UN High-Level Thematic Debate "Addressing Climate Change: The United Nations and the World at Work"', www.un.org/ga/president/62/ThematicDebates/ReducingCarbonFootprint.shtml (visited 23 December 2008).
19. KPMG Sustainability, 'Climate Neutral Group: Overview of Emission Reduction Projects for Carbon Offsetting', Report No. 2008-098 (June 2008), p. 15, www.klimaatneutraal.nl/Downloads/080701%20KPMG%20report%20CNG%20projects.pdf (visited 13 January 2009).
20. The Gold Standard, www.cdmgoldstandard.org/ (visited 23 December 2008).
21. The Gold Standard, 'Supporters', www.cdmgoldstandard.org/about_goldstandard.php?id=16 (visited 23 December 2008).
22. Gold Standard, 'The Gold Standard for Voluntary Offsets', www.cdmgoldstandard.org/how_does_it_work.php?id=44 (visited 23 December 2008).
23. The Gold Standard Registry, http://goldstandard.apx.com/ (visited 23 December 2008).
24. Gold Standard, 'What does the Gold Standard Foundation do?', www.cdmgoldstandard.org/faqs.php?type=What+is+the+Gold+Standard%3F (visited 22 February 2009).
25. Gold Standard, 'The Gold Standard for Voluntary Offsets', above n. 22. Normally the Gold Standard does not issue credits itself.
26. The Voluntary Carbon Standard, www.v-c-s.org (visited 23 December 2008).

27. The Climate Group, 'What We Do', www.theclimategroup.org/about/what_we_do (visited 23 December 2008). The Climate Group's members are leading global companies and several local and state/provincial governments from Australia, Canada, the UK and US.
28. Ibid.
29. Ibid.
30. See Voluntary Carbon Standard 2007.1, Specification for the project-level quantification, monitoring and reporting as well as validation and verification of greenhouse gas emission reductions or removals (18 November 2008), http://www.v-c-s.org/docs/Voluntary%20Carbon%20Standard%202007_1.pdf.
31. The Voluntary Carbon Standard, 'About the VCS', www.v-c-s.org/about.html (visited 23 December 2008).
32. Voluntary Carbon Standard, 'Projects and Registries', www.v-c-s.org/projects.html (visited 23 December 2008).
33. Others include VER+, the Voluntary Offset Standard (VOS), Chicago Climate Exchange (CCX), the Climate, Community & Biodiversity Standards (CCBS), the Plan Vivo System, ISO 14064 and 14065, and the GHG Protocol. See Kollmuss, et al., 2008; KPMG Sustainability, above note 19, pp. 15–16; and EcoBusinessLinks Environmental Directory, 'Carbon Offset Survey', www.ecobusinesslinks.com/carbon_offset_wind_credits_carbon_reduction.htm (visited 23 December 2008) for comparison of the various initiatives.
34. Klimaatcompensatie, www.klimaatcompensatie.nl (visited 23 December 2008).
35. The quoted price of CAD $98.75 was converted to Euros using the Bank of Canada's 10-year Currency Converter (cash rate for 26 September 2008), at www.bankofcanada.ca/en/rates/exchform.html (visited 23 December 2008).
36. Dutch Parliament, Second Chamber, 2007–2008, File 31 209, No. 32, p. 1 (16 June 2008) (only available in Dutch).
37. Ibid., p. 7.
38. Ibid.
39. Dutch Parliament, Second Chamber, 2007–2008, File 31 209, No. 12, p. 2 (15 February 2008) (only available in Dutch) (stating that this assessment would be done in due time).
40. KPMG Sustainability, above note 19.
41. Climate Neutral Group, press release, 'persbericht KPMG-studie naar duurzame energieprojecten voor klimaatcompensatie, Overgrote deel CO2-projecten Climate Neutral Group gecertificeerd' (9 July 2008), (in Dutch only; on file with author).
42. Dutch Parliament, Second Chamber, 2007–2008, File 31 209, No. 32, pp. 2–3 (16 June 2008) (only available in Dutch).
43. Dutch Parliament, Second Chamber, 2007–2008, File vra2008vrom-01, pp. 2–3 (26 March 2008) (only available in Dutch).
44. It is uncertain whether the general duty of care codified in Art. 1a of the Dutch Environmental Management Act can be applied or whether a civil law procedure would be possible. These possibilities have however not been discussed explicitly by the Minister.
45. Directive 2005/29/EC of the European Parliament and of the Council of 11 May 2005 concerning unfair business-to-consumer commercial practices in the internal market and amending Council Directive 84/450/EEC, Directives 97/7/EC, 98/27/EC and 2002/65/EC of the European Parliament and of the Council and Regulation (EC) No 2006/2004 of the European Parliament and of the Council (Unfair Commercial Practices Directive) [2005] OJL149/22.
46. Dutch Parliament, Second Chamber, 2007–2008, File 31 209, No. 32, p. 7 (16 June 2008) (only available in Dutch); Dutch Parliament, Second Chamber, 2007–2008, File vra2008vrom-01, p. 6 (26 March 2008).
47. Defra, 'Climate Change: Carbon Offsetting – Code of Best Practice' (Defra web page, visited 5 January 2009, removed shortly thereafter). The Defra web pages consulted in

the course of this research were removed in early 2009 after Defra was reorganised and responsibility for the carbon offset Code was transferred to the new Department of Energy and Climate Change (DECC). Copies of documents originally posted on those web pages (including Defra 2007a, 2007b, and 2008) are on file with the author. Limited information on carbon offsetting can now be found at Defra's new carbon offsetting web page, http://offsetting.defra.gov.uk/cms/ (visited 3 March 2009).

48. According to Defra, 'Offsetting is a complimentary [sic] interim measure for tackling the climate change impacts from aviation, which could be promoted to the wider travelling public to be taken up on a voluntary basis': Defra, 'Climate Change: Carbon offsetting – Frequently Asked Questions' (Defra web page, visited 13 January 2009, removed shortly thereafter).
49. Ibid. See also Defra (2007a), p. 11.
50. Hillary Benn, Secretary of State for Environment, Food and Rural Affairs, 'Open Letter on the case for the offsetting industry to establish a standard for Voluntary Emissions Reductions' (February 2008) (originally posted on Defra web page, visited 13 August 2008, removed early 2009), copy on file with author.
51. Ibid.
52. Information on the QAS can be found on Defra's new carbon offsetting website, http://offsetting.defra.gov.uk/cms/ (visited 3 March 2009).
53. Defra, 'Approved Offsets', at http://offsetting.defra.gov.uk/cms/approved-offsets/ (visited 3 March 2009).
54. De Boer, above note 15.
55. Defra, 'Carbon Offsetting: Government Emissions' (Defra web page, visited 13 January 2009, removed shortly thereafter). For the current policy see Defra, 'Carbon Offsetting: UK Government Emissions', at www.defra.gov.uk/environment/climatechange/uk/carbonoffset/govt.htm (visited 3 March 2009).
56. Defra, 'Climate Change: Carbon Offsetting' (Defra web page, visited 13 January 2009, removed shortly thereafter).
57. Ibid.
58. Defra, 'Carbon Offsetting: UK Government Emissions', at www.defra.gov.uk/environment/climatechange/uk/carbonoffset/govt.htm (visited 3 March 2009).
59. See ENDS Europe, 'MEPs Demand Green EU Airline Booking Rules' (5 September 2008), www.endseurope.com/15447 (visited 13 January 2009); European Parliament legislative resolution of 4 September 2008 on the proposal for a regulation of the European Parliament and of the Council on a Code of Conduct for computerised reservation systems (COM(2007)0709 – C6-0418/2007 – 2007/0243(COD)), www.europarl.europa.eu/sides/getDoc.do?pubRef=-//EP//TEXT+TA+P6-TA-2008-0402+0+DOC+XML+V0//EN&language=EN#BKMD-1 (visited 13 January 2009).
60. C. Jakubyszyn (2008), 'M. Borloo veut généraliser le "bonus-malus" écologique', *Le Monde* (20 August), at www.lemonde.fr (visited 13 January 2009).
61. KPMG Sustainability, above note 19, p. 19.
62. European Commission, COM(2008)16, Proposal for a Directive of the European Parliament and of the Council amending Directive 2003/87/EC so as to improve and extend the greenhouse gas emission allowance trading system of the Community, [2008] OJC118/3.
63. In this way citizens could even get JI or CDM credits under the Kyoto Protocol, as those mechanisms have been linked to the EU ETS, and the European Commission has proposed to go ahead with such mechanisms even if no international agreement is reached for new post-Kyoto commitments. Moreover, the European Commission has proposed to provide a mechanism for voluntary domestic offsets. This would mean that voluntary greenhouse gas reductions within the EU could be awarded tradable allowances, which of course could be equally bought by citizens: see COM(2008)16, Proposal for a Directive of the European Parliament and of the Council amending Directive 2003/87/EC so as to improve and extend the greenhouse gas emission allowance trading system of the Community, [2008] O.J. C 118/3, Art. 24a.

REFERENCES

Australian Competition and Consumer Commission (ACCC) (2008), *Carbon Claims and the Trade Practices Act* (Commonwealth of Australia).

Department of Environment, Food and Rural Affairs (Defra) (2007a), *Consultation on Establishing a Voluntary Code of Best Practice for the Provision of Carbon Offsetting to UK Customers* (Defra).

Department of Environment, Food and Rural Affairs (Defra) (2007b), *Summary of Responses to the Consultation on Establishing a Code of Best Practice for Selling Offsetting to Consumers, 18 January to 13 April 2007* (July), (Defra).

Department of Environment, Food and Rural Affairs (Defra) (2008), *Draft Code of Best Practice for Carbon Offset Providers: Accreditation Requirements and Procedures* (Defra).

Hahn, R.W. and G.L. Hester (1989), 'Where Did all the Markets Go? An Analysis of EPA's Emission Trading Program', *Yale Journal of Regulation* 6(1), 109.

House of Commons Environmental Audit Committee (2007), *The Voluntary Carbon Offset Market: Sixth Report of Session 2006–07* (HC 331) (HMSO).

Intergovernmental Panel on Climate Change (IPCC) (2007), *Fourth Assessment Report* (Cambridge University Press).

Kerr, T. (2007), 'Voluntary Climate Change Efforts', in M.B. Gerrard (ed.), *Global Climate Change and U.S. Law* (American Bar Association).

Kollmuss, A., et al. (2008), *Making Sense of the Voluntary Carbon Market* (WWF Germany).

Ogus, A.I., M.G. Faure and N.J. Philipsen (2006), *Best Practices for Consumer Policy: Report on the Effectiveness of Enforcement Regimes* (Organisation for Economic Co-operation and Development).

Peeters, M. (2007), 'EU Climate Change Policy: Critical Issues and Challenges for the Future', *Yearbook of International Environmental Law* 16, 179.

Peeters, M. (2008), 'Legislative Choices and Legal Values: Considerations on the Further Design of the European Greenhouse Gas Emissions Trading Scheme from a Viewpoint of Democratic Accountability', in M.G. Faure and M. Peeters (eds), *Climate Change and European Emissions Trading: Lessons for Theory and Practice* (Edward Elgar).

Ribón, L. and H. Scott (2007), *Carbon Offset Providers in Australia 2007* (Global Sustainability at RMIT University).

Richardson, B.J. (2008), *Socially Responsible Investment Law: Regulating the Unseen Polluters* (Oxford University Press).

Schneider, L. (2007), *Is the CDM Fulfilling its Environmental and Sustainable Development Objectives? An Evaluation of the CDM and Options for Improvement* (report prepared for WWF) (Öko-Institut e.V).

Tietenberg, T.H. (2006), *Emissions Trading. Principles and Practice* (2nd edn, Resources for the Future).

Van Boom, W. and M. Loos (eds) (2007), *Collective Enforcement of Consumer Law* (Europa Law).

Van den Bergh, R. (2007), 'Should Consumer Protection Law be Publicly Enforced? An Economic Perspective on EC Regulation 2006/2004 and its Implementation in the Consumer Protection Law of the Member States', in W. van Boom and M. Loos (eds), *Collective Enforcement of Consumer Law* (Europa Law).

15. Climate and trade in a divided world: can measures adopted in the North end up shaping climate change legislative frameworks in the South?

Francesco Sindico*

1. INTRODUCTION

Developing countries face a double challenge when dealing with climate change: they are asked to adopt mitigation and adaptation measures to start tackling this grave problem but, at the same time, development remains the number one priority for the vast majority of them. These two objectives may create tensions between sectors of a government wishing to prioritize environmental goals and those more concerned with economic development. Furthermore, this tension between addressing climate change and pursuing economic development may lead to disputes with other countries.[1] To foster development, many developing countries will strive to participate more actively in the global market, since one avenue to promote economic development is for producers to export freely their goods and services. It is therefore crucial that they be able to rely on international legal norms that facilitate exports to other developing countries and even more so to industrialized ones, as it is in the latter that most goods they produce are consumed.

This chapter seeks to explore to what extent climate change policies in the North may end up shaping policies in the South by obliging exporters from developing countries to comply with specific climate change requirements. Section 2 briefly highlights carbon leakage and competitiveness concerns as reasons why an industrialized country may consider targeting imports from developing countries. Section 3 focuses on current United States (US) legislative initiatives at the federal level proposing to link imports from developing countries to climate change policy. Section 4 examines whether the current international trade rules enshrined in the

World Trade Organization (WTO) Agreements allow for such domestic climate change measures and what options are open to developing countries in case of a climate and trade dispute. It seeks to determine, in particular, whether relying on the multilateral trading system would actually benefit these countries. The conclusion suggests a way forward.

2. CARBON LEAKAGE AND COMPETITIVENESS CONCERNS

Economic and environmental concerns may prompt an industrialized country to adopt trade measures against imports coming from developing countries. A company based in the North may worry that competitors in the South that are not bound by comparable climate change obligations will enjoy an unfair competitive advantage on the market. The company could lobby its government to adopt climate change related trade measures as a means to tackle the loss of competitiveness on the global market. The government adopting climate change related trade measures might also consider it necessary to deal with the problem of carbon leakage, which is the increase of greenhouse gas (GHG) emissions in one jurisdiction resulting from the adoption of climate change policies in another. Carbon leakage results, notably, from the relocation of production from countries with stronger climate change requirements, presumably in the North, to countries with less rigid ones, presumably in the South. To deal with carbon leakage, a government might decide to adopt a climate change related trade measure aimed at preventing companies from relocating production to jurisdictions with weak GHG standards.

Have companies in the North already started to complain about their position on the global market vis-à-vis companies in developing countries because of different climate change requirements? Have governments paid attention to these concerns when developing climate change policies and law? Certainly, a number of energy intensive industries in the European Union (EU) have warned that if the new European climate change legislation, and in particular the revised emissions trading scheme, does not deal comprehensively with 'cheap' imports from third parties, they may consider relocation as part of their business strategy.[2] The same is true in the US where the decision not to ratify the Kyoto Protocol was mainly a political decision based on competitiveness concerns.[3] Despite the fact that the debate on climate change has changed radically in that country in the last few years, most sectors of the economy are still of the view that accepting binding international climate change obligations when emerging developing countries are not is not a price worth paying (Houser, et al., 2008, pp. 1–13).[4]

Both the EU and the US are aware of concerns raised by industrial associations. Furthermore, the current economic situation heightens fears by industries in the North of loss of competitiveness in the global market. These concerns are reflected in the current revision of the EU Emissions Trading Scheme (European Commission, 2008a, p. 11; European Commission, 2008c, recitals 19 and 20), and in a number of climate proposals tabled in Congress in the US in recent years.[5]

Relocating to a country with less stringent climate change requirements may not be environmentally friendly, but ultimately companies respond to their shareholders. Profit-making – not reducing emissions – is their main goal. However, other considerations may deter a company from relocating. First, a company in the North may consider that, even if faced with tougher climate change requirements, it is good business strategy and consistent with its corporate social responsibility policy not to move (Bubna-Litic, 2007; Richardson, 2008).[6] Second, the loss of competitiveness on climate change grounds is confined to specific sectors. Research has shown that only a relatively limited, albeit important, part of the economy will face problems due to less stringent environmental requirements in other countries (Ratnayake, 1998). Climate change does not seem to be an exception (Demailly and Quirion, 2006).[7] If a country still relies primarily on fossil fuels, the amount of energy needed in the production process will determine whether a sector is at risk or not (Houser, et al., 2008, pp. 5–7). Petroleum refining, paper and pulp, steel, non-metallic mineral products, chemicals, cement, and ferrous and nonferrous metals sectors have been identified as energy intensive industries most at risk in the EU and the US (Houser, et al., 2008, p. 8; European Commission, 2008b, p. 11).

3. CAN THE CLIMATE CHANGE LAWS IN A DEVELOPING COUNTRY BE INFLUENCED BY MEASURES ADOPTED IN THE NORTH?

If a government adopts a measure for domestic producers and requires foreign producers of like goods destined for that domestic market to comply with comparable measures, this will inevitably have implications for the exporting country. In such a scenario foreign steelmakers, for example, would be allowed to enter the domestic market only if they have produced the steel in the same or in a comparable manner as domestic steelmakers. Should this be the case, and should it go unchallenged, the policy implication would be that the domestic steel production process would in effect be indirectly imposed on other countries. This section focuses mainly on whether the current climate change debate in the US is leading to this kind of scenario.

US Climate Change Laws

The US does not have a federal climate change law package that deals comprehensively with emission reductions across the entire country (Driesen, 2007; Szymanski and Stone, 2008). While this is not surprising, given the position of the Bush administration on climate change (especially during its first mandate), the situation may change under President Obama.[8] In addition, a recent decision of the US Supreme Court seems to have pushed climate change on the government agenda.[9]

It is in this context that an overarching federal climate change bill, America's Climate Security Act (CSA), proposing a cap-and-trade scheme, was introduced in Congress in 2007.[10] However it did not pass the two houses' scrutiny. Amendments were proposed in 2008 by Senator Boxer[11] but the proposed Act was defeated in a June cloture vote, and will have to be reintroduced in 2009. Despite the fact that CSA is not yet a US federal law, I refer to it throughout this chapter since it proposes to deal with carbon leakage and competitiveness concerns through, among other things, the adoption of border adjustment measures.

The US, especially at the state level,[12] appears to be rising to the challenges posed by the threat of climate change. Considering the importance of overall US emissions at the global level,[13] state and regional mitigation efforts are a very positive first step. The next one needs to cover the entire US territory through a federal climate change law package. The presence of a new president in the White House is likely to trigger action in this direction.[14] A fragmented US policy on climate change is unsustainable over the long term as Americans are asking for federal action to enhance legal certainty across the country in order for them to conduct their business accordingly (Driesen, 2007, p. 43). Therefore, time will come, probably sooner rather than later, when an overall federal climate change legislative package will be adopted.

The US Approach to Carbon Leakage and Competitiveness Concerns

In the same way that that some players in the industry are concerned with having to comply with a diversity of US states' regulations,[15] they are also lobbying the government to adopt measures meant to target imports from developing countries whose climate change policy is not as stringent as the one envisioned in the US.[16]. A White Paper on Competitiveness suggested three approaches (Subcommittee on Energy and Air Quality of the

US House of Representatives, 2008, p. 8): Border Adjustment Measures (BAMs), mandatory energy-intensity performance standards and specific carbon market design aimed at encouraging cleaner production in developing countries.

In this chapter, I will focus only on BAMs, since developing countries may also face these types of measures in their trade relations with the EU (European Commission, 2008a, p. 11).[17] These measures would be 'trade related policies that use tariffs, taxes, or other mechanisms such as requiring foreign goods imported into the US to be accompanied by emission allowances' (Subcommittee on Energy and Air Quality of the US House of Representatives, 2008, p. 8). An approach of this kind would mean, for example, that a developing country like China or India would have to buy allowances on a future US emissions trading scheme to gain market access to the US. According to the proposals on the table, the US government would determine whether a developing country has taken measures comparable to those in the US. This approach would apply to a limited number of energy intensive sectors. As a result, if an Indian producer wishes to export an energy intensive product such as paper or steel, and if the US government determines that the Indian government has not enacted climate change legislation comparable to that of the US, the Indian goods would be allowed into the US only if the Indian producer purchased allowances on the US emissions trading scheme.[18]

In addition to border adjustment measures requiring foreign exporters to participate in a cap and trade scheme through the mandatory purchase of emission allowances, other BAMs have been considered to deal with loss of competitiveness and carbon leakage. Some countries, France in particular, have argued that border tax adjustments could also be a feasible option to level the playing field (Wiers, 2008, pp. 18–19). According to this option, imports from countries with lower climate change standards would be taxed at the border at the same level as domestic products, therefore eliminating any competitive advantage arising from lower standards in the producing country. Border tax adjustments would also apply to products originating from countries with higher standards. Climate change taxes on these products would be rebated at the border if the products were exported to countries with lower climate change standards (Fauchald, 1998, pp. 164–93). Despite the ongoing debate on border tax adjustments, there does not seem to be currently much political interest in this option. This is why I will focus on BAMs linked to cap and trade schemes.

A developing country faced with US trade measures of this kind would have three policy options. First, it could decide to modify its legislation in order to harmonize it somewhat with the US legislation. Countries like China and India would probably not do so, and should not do so just to please the US economy. However, this would be a very difficult choice to make in cases where the US export market is particularly important for the developing country affected by the trade measures. Second, it could choose to redirect trade. Trade flows in energy intensive industries from China to the US, for example, are not as dominant as one may think. Except for cement, imports from China amount to less than 10 per cent of carbon intensive imports, Canada being the main importer (Houser, et al., 2008, pp. xviii and 45). China could, therefore, decide to export steel elsewhere and strengthen other commercial ties, including South–South trade. The last option, discussed in the next section, would be to lodge a complaint against the US measures before the Dispute Settlement Body (DSB) of the multilateral trading system.

4. COMPATIBILITY OF POTENTIAL US BORDER ADJUSTMENT MEASURES WITH WTO LAW

Before proceeding further it is worth remembering that this part of the chapter is obviously speculative in nature since no US climate change law has been adopted yet and, therefore, no dispute of this kind between the US and a developing country can arise at this stage.

The following section first explains the design of the climate change related trade measures included in the America's Climate Security Act 2008 and then it examines how an affected developing country may seek redress through the WTO dispute settlement system.

Climate Change Related Trade Measures in the 2008 Climate Security Act

The overarching goal of Title XIII of the Act was 'to promote a strong global effort to significantly reduce greenhouse gas emissions'.[19] The first and best route to achieve this is through international negotiations. Therefore, the Act took an explicit stand in favor of multilateralism.[20] In doing so, however, it also raised concerns about carbon leakage and the overall ineffectiveness of global action against climate change if other countries did not take comparable measures. Therefore, another overarching goal of this title was 'to ensure, to the maximum extent practicable, that greenhouse gas emissions occurring outside the United States do not undermine the objectives of the United States in addressing global climate change'.[21] This second goal was to be achieved through 'measures carried out by the United States that comply with applicable international agreements'.[22] Therefore, while multilateralism was stated as the best way forward, CSA also considered unilateralism explicitly as the second best means to deal with global climate change.

Section 1306 of the Act imposed the following requirement on US importers of goods from developing countries:

> Effective beginning January 1, 2014, a United States importer of any covered good shall, *as a condition of entry of the covered good into the United States*, submit to the Administrator and the Bureau of Immigration and Customs Enforcement a written declaration . . . [that will] include a statement certifying that the applicable covered good is (A) subject to the international reserve allowance requirements of this section . . . or (B) exempted from the international reserve allowance requirements of this section . . .. [T]he written declaration for the covered good shall include . . . an estimate of the number of international reserve allowances that are required for entry of the covered good into the United States . . . and the deposit of . . . international reserve allowances in a quantity equal to the estimated number required for entry.[23]

This climate related trade measure had several elements. Which goods were covered? Did the measure affect developing countries? In other words, where did developing countries fit in the list mentioned in the Act, and what were the criteria on which this list was based? And, finally, if imports from developing countries were covered, what did a climate change trade related measure like this actually require from the affected foreign country to allow its goods to enter the US market? In other words, what was the Act referring to when it mentioned 'international reserve allowances'?

Coverage

A developing country would be affected if it traded with the US in 'primary products or manufactured items for consumption'. Primary products are a limited number of carbon intensive goods. CSA 2008 explicitly mentioned 'iron, steel mill products, aluminium, cement, glass, pulp, paper, chemicals, or industrial ceramics'.[24] However, since the relevant section in CSA 2008 provided for cumulative conditions, not all carbon intensive goods were covered but only those that:

> generate[s], in the course of the manufacture of the good, a substantial quantity of, direct greenhouse gas emissions or indirect greenhouse gas emissions; *and* [are] closely related to a good the cost of production of which in the United States is affected by a requirement of this Act.[25]

The second category of covered goods was 'manufactured items for consumption'. These are not primary products and, according to CSA 2008, was a product that: 'generates, in the course of the manufacture, a substantial quantity of direct greenhouse gas emissions or indirect greenhouse gas emissions, *including emissions attributable to the inclusion of a primary product in the manufactured item for consumption*;'.[26] Therefore, CSA 2008 expanded the coverage to goods produced through a process resulting in significant GHG emissions. In other words, not only could steel imports (primary product) be targeted, but also cars the production of which depends on carbon intensive steel (manufactured item for consumption). However, this was the case only when the application of the measure was 'administratively feasible and necessary'.[27]

The criteria according to which a developing country would be targeted

How could developing countries be targeted? The climate change related trade measure provided for in CSA would only be imposed on foreign countries that would not take *comparable action* to the US in tackling climate change. The procedure called for an international climate change commission to evaluate whether a foreign country had taken comparable measures.[28] If it had, that country would be put on an 'excluded list', while others would fall into the 'covered list'.[29] Only those exporting into the US from countries on the 'covered list' would face the climate change related trade measure.[30] The key issue here is to understand what was meant by comparable measures. CSA 2008 amended the guidelines found in CSA 2007,[31] and provided for the following:

> [A] foreign country shall be considered to have taken comparable action if the Commission determines that the percentage change in greenhouse gas emissions in the foreign country during the relevant period is equal to or greater than the percentage change in greenhouse gas emissions of the United States during that period.[32]

CSA allowed some flexibility to a foreign country in designing a climate change policy, since it did not require the same cap and trade scheme.[33] Comparable measures were assessed against baseline emission levels,[34] defined as 'the total annual greenhouse gas emissions attributed to the category of the covered good of the foreign country'.[35] Best available technology was one of the criteria used in order to determine the abovementioned amount.[36] A strict application of this requirement meant that most emerging developing countries were included in the 'covered list' and, therefore, that their imports would be subject to the climate change related trade measure provided for in CSA. This brings us to discuss the design of the measure itself.

The implementation of the trade measure

What did the US require from a developing country in order to allow it to export its goods into the US territory? Countries found not to have taken comparable action on climate change would be obliged to purchase allowances from an international reserve allowance programme set up for this purpose. The measure would have begun in January 2014. A company exporting its products into the US would be required to submit a written document indicating if a product was manufactured or processed in a country included in the 'covered list'.[37] If that was the case, the company would estimate in the document the number of international reserve allowances that it would have to surrender to be granted access to the US market.[38] A final decision on the correct number of international reserve allowances would be made by the Administrator 180 days after the submission of the written document.[39] How many international reserve allowances would the developing country have to buy? CSA 2007 provided that methodology rules and formulas for this purpose had to be established and, therefore, there was no clear-cut answer to this question.[40] The amendments presented in 2008 dealt very thoroughly with this question. According to section 1306(d), entitled 'Quantity of Allowances Required', a company exporting its products into the US had to purchase a number of allowances equal to:

> the product obtained by multiplying . . . the national greenhouse gas intensity rate for each category of covered goods of each covered foreign country for the compliance year . . . ; the allowance adjustment factor for the industry sector of the covered foreign country that manufactured the covered goods . . . ; and the economic adjustment ratio for the covered foreign country,. . .[41]

Drafters allowed alternative options to the purchase of international reserve allowances. In fact it was clearly stated that a foreign country could rely on allowances issued under non-US cap and trade schemes (for instance, allowances from the European emissions trading scheme) to gain access to the US market.[42]

In sum, an analysis of goods covered, of the criteria according to which a foreign country could be targeted, and of the implementation of the measure provided for in CSA demonstrate that developing countries could be negatively affected by such a measure.

WTO Compatibility of Potential US Border Adjustment Measures Concerning Developing Countries

Could a developing country targeted by such a measure have successfully questioned its legality before the WTO DSB? It seems that the drafters of the Act were well aware of this possibility since a sub-section of the provision that established the climate change related trade measures in CSA 2007 was entitled 'Consistency with international agreements' and provided: '[t]he Administrator . . . shall adjust the international reserve allowance requirements established under this section . . . as the

Administrator determines to be necessary to ensure that the United States complies with all applicable international agreements'.[43] While this subsection was not replicated in CSA 2008, the latter did provide that any other action deemed necessary 'to address GHG emissions attributable to the production of covered goods in covered foreign countries' had to be adopted 'in compliance with all applicable international agreements'.[44] It was reasonable to expect that the application of the climate related trade measure upon companies wishing to gain access to the US market must also follow this requirement. Interestingly, under the definition of 'international agreements' CSA 2008 specifically included the Marrakesh agreement establishing the WTO.[45]

The US House of Representatives Subcommittee on Energy and Air Quality White Paper was also concerned that unilateral climate change measures be compatible with WTO obligations. It expressed the following view:

> [s]ince the US cannot unilaterally bind other countries, our goal will be to craft legislation that also induces developing countries to limit their emissions growth . . . in a manner that is reasonably certain to withstand challenge before the World Trade Organization (Subcommittee on Energy and Air Quality of the US House of Representatives, 2008, p. 2).

It is not my goal here to provide a detailed account of the technicalities of the climate change and trade debate (see, e.g., Pauwelyn and Sindico, 2008; Green, 2005; Zhang and Assunção, 2004; Pauwelyn, 2007), but instead to draw attention to those points in the CSA that a developing country could have focused on, had a US measure been brought before the WTO DSB. I will also assess whether bringing the multilateral trading system into the carbon leakage debate is in the interests of a developing country. If a climate policy from the North can end up shaping the climate and non-climate policies in the South, can a referral to the WTO prevent this, or does it make the situation worse for the developing country?

A developing country's position in a possible WTO dispute

Imagine a scenario in which imports from a developing country steelmaker, for instance India, were caught by a US climate change related trade measure because the latter considered that India had not taken comparable climate change measures. Could India argue that the US requirements either to purchase a specific number of international reserve allowances or to rely on allowances from other cap and trade schemes constitute a breach of WTO law? This is not a far fetched scenario. The Confederation of Indian Industry has already stated, in relation to the EU proposal for the reform of the directive on the European trading scheme that, '[p]rincipally, we do not agree with the proposal. It seems a new tariff barrier is being set up and we are preparing inputs for the government to fight it at the WTO.'[46]

Despite the fact that the EU proposal does not include a specific climate change related trade measure such as the one found in CSA, it does put forward a 'carbon equalization system' (European Commission, 2008c, recital 20; Sindico, 2008), which may be, in the end, similar to the climate change related trade measure found in CSA. Therefore, speculating on how a developing country could challenge the latter is not a mere academic exercise.

A developing country could focus its attention on three different sets of issues before the WTO DSB: it could look at the nature of the goods covered by the import requirement scheme; it could consider the criteria under which the developing country was targeted; and, finally, it could focus on the design and implementation of the climate change related trade measure.

Coverage CSA 2008 stated that the import requirement scheme could also apply to manufactured items for consumption on the basis of the amount of emissions released by the primary products during the manufacturing process. This meant that the US could treat domestic low carbon steel differently from carbon intensive products imported from a developing country. India could then argue that the US treated two products differently solely on the basis of production and process methods (PPMs). WTO law does not allow a Party to impose requirements that result in imported like products being given 'less favorable treatment' than domestic ones.[47] A developing country could raise two questions on this issue.

First, it could argue that low carbon and carbon intensive goods are like products and, therefore, that they should be treated similarly. If, from an environmental point of view, the differentiation of two products based on their climate friendliness makes sense, it can be argued that this is not the case (yet) from an international trade law perspective. However, while the GATT case law on trade and environment focused on the final product, notwithstanding how it was produced,[48] the case law under the WTO DSB has taken other elements into account to determine whether two products are 'like': end uses, consumer habits and tariff classification.[49] This could open up some space for the adoption of measures based on the process and production methods of a product. Indeed, the WTO Appellate Body has stated that two products may not be 'like' based on their PPMs if the way a product has been manufactured is recognized by a consumer in a specific marketplace as a criterion to differentiate it from another otherwise 'like' product.[50] However, consumer differentiation can be assessed only on a case by case basis. Furthermore, it is arguably not very likely that consumers will distinguish between energy intensive imports based on their carbon intensity, despite the fact that this could change if more information is conveyed to consumers. In sum, despite the fact that WTO case law on trade and environment has widened the scope of 'like' products, a developing country can make a strong case that two products are 'like', despite their different carbon intensity.

Second, a developing country could dispute the nature of the cap and trade scheme linked to the measure. If it was interpreted as domestic regulation, then GATT Article III: 4 would apply and, if the two products are 'like', no less favorable treatment could have been accorded to the developing country import by the US administration. On the other hand, if the cap and trade scheme was understood as a tax under WTO law,[51] then GATT Article III: 2 would come into play. In that case there would be even less policy space for the US Administrator since any measure adopted against imports 'in excess of those applied, directly or indirectly, to like domestic products' would constitute a breach of WTO law.[52] From a developing country point of view it is preferable to consider the cap and trade scheme to which the climate related trade measure is attached as a tax, and not as a domestic regulation, because in the former case 'any' difference in treatment between the US domestic product and the like product being imported from a developing country will make the trade restrictive measure WTO incompatible.

Therefore, from a developing country perspective the best option would be to maintain that the two products, the domestic low carbon product and its high carbon imported product, are 'like' products, and that the cap and trade scheme to which the trade measure was linked amounted to a tax under WTO law. If either of these arguments were accepted by the WTO DSB, any discriminatory measure would constitute a breach of WTO law. A developing country has to be aware that the nuances of the 'like' products concept may lead to different conclusions in the future, and that a cap and trade scheme may also be considered a domestic regulation, in which case the US would have wider policy space to treat imported products differently.

The criteria according to which a developing country could have been targeted The Act took a list approach by providing that foreign countries were to be divided into countries with and without comparable climate change action. The adoption of a climate change related trade measure depended on the foreign exporter being based in a country listed among the climate change 'unfriendly'. Measures taken according to this approach were country based measures. Since the import requirement scheme applied on all exporters coming from a country included in the list of climate 'unfriendly' states, a developing country could have easily attacked the US measure as a violation of the most favored nation principle (MFN). The MFN principle is one of the cornerstones of the multilateral trading regime. It provides that if a WTO party confers special treatment to another WTO party, it will have immediately and unconditionally to do the same for all other WTO parties.[53]

Furthermore, the criteria by which a country was deemed to have a comparable climate policy could be questioned, and this despite the fact that the wording of the Act provided for some flexibility. This could be raised in the second phase of a possible dispute. If a developing country's claim of the non-legality of the US climate change related trade measure with WTO law per se was to be accepted by the DSB, the US could argue that the measure was, nevertheless, valid by virtue of one of the exceptions under GATT Article XX. This Article allows for otherwise unlawful WTO measure if they are necessary, among other things, to protect human health,[54] or they relate to the conservation of exhaustible natural resources.[55] It would be difficult for a developing country to argue that these exceptions are not applicable, given that climate change science supports climate measures.[56] Instead, in our hypothetical scenario, India should focus on the *chapeau* of Article XX, which provides that a country can adopt an otherwise unlawful WTO measure 'subject to the requirement that such measures are not applied in a manner which would constitute a means of arbitrary or unjustifiable discrimination between countries where the same conditions prevail, or a disguised restriction on international trade'.

One of the criteria that the WTO trade and environment case law has established to determine whether the application of a measure is actually an unlawful disguised restriction on international trade is its 'flexibility'.[57] According to the DSB, flexibility entails the possibility of taking into account policies from other countries which, without being exactly identical with the domestic measures, achieve similar goals. CSA provided that 'any' foreign climate measure would be considered comparable if it achieved similar reductions.[58] Therefore, if a developing country adopts a tax or a command and control scheme to improve the energy efficiency of domestic steel makers, this could be considered 'comparable' by the US. The application of the measure in this way would favor the US in a dispute against a developing country. However, how would the US decide whether the developing country has achieved results 'comparable in effect to actions carried out by the United States'?[59] According to the CSA, comparable effect would be assessed by reference to the baseline emissions level. The latter, as mentioned earlier, amounts to 'the total annual greenhouse gas emissions attributed to the category of the covered good of the foreign country'.[60] in a specific time frame. The climate record of a developing country would be based on best available information, described as follow:

> all relevant data available that are available for a particular period; and to the extent necessary economic and engineering models; best available information on technology performance levels; and any other useful measure or technique for estimating the emissions from emission activities.[61]

Criteria such as 'best available technology' are still not clear-cut and, more importantly, are not harmonized internationally. India could, therefore, argue that the implementation of a climate change related measure based on this criterion lacks international recognition and, therefore, that it would lead to an arbitrary application of the baseline emissions level criterion. Furthermore, the CSA's wording did not allow the possibility for a foreign country to provide the relevant and necessary information to assess the average GHG emissions arising from its production of goods under a covered sector. As a result India could also argue that the imposition of US methodologies to determine the climate 'friendliness' of Indian industries, without taking into account Indian data, demonstrate a considerable lack of flexibility in the application of the measure and, therefore, amount to a disguised restriction on international trade.

The implementation of the measure A developing country could also raise at least two arguments on the implementation of the climate change trade related measure. First, despite the fact that the wording secured equal treatment for the price of the allowances,[62] the actual implementation could be scrutinized. If an importer from a developing country purchased international reserve allowances at a higher price than domestic producers, the foreign importer could argue that he had been treated 'less favourably'.[63] Furthermore, this line of argument could also be invoked by a developing country within GATT Article XX, since one of the decisive criteria to determine whether the implementation of a measure constitutes a disguised restriction on international trade is the measure's design and architecture.[64]

Second, CSA 2008 dealt with the competitiveness concerns of domestic industries by allocating to them, at no cost, a specific number of allowances under the domestic cap and trade scheme.[65] In other words, the US would grandfather allowances to carbon intensive domestic manufactures until 2030 to facilitate their transition to a low carbon economy. Despite the fact that the import allowances requirement to companies wishing to export their products to the US did take this factor into account,[66] a developing country could still claim that its goods were, contrary to WTO law, treated differently from domestic 'like' products.

Third, a developing country exporter could claim that the final decision on the amount of international reserve allowances that it has to surrender to gain market access was ultimately arbitrary. I have mentioned that the CSA 2008 provided for a fairly complex formula according to which the total amount of allowances that a US importer had to purchase is equal:

> the product obtained by multiplying . . . the national greenhouse gas intensity rate for each category of *covered goods* of each covered foreign country for the compliance year . . .; the allowance adjustment factor for the *industry sector* of the covered foreign country that manufactured the *covered goods*. . .; and the economic adjustment ratio for the *covered foreign country*,. . . .[67]

While the first two elements of this formula, national GHG intensity rate[68] and allowance adjustment factor,[69] related to the goods manufactured in a specific country, the last element, the economic adjustment ratio,[70] referred directly to the country itself. In other words, despite the methodological problems at the time of applying the climate change related trade measures,[71] the number of international reserve allowances could be tracked down to objective criteria under the first two elements of the formula. The third element, on the contrary, was per se subjective and reveals the country based nature of the climate related trade measure since it refers to the appreciation of the domestic climate efforts occurring in the country wishing to export its products to the US. The developing country could, therefore, focus on this third element of the formula to establish that the measure was in violation of the MFN rule and/or that the measure constituted a disguised restriction on international trade under GATT Article XX.

There is one implementation related argument on which a developing country could probably not rely. A trade related measure will not fall within the *chapeau* of GATT Article XX if the country adopting the measure fails to meet the prior negotiation effort criterion.[72] Indeed, a trade restrictive measure must be a last resort after serious negotiations have been initiated with the country that may be affected by the measure. The CSA 2008 clearly stated that international negotiations are to be preferred, and it reaffirmed its commitment to the United Nations Framework Convention on Climate Change.[73] Furthermore, it strategically maintained that, should the multilateral or bilateral agreements negotiated with foreign countries 'involve measures that will affect international trade in any goods or service', the objective of the negotiations would be environmental (i.e., 'the reduction of greenhouse gas emissions').[74] If the future US policy on climate change follows these requirements it will be difficult for a developing country to claim that a trade related measure has been adopted only to deal with US domestic competitiveness concerns.[75]

Finally, the CSA allowed foreign countries to rely on alternative compliance options: foreign allowances and credits, but these were not available to all US trade partners. If a developing country had not established a cap and trade scheme as a measure to tackle climate change within its domestic climate policy, how could it purchase a 'commensurate' foreign allowance? Furthermore, developing countries were not allowed to accrue credits from clean development projects since they are the beneficiaries of these projects. Where would they have been able to accrue credits from international offset projects? Given the above, a developing country could argue that industrialized countries that could rely on foreign allowances or on credits were given special treatment compared to a country like India. There would have been, therefore, scope to explore the possibility, once again, of a breach of the MFN rule.

As mentioned earlier, the goal of this chapter is not to provide a definitive answer to the WTO compatibility of a climate change trade related measure like the one that was included in CSA. This would be a counterproductive exercise since the Act has not been enacted into law and because WTO jurisprudence on trade and environment is still evolving. Yet, authors tend to agree that a climate change related measure of the kind provided for in the CSA 2007, which is roughly similar to the one provided for a year later, would be lawful only if careful attention were given to its implementation (Janzen, 2008, p. 24; Morgenstern, 2007, p. 116).[76] Only if procedural requirements such as previous notification of the measure to affected trading partners were followed and, in particular, only if a real degree of flexibility were granted to the developing country's climate change policies would it be possible to maintain that such a measure could withstand a WTO dispute as an environmental exception under Article XX of the GATT. The CSA 2008 has improved some of the pitfalls in the previous bill, but a developing country could probably still raise worthwhile arguments against the nature and, in particular, the implementation of the measure.

Developing country strategies to deal with the climate and trade tension

The second question to address with regard to a climate change related trade measure and the WTO is whether a developing country 'should' bring a dispute before the WTO. In this context it is worth exploring whether a developing country has any other strategic options, in particular within multilateral forums.

Countermeasures at the WTO: should a developing country consider them? Would it be beneficial for a developing country to raise the tension with the US to such a point that a trade dispute is inevitable and a dispute arises at the WTO? The same question is also asked by policymakers in the US. For instance, the White Paper posits that:

> [s]ince the US cannot unilaterally bind other countries, our goal will be to craft legislation that also induces developing countries to limit their emissions growth . . . on terms that pose acceptable risks to U.S. interests in the event of a negative WTO determination (Subcommittee on Energy and Air Quality of the US House of Representatives, 2008, p. 2).

This passage seems to suggest that the US is ready to take WTO unlawful actions if the consequences thereof are acceptable for its economy. This approach is self-explanatory when one considers the retaliatory effect of final decisions by the WTO DSB. The DSB does not have competence to award any damages. It can only recommend to the party in non-compliance with WTO law that it modify its laws accordingly. At this point the WTO party that has adopted the unlawful measure has two choices: it can either comply by modifying its domestic legislation or it can do nothing and suffer countermeasures from the other party in the dispute. These countermeasures will be of a retaliatory nature. In fact, the WTO Dispute Settlement Understanding allows the party that has won the dispute to suspend its trade obligations under any of the WTO Agreements until the other party to the dispute modifies its domestic legislation according to the recommendations of the DSB.[77]

We must assess the impact on developing countries of the possibility that a party may willingly decide not to take into account a final decision by the DSB. Would the US be concerned about suffering retaliatory trade measures from a developing country whose carbon intensive imports had been subjected to a restrictive trade measure for climate change related reasons, if its measure were not upheld by the WTO DSB? The answer to this question belongs to the realm of politics and current international trade power relations rather than law. While I do not wish to move too far into this field, the obvious however can be stated. If a developing country plays an important role on the global market, then it will be able to impose countervailing measures that may harm the US economy.[78] Whether these are more or less harmful to the US than entry of the banned imports would be is a question that can be answered only on a case-by-case basis. However, if the developing country has a small economy, the US exports would easily find other replacement markets in response to validly imposed countermeasures.[79] In any event, this scenario is not likely, since the goods covered by a climate change related trade measure will mostly be carbon-intensive goods produced in big developing countries. Therefore, it is likely that trade disputes will be between the US and economically important developing countries.

Should a developing country decide to use countermeasures in the context of a climate change trade dispute against the US, it will have to decide which US imports would be the object of countermeasures and how these would be implemented.

Given that the WTO dispute settlement system allows countermeasures as a pressure mechanism and that there is an urgency to deal with climate change by shifting toward alternative energy sources, it seems undesirable to resolve a future climate and trade dispute between the US and a developing country through the DSB. What may begin as climate change dispute may end up involving a wide range of sensitive economic areas, increasing the political tension between industrialized and developing countries.[80] Furthermore, given that countermeasures under the WTO DSB can slow down mitigation action necessary to deal with the urgent need to address climate change, it appears self-defeating to rely on the multilateral trading system to deal with a climate and trade dispute.

Alternative multilateral options to disentangle the climate and trade deadlock The WTO Director General argued recently that this deadlock could be solved through international negotiations within the current Doha Development Agenda.[81] There are however serious reasons to doubt such optimism. Paragraph 31 of the Doha Mandate sets out the terms for a negotiation the goal of which is to promote the liberalization of environmental products and services, '[w]ith a view to enhancing the mutual supportiveness of trade and environment, we agree to negotiations, without prejudging their outcome, on: the reduction or, as appropriate, elimination of tariff and non-tariff barriers to environmental goods and services'.[82]

While a first reading of this part of the mandate may seem to lead to a win-win solution in both the fight against climate change and the promotion of economic development, the outcome of the negotiations until now is not promising. These negotiations are driven by industrialized countries. Most developing countries have other priorities within the Doha mandate, focusing in particular on removing barriers to agricultural trade.[83] In addition, developing countries should not rely too much on the trade and environment mandate within Doha because trade and environment negotiations are linked to the overall trade negotiations. Countries cannot just settle specific trade issues; they need to reach a global compromise on the entire Doha talks. Unfortunately, the round is currently in an impasse, and no positive result is foreseeable in the near future (Drache and Froese, 2007).

5. CONCLUSION: FROM UNILATERAL CLIMATE CHANGE RELATED TRADE MEASURES TO A MULTILATERAL PRODUCTION AND CONSUMPTION DIALOGUE

This chapter aimed to demonstrate that measures that an industrialized country adopts in its struggle to cope with competitiveness and carbon leakage might end up shaping climate (and non-climate) policies in the global South. Using proposals in the US climate debate as a case study, it explained how a developing country could challenge these measures before the WTO's DSB. Questions related to the coverage of goods, to the criteria according to which a developing country would be targeted and to the actual implementation of the climate related trade measure could be argued before a WTO Panel. Despite some obvious uncertainty as to whether the final decision would be in favor of or against a developing country affected by a climate change related measure, the important point is that a WTO dispute may by no means be the end of the story. The possibility for the winning party to adopt retaliatory countermeasures may actually exacerbate tensions and be counterproductive, from both an environmental and an economic point of view. Unfortunately, current talks at the WTO to find win-win avenues on climate and trade by liberalizing environmental goods and services are not promising.

Nevertheless, despite the current stalemate, the Doha Round points in the right direction. The climate and trade deadlock could be resolved if unilateralism were replaced by genuine multilateralism, if countries understood that climate change is not only an environmental or an economic problem, but first and foremost a global challenge to energy consumption and production patterns. The international community will be moving forward only when states decide to engage in serious efforts to promote sustainable production policies that favor low carbon energy sources and, far more challenging, to devote time and funding to adopting and implementing sustainable consumption policies (Bradbrook and Wahnschafft, 2001). Only the latter will move society towards a low carbon economy, which must be the aim of the international community when dealing with climate change.

NOTES

* Lecturer in Law, School of Law, University of Surrey, Guildford, UK and Deputy Director, Environmental Regulatory Research Group. Research for this chapter has been facilitated by funding from the School of Law of the University of Surrey

and from the URSF Conference support scheme. The author wishes to thank Rosa Fernández Egea, Harro Van Asselt and Michael Mehling for comments on earlier drafts of this article. Errors remain the author's sole responsibility.

1. As seen later in this chapter, what the European Commission perceives as a legitimate environmental concern amounts to protectionism for India. See A. Bhatta (2008), 'Europe Threatens Carbon Tax on Third World', *Down to Earth: Science and Environment online*, 16, 6 May 2008, www.downtoearth.org.in/full6.asp?foldername=20080515&filename=news&sec_id=4&sid=11 (verified 28 January 2009).
2. For example, in June 2008 the spokesman for Voestalpine, an Austrian steel firm, indicated that if the company did not get a clear signal from the Commission on how it plans to deal with competitiveness concerns arising from energy intensive sectors, such as the steel industry, it would consider investing in a new plant outside the EU. See Voestalpine (2008), *Annual Report 2007–08* (Voestalpine).
3. See Byrd-Hagel Resolution (US Senate Resolution, 98), July 1997.
4. However, others are of the view that the US should be setting ambitious climate targets for itself irrespective of developing country participation. This is for instance the position taken by a coalition of NGOs and businesses. See USCAP (2008), 'A Call for Action – Consensus Principles and Recommendations from the U.S. Climate Action Partnership: A Business and NGO Partnership', www.us-cap.org/ (visited 29 January 2009).
5. See infra note 14.
6. This could end up being a prudent business strategy if consumers in a given market are climate conscious and reward the company by buying its products, despite the fact that they may be more expensive than 'dirty' products. In other words, climate friendly consumers would not consider the two products to be 'alike'.
7. The cement sector is one of the energy intensive industries that will most likely suffer losses from the adoption of higher climate standards (OECD, 2005). It must be acknowledged that research on whether the economy can suffer negative effects from lower climate standards elsewhere has been undertaken thoroughly only in the framework of the first phase of the European emission trading scheme in which generous allocation of allowances prevented carbon leakage from happening. It will be interesting to see if the situation will change now that the trend will be to move towards further auctioning and less grandfathering of allowances (Reinaud, 2008a and 2008b).
8. President Obama has been supportive of an emission trading scheme that will help reduce emissions by 80 per cent below 1990 levels by 2050. He also vows to bring the US back into multilateral forums on climate change. See B. Obama and J. Biden (2008), 'New Energy for America', www.barackobama.com/pdf/factsheet_energy_speech_080308.pdf (visited 6 December 2008).
9. *Massachusetts v. EPA*, 127 SCt 1438 (2007). The Court ruled that the Environment Protection Agency (EPA) should make a so-called 'endangerment finding' on whether CO_2 constitutes a pollutant under the Clean Air Act and thus needs to be regulated. The EPA has only published an advance notice of regulatory intent implying that EPA regulation of GHGs would not be the best approach. Nevertheless, this case is generally understood as confirming that the federal EPA has jurisdiction to regulate GHG emissions.
10. Climate Security Act (2007), S. 2191, 110th Cong. 1st Sess.
11. Lieberman–Warner Climate Security Act (2008), S. 3036, 110th Cong. 2nd Sess.
12. However, while the US federal government has been considered a climate laggard for many years, state governmental and regional initiatives have been well received by the climate community. Three initiatives merit particular mention due to their scope and the current extent of implementation: vehicle emission standards in California, the cap and trade scheme established by the Northeast States' Regional Greenhouse Gas Initiative (RGGI) and the Renewable Portfolio Standards.
13. Ecofys (2008), 'Factors Underpinning Future Action – Country Fact Sheets 2008 Update', www.ecofys.com/com/publications/reports_books.asp (visited 6 December 2008).

14. Two initiatives from President Obama, before he assumed the presidency, suggested a change of direction. He appointed the Nobel-prize winning scientist Steven Chu, well known for his stand in favor of promoting renewable energies, as Secretary of Energy. In addition, he created a new position – Assistant to the President for Energy and Climate Change – which will be held by former EPA administrator C.M. Browner. See S. Goldenberg (2008), 'Obama's New Team Raises Hope for US Environment', *The Guardian*, 11 December.
15. President Obama's stand on climate change will not reverse this trend. In fact, one of his first announcements was to let states set their own vehicle emissions standards, which will surely lead to complaints from the auto industries. See P. Walker (2009), 'Obama to Put Bush Car Pollution Policies into Reverse', *The Guardian*, 26 January.
16. See the Low Carbon Economy Act, S. 1766, 11 July 2007, the proposal put forward by the International Brotherhood of Electrical Workers-American Electric Power IBEW/AEP, which then inspired the America's Climate Security Act.
17. Supra note 1.
18. This kind of BAM was provided for in the Climate Security Act (not enacted into law), which will be analysed in depth in section 3 infra.
19. CSA 2008, supra note 11, s. 1302(1).
20. Ibid., s. 1303(a): 'Congress finds that the purpose described in section 1302 can be most effectively addressed through agreements negotiated between the United States and foreign countries'.
21. Ibid., s. 1302(2).
22. Ibid., s. 1302(3)(B).
23. Ibid., s. 1306(c)(1), (c)(2)(a), (c)(3)(C) and (c)(3)(D)(i).
24. Ibid., s. 1301(15)(A). We have already underlined that, especially for China, trade flows with the US in these kinds of goods are much less significant than one would expect (Houser, et al., 2008, pp. xviii and 45).
25. Ibid., s. 1301(7), emphasis added.
26. Ibid., s. 1301(13)(B), emphasis added.
27. Ibid., s. 1301(13)(C).
28. Ibid., s. 1305(a).
29. Ibid., s. 1306(b)(2) and (b)(3).
30. Ibid., s. 1306(d)(1)(A). This was much clearer in CSA 2007, s. 6006 (b)(3)(A): 'the President shall identify and publish in a list, to be known as the "covered list", each foreign country the covered goods of which are subject to the requirements of this section'.
31. CSA 2008, s. 6001(2).
32. Ibid., s. 1301(4)(b)(i).
33. Ibid., s. 1301(4)(A). The CSA 2007 did consider also the level of economic development: see CSA 2007, supra note 10, s. 6001(2).
34. CSA 2008, s. 1301(14).
35. Ibid., s. 1301(1)(A).
36. Ibid., s. 1301(2).
37. Ibid., s. 1306(c)(2).
38. Ibid., s. 1306(c)(3)(A).
39. Ibid., s. 1306(c)(4).
40. Ibid., s. 1306(d)(1).
41. Ibid., s. 1306(d)(2). The implications of the definitions of national greenhouse gas intensity rate, of allowance adjustment factor and of economic adjustment ratio will be discussed later in this chapter.
42. Ibid., s. 1306(e).
43. CSA 2007, s. 6006(g).
44. CSA 2008, s. 1307(b)(3).
45. Ibid., s. 1301(11).
46. Supra note 1.

47. General Agreement on Tariffs and Trade (GATT), 1994, Marrakesh Agreement Establishing the World Trade Organization, Annex 1A, 1867 UNTS 187, (1994) ILM 33, 31, Art. III:4: '[t]he products of the territory of any contracting party imported into the territory of any other contracting party shall be accorded treatment no less favourable than that accorded to like products of national origin in respect of all laws, regulations and requirements affecting their internal sale, offering for sale, purchase, transportation, distribution or use'.
48. *United States – Restrictions on Imports of Tuna*, Doc. DS21/R, 3 September 1991, para. 5.15.
49. *European Communities – Measures Affecting Asbestos and Asbestos-Containing Products (EC – Asbestos)*, Appellate Body Report, Doc. WT/DS135/AB/R, 12 March 2001, s. 101.
50. Ibid., ss. 113, 114.
51. Several authors argue that an emissions trading scheme can be considered a tax for WTO purposes (De Cendra, 2006, pp. 135–36; Ismer and Neuhoff, 2004, p. 11), but others do not share the same view (Wiers, 2008, p. 22).
52. GATT, Art. III:2.
53. Ibid., Art. I.
54. Ibid., Art. XX(b).
55. Ibid., Art. XX(g).
56. The current case law on trade and environment also makes it easier for the country adopting the trade measure on environmental grounds to prove that the measure is actually *necessary* to protect human health: *Korea – Measures Affecting Imports of Fresh, Chilled and Frozen Beef*, Appellate Body Report, Doc.WT/DS161/AB/R, 11 December 2000, s. 164; *Brazil – Measures Affecting Imports of Retreated Tyres (Brazil – Tyres)*, Appellate Body Report, Doc. WT/DS332/AB/R, 3 December 2007, s. 210, or that it really *relates to* the conservation of exhaustible natural resources: *United States – Standards for Reformulated and Conventional Gasoline (US – Gasoline)*, Appellate Body Report, Doc. WT/DS2/AB/R, 29 April 1996, section III.B).
57. *United States – Import Prohibition of Certain Shrimp and Shrimp Products (US – Shrimp)*, Appellate Body Report (WT/DS58/AB/R, 12 October 1998), ss. 173-75.
58. CSA, supra note 11, s. 1301 (4)(A).
59. Ibid.
60. Ibid., s. 1301(1).
61. Ibid., s. 1301(2).
62. Ibid., s. 1306(a)(3).
63. See reference to GATT, supra note 47, Art. III:4. Once again, the question would have been whether the national treatment principle had been violated.
64. *European Communities – Measures Affecting Asbestos and Asbestos-Containing Products (EC – Asbestos)*, Panel Report, Doc. WT/DS135/R, 18 September 2000, s. 8.236.
65. CSA 2008, supra note 11, Title V, sub-titles E, F, G and H.
66. Ibid., s. 1306(d)(4)(A)(i) and (4)(B).
67. Ibid., s. 1306(d)(2).
68. Ibid., s. 1306(d)(3).
69. Ibid., s. 1306(d)(4).
70. Ibid., s. 1306(d)(5).
71. The main problem will be to calculate correctly the average greenhouse direct and indirect emissions accruing from covered goods in a specific country where data may not be available or the foreign country may not be willing to provide them. Furthermore, should the number of international reserve allowances be decided on a per-unit basis (this would be preferable from an international trade law perspective), on a covered goods basis or on an industry sector basis. CSA 2008 does not clarify this in any detail since it uses the three expressions within the same formula (per-unit base for the greenhouse gas intensity sector and industry sector and covered goods for the allowance adjustment factor).

72. Supra note 56, s. 172.
73. CSA 2008, supra note 11, s. 1303(a) and (b).
74. Ibid., s. 1303 (b)(2)(A).
75. Furthermore, the Act requires speedy notification to foreign countries of the negotiating objective: see ibid., s. 1303(c).
76. Supra note 45, p. 41.
77. 'Understanding on Rules and Procedures Governing the Settlement of Disputes (DSU)', printed in WTO, *The Legal Texts. The Results of the Uruguay Round of Multilateral Trade Negotiations*, 1999, art. 22.
78. *United States – Subsidies on Upland Cotton Request for Consultations by Brazil (US – Upland Cotton)*, Doc. WT/DS267/1, 3 October 2002.
79. *United States – Measures Affecting the Cross-Border Supply of Gambling and Betting Services – Request for Consultations by Antigua and Barbuda (US – Gambling)*, Doc. WT/DS285/1, 27 March 2003.
80. It should be recalled that we are currently in the midst of key global negotiations that will shape the future of the international climate and international trade regimes. The Bali Road Map and the Doha Round already stand on very unstable feet. Increased tension between key countries – such as the US and China – over the adoption of a climate related trade measure may undermine ongoing negotiations further. These observations suggest that, at least until global climate and trade negotiations have been addressed properly, the WTO DSB, despite how attractive it may seem, should not be considered as the correct forum to deal with a difficult climate and trade relationship.
81. P. Lamy (2007), '*Doha* Could Deliver Double-Win for Environment and Trade', Informal Trade Ministers' Dialogue on climate change – Bali, Indonesia, 9 December 2007, www.wto.org/english/news_e/sppl_e/sppl83_e.htm (visited 15 March 2009).
82. Doha Ministerial Declaration (2001), WT/MIN(01)/DEC/120, 20 November 2001, para. 31(iii).
83. Since the early 2000s, divisions have persisted between a list-based approach promoted by industrialized countries and a project-based approach favored by some developing countries. According to the former approach, countries should agree on specific goods and services which are per se environmentally friendly, and then promote trade therein amongst them. According to the project-based approach, goods and services to be further liberalised should be those that aid in the development of environmentally friendly projects. See TN/TE/W/6 2007 and TN/TE/W/67 2006. It is frustrating that after so many years there is no agreement yet on the definition of environmental goods and services (Vikhlyaev, 2004).

REFERENCES

Bausch, C. and S. Cavalieri (2007), 'Allocation of Greenhouse Gas Allowances in the United States – A Northeastern Example', *Carbon and Climate Law Review* 1, 129.

Bradbrook, A.J. and R.D. Wahnschafft (2001), 'A Statement of Principles for a Global Consensus on Sustainable Energy Production and Consumption', *Journal of Energy and Natural Resources Law* 19, 143.

Bubna-Litic, K. (2007) 'Corporate Social Responsibility: Using Climate Change to Illustrate the Intersection between Corporate Law and Environmental Law', *Environmental and Planning Law Journal* 24(4), 253.

De Cendra, J. (2006), 'Can Emissions Trading Schemes be Coupled with Border Tax Adjustments? An Analysis vis-à-vis WTO Law', *Review of European Community and International Environmental Law* 15, 131.

Demailly, D. and P. Quirion (2006), 'CO2 Abatement, Competitiveness and Leakage in the European Cement Industry under the EU ETS: Grandfathering Versus output-based Allocation', *Climate Policy* 6, 93.

Driesen, D.M. (2007), 'The Changing Climate for United States Law', *Carbon Climate Law Review* 1, 35.

Drache, D. and D.M. Froese (2007), 'Deadlock in the Doha Round: The Long Slow Decline of Trade Multilateralism' (SSRN Paper).

European Commission (2008a), Communication from the Commission to the European Parliament, the Council, the European Economic and Social Committee and the Committee of the Regions – *20 20 by 2020 – Europe's Climate Change Opportunity*, COM(2008)0030 final.

European Commission (2008b), *Commission staff working document accompanying document to the Proposal for a Directive of the European Parliament and of the Council amending Directive 2003/87/EC so as to improve and extend the EU greenhouse gas emission allowance trading system*, SEC(2008)52.

European Commission (2008c), *Proposal for a Directive amending Directive 2003/87/EC so as to improve and extend the greenhouse gas emission allowance trading system of the Community*, COM(2008)16 final, 23 January 2008.

Fauchald, O.K. (1998), *Environmental Taxes and Trade Discrimination* (Kluwer Law International).

Goldenberg, S. (2008), 'Obama's New Team Raises Hope for US Environment', *The Guardian*, 11 December.

Green, A. (2005), 'Climate Change, Regulatory Policy and the WTO', *Journal of International Economic Law* 8(1), 143.

Houser, T., et al. (2008), *Leveling the Carbon Playing Field: International Competition and U.S. Climate Policy Design* (Peterson Institute for International Economics).

Ismer, R. and K. Neuhoff (2004), 'Border Tax Adjustments: A Feasible Way to Address Non-participation in Emission Trading', *CMI Working Paper* (MIT Institute Electricity Project).

Janzen, B.G. (2008), 'International Trade Law and the "Carbon Leakage" Problem: Are Unilateral US Import Restrictions the Solution?', *Sustainable Development Law and Policy* 8, 22.

Lamy, P. (2007), 'Doha Could Deliver Double-Win for Environment and Trade', Informal Trade Ministers' Dialogue on climate change – Bali, Indonesia, 9 December 2007, www.wto.org/english/news_e/sppl83_e.htm (visited 15 March 2009).

Morgenstern, R.D. (2007), 'Addressing Competitiveness Concerns in the Context of a Mandatory Policy for Reducing US Greenhouse Gas Emissions', in R.J. Kopp and W.A. Pizer (eds), *Assessing U.S. Climate Policy Options: A Report Summarizing Work at RFF as Part of the Inter-industry US Climate Policy Forum* (Resources for the Future).

Organisation for Economic Co-operation and Development (OECD) (2005), *The Competitiveness Impacts of CO_2 Emissions Reductions in the Cement Sector* (OECD).

Pauwelyn, J. (2007), 'U.S. Federal Climate Policy and Competitiveness Concerns: The Limits and Options of International Trade Law' (Nicholas Institute for Environmental Policy Solutions, Duke University).

Pauwelyn, J. and F. Sindico (2008), 'Climate Change in a Global Economy', *Carbon and Climate Law Review* 2, 1.

Rabe, B. (2007), 'Race to the Top: The Expanding Role of State Renewable Energy Standards', *Sustainable Development Law and Policy* 7(3), 10.
Ratnayake, R. (1998), 'Do Stringent Environmental Regulations Reduce International Competitiveness? Evidence from an Inter-industry Analysis', *International Journal of the Economics of Business* 5, 77.
Reinaud, J. (2008a), 'Climate Policy and Carbon Leakage – Impacts of the European Emissions Trading Scheme on Aluminium' (International Energy Agency Information Paper).
Reinaud, J. (2008b), 'Issues behind Competitiveness and Carbon Leakage – Focus on Heavy Industry' (International Energy Agency Information Paper).
Richardson, B.J. (2008), *Socially Responsible Investment Law: Regulating the Unseen Polluters* (Oxford University Press).
Sindico, F. (2008), 'The EU and Carbon Leakage: How to Reconcile Border Adjustments with the WTO?', *European Energy and Environmental Law Review* 17, 328.
Subcommittee on Energy and Air Quality of the US House of Representatives (2008), *Climate Change Legislation Design White Paper* (US House of Representatives).
Szymanski, T. and S. Stone (2008), 'U.S. Efforts on Climate Change: New Beginnings and Enduring Realities', *Carbon and Climate Law Review* 2, 134.
USCAP (2008), 'A Call for Action – Consensus Principles and Recommendations from the U.S. Climate Action Partnership: A Business and NGO Partnership', www.us-cap.org/ (visited 29 January 2009).
Verrill, C.O. (2008), 'Maximum Carbon Intensity Limitations and the Agreement on Technical Barriers to Trade', *Carbon and Climate Law Review* 1, 43.
Vikhlyaev, A. (2004), 'Environmental Goods and Services: Defining Negotiations or Negotiating Definitions', *Journal of World Trade* 38(1), 93.
Voestalpine (2008), *Annual Report 2007–08* (Voestalpine).
Walker, P. (2009), 'Obama To Put Bush Car Pollution Policies into Reverse', *The Guardian*, 26 January.
Wiers, J. (2008), 'French Ideas on Climate and Trade Policies', *Carbon and Climate Law Review* 2, 18.
Zhang, Z. and L. Assunção (2004), 'Domestic Climate Policies and the WTO', *The World Economy* 27, 359.

16. Climate change in the European Union development cooperation policy

Marie-Pierre Lanfranchi* and Sandrine Maljean-Dubois**

1. INTRODUCTION

The climate policy of the European Union (EU) towards developing countries deserves our attention for at least three reasons. First, development is a traditional as well as a central element of European international relations. Already the main donor in the world with nearly 57 per cent of the total volume of public development aid (PDA), the EU will provide, until 2015, 90 per cent of the expected increase in aid to meet the Millennium Development Goals.[1] Second, the fight against climate change has become another one of its core priorities, at both the internal and international levels. Finally, the European policy regime on climate change, implemented early on and strengthened later in its objectives and means, provides a useful laboratory to reflect the 'post-2012' international regime.

In 1999, before the coming into force of the Kyoto Protocol,[2] both issues of climate change and development were linked for the first time in a European policy working paper (European Commission, 2003, pp. 54–55). It was used in support of the conclusions adopted by the European Council on 11 November 1999 (Council Conclusions, 1999), in which it reaffirmed that addressing climate change is a priority for the EU. The Council invited the European Commission to propose an action plan to integrate climate change into the EU's development cooperation policy. As a result, the political impulse was generated to add a new sectoral dimension to this cooperation policy. The Commission answered in 2003 with the adoption of a key document which laid the foundation for 'an integrated strategy for addressing climate change and poverty reduction concerns' and defined an 'action plan' for 2004–2008 (European Commission, 2003).

The Commission noted the lack of centralised funding for development actions on climate change, as it was financed in some instances by the

European Development Fund, on other occasions by the Community's general budget, and sometimes by the research budget for the Fifth Framework Programme.[3] The situation was even more complex and unsatisfactory, given that additional activities were undertaken by individual EU Member States with other sources of funding: bilateral or through the Global Environmental Facility or other multilateral channels. There was therefore an urgent need to rationalise EU policy. To increase the effectiveness of all actions it became important to define a coherent framework that would assist in the coordination of interventions of the EC and its Member States (European Commission, 2003, p. 55). As a result, the Action Plan was adopted by the Council on 22 November 2004 (Council of the European Union, 2004; European Commission, 2007d). In 2007, a second key document was adopted by the Council. Its aim is to build 'a Global Climate Change Alliance (GCCA) between the European Union and poor developing countries most vulnerable to climate change' (Council of the European Union, 2007). Together these texts put forward a very ambitious discourse. Our objective in this chapter is first to decode its meaning and subsequently to examine the practical implications of this discourse. We ask, among other things, whether the EU has taken the steps necessary to achieve its ambitious policy goals. As we will see, the post-Kyoto negotiations point to the pitfalls the EU faces in the implementation of its strategy.

2. DECODING AN AMBITIOUS DISCOURSE

The following section examines the bases for the discourse, focusing on legality and legitimation issues. It then considers its content as derived from successive texts.

Foundations for the Discourse

Legality: what legal bases?

The legal bases for these policy initiatives must emanate from the Treaty of Rome[4] establishing the EC. Various legal grounds can be invoked to support a climate policy towards developing countries. Together they result in a complex scheme which inspired the 'guiding principles' outlined in the Commission's 2003 strategic document.

At the institutional level, the competence of the EU to adopt a 'climate policy' towards developing countries is shared with Member States. Article 177 of the EC Treaty provides that the development cooperation policy of the EU, which is the framework into which climate actions must be defined, is complementary to the policy of its Member States (Balleix,

2008, pp. 377–78). This arrangement results in the superposition of two levels of decision-making.

At the diplomatic level, the legal bases that can be mobilised vary from case to case, depending on the closeness of the relationship with the partner countries: association agreements with African, Caribbean and Pacific countries (ACP countries);[5] development cooperation with non-ACP countries;[6] and financial and technical cooperation with middle-income countries from Latin America, Asia and Eastern Europe.[7] In the first case, the decision-making process is clearly more intergovernmental, as unanimity is required for approving association agreements. Only a qualified majority is needed for approving the other relationships.

Finally, as for substantive issues, the overall objective assigned to the development cooperation policy in Article 177 of the Treaty is neither original nor specific. It refers to 'the campaign against poverty in the developing countries' and to their 'economically and socially sustainable development'. This goal was clarified on a number of occasions and the evolving Community texts clearly point to an increased mainstreaming of environmental concerns in the design of development cooperation policy. In this regard, the political declaration adopted in December 2005 by the European Council is of particular importance (Council of the European Union, 2005). It sealed a European consensus on development, which means 'for the first time, a common vision' is shared by EU and its Member States which is consistent with the Millennium Development Goals (notably, to 'end poverty and hunger', and achieve 'environmental sustainability'). This consensus is in line with the approach previously announced by the Commission in the integrated strategy adopted in 2003. Two of the six 'guiding principles' in this strategy are to contribute to the overarching objective of poverty reduction as stated in the EC development policy and to contribute to the Millennium Development Goals and the outcome of the World Summit on Sustainable Development.

The consensus, in turn, set the scene for the conclusions of November 2007 on a Global Climate Change Alliance. In this document the Council, in a sombre tone, notes that 'Climate Change is becoming a major threat to achieving the Millennium Development Goals (MDGs) and may have a considerable impact on international security issues' (Council of the European Union, 2007, p. 3). Therefore, a solid legal basis and a strong political consensus exist to incorporate climate change into EU development cooperation policy. This is consistent with the fundamental principle of mainstreaming environmental goals found in Article 6 of the Treaty of Rome: 'environmental protection requirements must be integrated into the definition and implementation of the Community policies and activities . . . in particular with a view to promoting sustainable development'.

Beyond the legal bases found in the EU system, the international legal context fully justifies the integration of climate change into the EU cooperation policy. The 'common but differentiated responsibilities' principle and the commitment in the United Nations Framework Convention on Climate Change (UNFCCC)[8] and the Kyoto Protocol to supply new and additional resources provide impetus for the EU and Member States to act (European Commission, 2003, p. 3). Furthermore, the EU is also bound by commitments in the field of development cooperation, whether in terms of aid's volume or effectiveness, as evidenced, for example, by its contribution to the Millennium Development Goals as a result of the Monterrey Conference in 2002 on the financing of development or the 2005 Paris Declaration[9] (European Commission, 2008b, p. 3).

Legitimation: which justifications?

The EU action is not founded only on a legal basis. It is also supported by a discourse based itself mainly on two arguments: equity and the principle of common but differentiated responsibilities, which are both legitimising the policy. Equity obliges one to take into account climate change in development cooperation. In 2007–08, the United Nations Development Programme report on the fight against climate reasserted the idea which underlies the UNFCCC and the Kyoto Protocol: the challenges posed by climate change are of a different nature and order in the North and South (United Nations Development Programme, 2007).

Industrialised countries can mobilise needed resources to react to climate change while assuming historical responsibility for the problem. On the other hand, developing countries have a very low capacity to mitigate their greenhouse gas (GHG) emissions. At the same time, their need for development is very urgent and legitimate, bringing with it an increase in GHG emissions in the medium term, at least for some of them. In its 2003 strategy (European Commission, 2003, p. 51), the EU opted for a strategy of conciliation: the struggle against climate change should not preclude development and the North has to assist the South to adapt. Afterwards, the EU often relied on the much publicised Stern Review (Stern, 2006). This report posits that the choice is not between safeguarding the global climate and development, and that lack of action will prove much more expensive for the global economy than mitigation and adaptation measures (European Commission, 2007b, p. 104). The Stern Review also underlines the high costs of adaptation for the poorest countries, calling on the international community to support them. The equity argument is even more compelling for these countries: they are not responsible for global warming, but are particularly exposed and vulnerable to its risks, as explained in the 2007 report of the Intergovernmental Panel on Climate Change (IPCC, 2007).

The almost incantatory invocation of the Millennium Development Goals in the Community texts and the desire to link climate change with the fight against poverty can be seen as an effort to provide assurances to developing countries while encouraging them to confer equal priority on climate and on the MDGs. In its communication in 2003, the Commission endorsed this policy as one of its four strategic priorities: the partnership countries had to give more political importance to climate change (European Commission, 2003, p. 60).

The actions of the EU are motivated as much by its own interests as by a sense of international solidarity: '[i]t is . . . in *all parties' interest* to promote sustainable trends in GHG emissions also in partner countries' (European Commission, 2003, p. 3, our emphasis). The potentially dramatic consequences of climate change in developing countries could also threaten the industrialised world. European documents predict new migratory pressures from South to North with a massive displacement of population (European Commission, 2007b, pp. 1, 5–6). Potential risks of regional conflicts that may threaten international peace and security are also mentioned, particularly as a result of water scarcity. The UN Security Council (2007) has also discussed this issue. These analyses are consistent with the conclusions of the *Schellnhuber* report issued by the German Advisory Council on Global Change, a report endorsed by the United Nations Environment Programme in December 2007 (German Advisory Council on Global Change, 2008). These are recurring themes in Community documents adopted since 2007 (Council of the European Union, 2007). Faced with the spectre of millions of climate refugees and the risk of destabilisation,[10] the North has an interest in promoting preventive strategies by helping the South to adapt.

Migratory pressures, security and development: this rhetoric is not new, but here finds a new expression. The European Commission recognises that the fight against climate changes requires addressing as one both environmental issues and development cooperation: 'climate change is not only an environmental problem. It is also clearly a development problem' (European Commission, 2003, p. 3).

Content of the Discourse: From the Action Plan to the Global Climate Change Alliance (GCCA)

The integrated strategy and the Action Plan of 2003

The strategy is based on six 'guiding principles' and defines five 'strategic priorities' which provide a framework for the Action Plan. The guiding principles established by the European Commission (2003, pp. 16–17) are as follows:

> (i) Contribution to the overarching objective of poverty reduction as stated in the EC development policy and, where appropriate, its six core areas; (ii) Contribution to the Millennium Development Goals and the outcome of WSSD; (iii) Coherence, both internally and externally, and on the following levels: (a) coherence with other EC and MS policies, (b) coherence with other development sector/thematic policies/strategies, (c) and coherence/synergies with support action for other Multilateral Environmental Agreements; (iv) Coordination and complementarity between the Community, Member States and other donors; (v) Primacy of national ownership of development strategies and processes; (vi) Broad stakeholder participation in the implementation process.

Of these principles, the last two – ownership and participation – deserve a brief comment. They are part of the now very frequent rhetoric in the EU discourse on its relations with the developing world. As a political message to partners, they aim to provide assurances that the approach is egalitarian and respectful of sovereignty: 'the formulation of development priorities should be country-driven in order to further national ownership of the development process' (European Commission, 2003, p. 19). But are such pronouncements sincere or simply rhetoric?

The question is worth asking because, while the proposals of the EU only 'provide guidance', the Action Plan truly constitutes a strong incentive to follow an approach defined at the European level. Its stated 'strategic priorities' are to: (i) raise the policy profile of climate change; (ii) support adaptation to climate change; (iii) support mitigation of climate change; and (iv) build capacity (European Commission, 2003, p. 17). The first priority is essentially a political goal aimed at generating awareness as much within partnership countries as within the EC. The fourth priority is meant to reinforce the institutional and human capacities needed to implement the UNFCCC and the Kyoto Protocol, and to participate fully in future international negotiations. However, the heart of the proposed actions is captured in points (ii) and (iii): relying on a mix of mitigation actions and adaptation measures (European Commission, 2003, p. 19).

The adaptation actions (reducing vulnerability) outlined by the Commission (2003, pp. 29–31) target all the developing country partners, and aim for: (i) improving the robust design of infrastructure and long-term investments; (ii) increasing the flexibility of vulnerable managed systems (e.g., changing activity or location); (iii) enhancing the adaptability of vulnerable natural systems (e.g., reducing non-climatic stresses); (iv) reversing trends that increase vulnerability (e.g., slowing development in vulnerable areas such as flood plains and coastal zones); (v) improving the preparedness and awareness of society. The targeted areas are: natural resources management (water resources, coastal resources, forest resources), related productive sectors (agriculture, forestry, fisheries), infrastructures and human settlements.

The mitigation measures (reducing emissions) envisaged, however, are reserved mostly for the so-called emerging economy countries. Twenty states are identified as having high GHG emissions (in absolute or per capita terms) and are designated in Annex II: South Africa, Argentina, Azerbaijan, Belarus, Brazil, Chile, China, North Korea, India, Indonesia, Jamaica, Kazakhstan, Lebanon, Malaysia, Mexico, Uzbekistan, Trinity and Tobago, Turkmenistan, Thailand, Venezuela. The mitigation options outlined by the Action Plan include: price rationalisation, increased access to data and information, availability of advanced technologies, financial resources, and training and capacity building (European Commission, 2003, pp. 31–34). The primary economic sectors targeted for such measures are: energy supply, energy use and transport, land use, land use change and forestry options.

Based on these four strategic priorities, the Action Plan lists no fewer than 84 possible actions, drawing in particular on the methods supported by UNEP (adaptation) or on the European experience in reducing GHGs. Ultimately, the Plan leaves a nuanced impression: the observer is struck by the ambition of the discourse and the expertise it reflects. However, the plan also resembles a smorgasbord of miscellaneous ideas rather than a programme with clear, prioritised and achievable goals.

As the Bali Conference to pursue a post-Kyoto agreement approached, the Council took decisive steps in 2007 to reshape its strategy.

Reframing: the GCCA

Between January and November 2007, many important texts were adopted as the discourse on addressing climate change changed, leading to the GCCA initiative (European Commission, 2007d). A change in tone became evident: the emphasis shifted to the urgency of the situation and the worsening prognosis. This discourse probably made it easier for the EU to legitimise a more clearly differentiated approach among partner countries. The main objective became to create conditions favourable to the conclusion of a post-2012 agreement and for the EU to assume a leadership role in pursuing this objective.

As a result, the discourse adjusted to reflect a more differentiated approach. In a communication in January 2007, the European Commission stated that '[d]eveloping country emissions are projected to surpass those of developed countries by 2020' (European Commission, 2007c). Priority is given to mitigation over adaptation in order to prevent a worsening of the climate change prognosis. Adaptation actions are needed because global warming is already a reality, but they complement mitigation measures. The new discourse creates distinct categories among developing countries: emerging countries, on one side, and 'poor developing countries and most vulnerable to climate change', on the other side. With respect to the former, the EU seeks to deepen its mitigation strategy, while with the latter its main aim is to promote adaptation to climate change. The EU intends to encourage the emerging economy countries to limit their growth of GHG emissions and eventually to reduce them (European Commission, 2007c). Its aim is to convince these countries to accept quantified and binding commitments. The European Council approved this approach in its conclusions of March 2007:

> The European Council notes the increasing share of greenhouse gas emissions from developing countries and the need for these countries to address the increase in these emissions by reducing the emission intensity of their economic development, in line with the general principle of common but differentiated responsibilities and respective capabilities. The European Council stands ready to continue and further strengthen its support for developing countries in lessening their vulnerability and adapting to climate change.[11]

The Commission's document of January 2007 identified various options, such as extending the Clean Development Mechanism (CDM) to entire national sectors, generating greater flows of capital and technology, improving access to financing to invest in technologies and processes for clean energy production and creating carbon markets in sectors where monitoring is possible, especially among the major energy sectors such as aluminium, iron, steel, cement, refineries and pulp and paper. The preferred method is not to rely on the usual tools of cooperation, but to conclude bilateral partnerships like those already concluded with China, India and Brazil, respectively in 2005, 2006 and 2007. Concerning the other developing countries, the aim of the GCCA, as mentioned earlier, is to deepen the strategy of adaptation. As expected, the Alliance does not foresee a reduction commitment for these partners. However it defines five priorities for strengthening the capacity to adapt:

> (a) Adaptation to climate change:
> Objective: To help developing countries improve their knowledge base on the effects of climate change, to develop and implement adaptation strategies.
> (b) Reducing emissions from deforestation:
> Objective: To decrease CO_2 emissions from deforestation in developing countries, by creating economic incentives for forest protection, while preserving livelihoods and ecosystems depending on forests.
> (c) Enhancing participation in the Clean Development Mechanism:
> Objective: To help developing countries to participate in and benefit from the global carbon market, through the Clean Development Mechanism.

(d) Promoting Disaster Risk Reduction (DRR):
Objective: To improve the preparedness of developing countries and societies for climate-related natural disasters, and to mitigate the risks and limit their impact.
(e) Integrating climate change into poverty reduction efforts:
Objective: To assist developing countries in systematically integrating climate change into development strategies and investments, and to systematically integrate climate change into development cooperation (European Commission, 2007d, pp. 4–8).

The means *proposed* to achieve these aims are essentially sustained funding by the EU and a deepening of the dialogue in order to contribute to a convergence of views between Europe and developing countries in the negotiation of a 'post-2012' agreement (European Commission, 2007d, p. 4).

The discourse is thus obviously supporting the EU objective to position itself to play a leading role in the conclusion of a new international agreement on climate change. The initiatives adopted throughout 2007 and in particular the commitment by the EU to reduce GHGs by 20 per cent by 2020 (Council of the European Union, 2007; European Commission, 2007d) demonstrate a desire to lead. The GCCA is a central piece in this strategy. The text establishing the Alliance states that 'the European Union (EU) has taken a leadership role in promoting international action to tackle climate change' and that 'the GCCA will provide the EU with a unique opportunity to show international leadership and re-affirm the principles of multilateralism and global responsibility that underpin its international relations' (European Commission, 2007d, pp. 2, 10). Other documents adopted in 2007 (European Commission, 2007d) and in 2008 (High Representative and the EC, 2008) are in the same vein. The Poznan Conference reaffirmed this ambition, but also the failure of the Union to convince the developing world to accept its approach (Kempf, 2008).

The desire of the EU to take the lead is not new: it has nearly always wanted to be a major player on matters of international environmental governance. This stance is particularly true for the topic of climate change, as demonstrated by the early establishment of a climate regime, the initiatives to save the Kyoto Protocol and the actions launched in 2007 and 2008. By concluding partnerships with emerging countries and strengthening the dialogue with the entire developing world through sustained funding, the EU can favour the convergence of views between states in the post-2012 negotiations and, as a result, increase further its political influence.

While this discourse is both generous and ambitious, one wonders whether the EU has invested the means needed to implement it.

3. IS THE EU TAKING THE STEPS NECESSARY TO ACHIEVE ITS OBJECTIVES?

Recently the European Commission stated that '[w]e do not need new promises. Instead we should translate existing commitments into tangible results' (European Commission, 2008b). The effectiveness of the European Policy can be assessed in quantitative terms – means, financing and volume of aid – and in qualitative terms – the aid's efficiency.

Financing Issues

The context: The decline in the volume of development aid

The EU is the largest aid donor in the world, providing the equivalent of 93 euros per European citizen per year. In 2006, according to OECD figures,[12] it contributed 56.67 per cent (46.9 billion euros) of the entire world public development aid (PDA). The EU's collective target is to reach 0.56 per cent of gross national income (GNI) for development aid in 2010 (already representing an increase of 20 billion euros), and to reach the 2015 target of 0.7 per cent GNI. This target was set by the United Nations in the 1970s and reaffirmed many times since at the international level (Monterrey Conference,[13] Millennium Summit[14]) as well as at the European level (European Council, 2002 and 2005).[15]

However, in 2007 there was a reduction of EU budgetary aid. The trend is more general: for the second year in a row there has been a net drop in the volume of development aid from the international community. For the first time, the EU contributed to this decrease: '[e]xpressed in euro and as a percentage of gross national income, European aid stood at 0.38 per cent in 2007 compared with 0.41 per cent in 2006' (European Commission, 2008b). In 2007, the increase in budgetary aid allocated by the Commission was only of 30 per cent, when it should have been doubled to meet its professed commitments. Thus, an increase in development aid will have to come from the EU Member States whose contributions have been unequal (for example, Sweden has contributed 1 per cent GNI compared to Latvia, which contributed just 0.06 per cent of its GNI) (Balleix, 2008, p. 384).

Given the worsening global economic recession in 2009, the objective of reaching 0.7 per cent GNI in 2015 seems to be largely unrealistic at this point. Indeed, OECD figures over a longer time-frame show that the EU performance has stagnated: 0.37 per cent in 1996 against 0.38 per cent in 2007, figures which were below the average performance of the OECD Development Assistance Committee countries in 2007 (at 0.45 per cent).

Insufficient budget assigned to climate change

According to the World Bank, between US$10 and 40 billion per year will be necessary to fund climate change adaptation measures in the poorest countries. Meanwhile, the contributions to the Adaptation Fund under the UNFCCC presently do not amount to more than US$150 to 300 millions per year (European Parliament, 2007).

As for EU Member States, the '[c]ommitments by the EU-15 to activities marked with climate change as a principal or significant objective totaled US$554 million in 2003 and US$660 million in 2004' (External consultant, 2006, p. 105). This is in keeping with the commitment made in Bonn in 2001 (410 to 450 million euros per year starting in 2005). However, the 50 million euro budget allocated to the Alliance for 2008–2010 is not in line with the policy discourse making action on climate change a priority. In total, the European Commission plans to allocate approximately 345 million euros to climate change within the framework of its development cooperation policy (including 200 million euros from the European Development Fund), a small sum compared, for instance, to the 8.2 billion euros it contributed to public development aid in 2006. A report from the European Parliament found the amount to be 'woefully inadequate' and it 'calls on the Commission to establish a long-term financing goal for the GCCA of at least 2 billion euros annually by 2010 and 5-10 billion annually by 2020' (European Parliament, 2008a).

The European Commission is considering the creation of a joint financing mechanism with the Alliance, which would be managed by the Commission and would receive subsidies from the Commission and Member States. This funding would support and complete other initiatives and bilateral and multilateral funds, such as the Global Environment Facility or other mechanisms within the UNFCCC which finance projects to fight climate change. To our knowledge, this mechanism has not yet been established.

The need to imagine new ways of financing

Given the limits of traditional sources of financing, the necessity to find 'innovative sources' is a *Leitmotiv* in most of the Commission's documents. Several options have been explored, with some in a more advance stage of development than others. However, much remains to be done.

The Commission planned, with the 'Energy-Climate Package', progressively to replace free allocation of emission allowances in the EU cap and trade scheme with an auction of allowances (from 20 per cent in 2013 to 100 per cent in 2020). In its initial proposal, the Commission expected the auctioning process to generate significant revenues, up to 40 to 50 billion euros per year in 2020. It suggested that part of these revenues could be used to assist developing countries to adapt to the effects of climate change: 'a certain percentage of the proceeds from the auctioning of allowances should be used to (. . .) contribute to the Global Energy Efficiency and Renewable Energy Fund, for measures to avoid deforestation and facilitate adaptation in developing countries' (European Commission, 2008a). However, this 'certain percentage' was not defined. The European Parliament proposed that at least 50 per cent of the revenues generated from the auctioning of allowances should be used in a dedicated international fund as follows:

> a) one quarter for measures to contribute to funds to avoid deforestation and increase afforestation and reforestation in developing countries that have ratified the future international agreement, taking into account: the rights and needs of indigenous peoples; the preservation of biodiversity; and the sustainable use of forest resources;
> b) one quarter to reduce emissions in developing countries that have ratified the future international agreement, and to transfer technology to those countries, e.g. through the Global Energy Efficiency and Renewable Energy Fund;
> c) one half to facilitate adaptation to the adverse effects of climate change in developing countries that have ratified the future international agreement on climate change (European Parliament 2008c).

Unfortunately, the recent European Council of 11–12 December 2008 has lowered expectations considerably, essentially due to the risk of carbon leakage. On the one hand, the period for auctioning is lengthened (20 per cent in 2013 with a view to reaching 100 per cent in 2027) and numerous derogations allowed. This will result in lower revenues than expected – no more than 30 billion euros a year. On the other hand, the key issue of how to use the revenues was not tackled. The EU delayed the decision as to the proportion of the package funds that will be dedicated to developing countries to March 2009. Not surprisingly, this news was not well received in Poznan (Kempf, 2008).

The Commission also created a new tool, the Global Energy Efficiency and Renewable Energy Fund (GEEREF). It is an innovative risk capital fund to attract private investment in small energy efficiency and renewable energy projects in developing countries and economies in transition. Essentially, the fund will protect private investors' investments against risks. Announced in Bali in December 2008, the Fund is now operational. The European Commission will inject 80 million euros into the GEEREF between now and 2010.[16]

The CDM can of course also be used as a lever for a win-win strategy. However, it is a somewhat controversial tool for cooperation (Moliner-Dubost, 2004, p. 963). Its operations and results are currently being discussed and assessed, and this will be likely to lead to an in-depth reform. It quickly attracted many takers[17] (Wara and Victor, 2008), but has come under criticism on govervance issues. Even more worrying is some evidence that it has sometimes led to fake reductions at a high cost. The additionality of projects – an innovative feature – is questioned in some of those projects (Wara, 2008, p. 410). Moreover, the CDM cannot remain the only tool to reduce emissions in emerging countries. As Michel Colombier points out, the number of credits that would need to be purchased would be too important, both for industrial carbon markets which may not be able to achieve their goal, and for Annex I Parties who would have to agree to purchase an artificially high number of these credits while proceeding with domestic reductions (Colombier et al., 2007, p. 4). Finally, because it is an incentive and voluntary tool, projects are concentrated in countries providing the best opportunities and the best security for investments. The poorest ones, in particular in Africa, are almost excluded from it (only 2 per cent of the projects are located in Africa). All of these factors lead one to conclude that the CDM cannot be the only international mitigation policy in developing countries. However, one helpful component of the CDM is that, unlike other funds within the UNFCCC, the Adaptation Fund is financed by a share of proceeds from CDM project activities (2 per cent of certified emission reductions). In Poznan, developing countries asked that an equivalent percentage also be levied from carbon markets. This would increase the Adaptation Fund substantially. However, no agreement was reached on this proposal.

The Commission has also indicated that, in consultation with the World Bank, it is considering the idea of a world loan drawing on the resources raised by auctioning emission rights on a future carbon market (European Commission, 2008b, p. 8).

For its part, the European Parliament proposed the establishment of an adaptation financing mechanism modelled on the International Finance Facility for Immunization. This mechanism raises funds by issuing bonds on capital markets and converting long-term government pledges into immediately available cash resources (European Parliament, 2008a). Another model is UNITAID, a tax on air tickets which allows an international facility to purchase drugs to fight diseases caused by poverty.[18] Finally, another idea, mentioned in a report of the European Parliament, is to encourage public–private partnerships (European Parliament, 2008a).

Effectiveness Issues

The effectiveness of the European development cooperation policy has come under heavy criticism.[19] Some have talked of the existing gap between the amount of aid and the low visibility of the policy (Balleix, 2008, p. 376).

The debate about its effectiveness has spanned the following issues: the consistency of the aid, its relevance, its predictability and its monitoring.

Consistency

Consistency is probably the first challenge to address in order to improve aid's effectiveness. Such consistency must exist both within and outside the EU. At the internal level, within the EU, consistency can be analysed vertically (the relationship between Member States and the EU) and horizontally (consistency between sectoral policies of the EU).

In relation to vertical consistency, internal unity is seen as a precondition of European influence internationally (Vogler and Hannes, 2007, p. 408). But the necessary cooperation for development policy is a responsibility shared by the Community with its Member States, which raises consistency – not to speak of synergy – issues between Member States and the EU. Article 180 of the Rome Treaty provides that:

> 1 The Community and the Member States shall coordinate their policies on development cooperation and shall consult each other on their aid programmes, including in international organizations and during international conferences. They may undertake joint action. Member States shall contribute if necessary to the implementation of Community aid programmes.
> 2. The Commission may take any useful initiative to promote the coordination referred to in paragraph 1.

However, overlaps, duplications, omissions and even inconsistencies in policy-making are frequent. The development cooperation policy remains a poorly coordinated assembly of 27 +1 European development policies (Balleix, 2008, p. 377).

The Commission recently proposed a voluntary code of conduct to share responsibilities, but it is too early to make a proper assessment of it. Despite guidelines, pilot experiences, action plans and codes of conduct, the Member States continue to try to maintain their own policy visibility. Given the political dimension of aid, they press in the opposite direction (Balleix, 2008, p. 386). To this day, '[r]esponsibilities between Member States or General Directorates of the European Commission are not clearly established' (External consultant, 2006, p. 65).

In relation to horizontal consistency between sectoral European policies, the Rome Treaty tasks the Community with the promotion of a 'balanced and sustainable development of economic activities' (Article 2). In addition, as mentioned before, it affirms that '[e]nvironmental protection requirements must be integrated into the definition and implementation of the Community policies and activities referred to in article 3, in particular with a view to promoting sustainable development' (Article 6).

The potential consequences of this principle are tremendous – as are the difficulties – and the European Commission made it its hobbyhorse within the 'Cardiff Process' (the label given to its efforts to introduce a horizontal approach to environment policy by incorporating it into all Community policies). It defined some sectoral strategies to implement this principle (European Commission, 1998): following a communication in May 2000 (European Commission, 2000), the European Council on 31 May 2001 adopted a strategy for integrating environmental requirements into the development cooperation policy to promote a sustainable development (European Commission, 2004).

The EU is not the only one faced with some difficulties in achieving horizontal consistency: every state is confronted with the same difficulties. But the issue is more acute at the European level; as Bretherton and Vogler (2008, p. 405), explain, '[t]he "overarching objective" of sustainability would challenge the Policy coherence mechanisms of any political system, and the complex and fragmented nature of the EU generates unique coordination problems'.

In addition, the European Commission is conducting impact assessments of the sustainability of all its substantial proposals within the framework of its 'Best Regulation' initiative (European Commission, 2002 and 2008c). The EU wants to ensure the consistency of its policies with development goals. Policies in various fields – trade, security, migration, environment – are now expected to be analysed to assess their support for the Millennium Development Goals (MDGs) (European Commission, 2008b). However, some studies show that in practice the impact assessment process is not an efficient tool for implementing European commitments to promote sustainable development in developing countries (Adelle, Hertin and Jordan, 2006, p. 57). The Commission also conducted a process of internal reorganisation to achieve better consistency (Dearden, 2007, p. 5). Again, despite these good intentions, administrative and political difficulties and obstacles are real and persistent. Risks of inconsistency are numerous, in particular in trade, agricultural, fishing and even migratory policies (Balleix, 2008, p. 378; Zito, 2005, p. 363).

In 2006, the European Commission adopted a set of concrete measures to improve consistency, creating in particular the Commission's internal Interservice Group on Environmental Mainstreaming in Development Co-operation (European Commission, 2008b). A report at the time highlighted many deficiencies to be overcome:

> Policies, programmes and projects have not been systematically checked for consistency with climate change goals. International resolutions towards

> the integration of climate change into development cooperation provide broad frameworks and demonstrate commitments at the policy-making levels. However, the operational implications of these commitments in terms of resource allocation and institutional reforms are yet to be systematically assessed to put into practice climate change integration into development policies . . . the overall picture remains that of development activities and projects planned and implemented without due consideration for their impacts on climate change or the impacts of climate change on their proposed contributions (External consultant, 2006, p. 34).

This negative assessment was echoed by a special report from the European Court of Auditors, which found that the process of environmental mainstreaming in projects is not systematically implemented (European Court of Auditors, 2006, p. 8). In October 2007, the European Commission published, with little media coverage, its first biennial report on the consistency of the development cooperation policy (Balleix, 2008, p. 382). Climate change is one of the 12 priority fields defined in this approach.

In relation to 'external consistency', Article 181 of the Rome Treaty on the development cooperation policy states that '[w]ithin their respective spheres of competence, the Community and the Member States shall cooperate with third countries and with the competent international organizations'. Major UN conferences have facilitated the definition of global frames of reference (e.g., the Monterrey consensus and the Millennium Development Goals), thereby contributing to building consistency among the policies of various development cooperation actors, at least at the strategic level, if not at the operational level (Bretherton and Vogler, 2008, p. 407). The 'European consensus' is also based on instruments issued from these initiatives, with some specificity such as the support for European integration. Of course, multilateral agreements in the environmental field also contribute to the building of a common discourse. This is true in particular of the international climate change regime (UNFCCC and Kyoto Protocol), given the number of state parties to these treaties, their importance and their coverage in the media. But, there again, 'extensive scholarship has noted how the EU role as an external trade actor and a promoter of environmental protection values has led to internal conflict, often resolved in favour of others interests like the trade interests' (Bretherton and Vogler, 2008, p. 407).

Relevance

One can also question the relevance of the EU contribution. In theory, the multilateral financing it brings could offer more objective and consistent aid, freer from the trade or political interests which often characterise national development cooperation policies.[20] Indeed, the European strategy and its objectives are becoming clearer with time. These objectives were not even articulated before 1985.[21] Nevertheless, according to the European Court of Auditors' report, while projects are often 'relevant', they often achieve only half of their objectives (European Court of Auditors, 2006, p. 15). Even the European Commission recognises that '[f]urther improvements are required regarding the integration of climate change concerns into the policy dialogue with developing countries as well as into development cooperation programmes' (European Commission, 2007a).

Predictability

The issue of predictability is also well identified, including in the 'European Consensus'. Predictability can be assessed by looking at the percentage of payments made according to schedules agreed on an annual or multi-year framework.[22] The Commission identified three ways to reinforce it: (i) to improve the predictability of financial flows by using multi-year schedules; (ii) to establish joint multi-year programmes; and (iii) to ensure the predictability of disbursements (European Commission, 2008b, p. 7). The Commission has also alluded to the possibility of creating MDGs contracts, under which some countries could receive aid for six years to support their national budgets, in exchange for close monitoring of their results in achieving the MDGs.

Monitoring

Monitoring of development cooperation policies is notoriously inadequate. A follow-up of the Action Plan was planned with the drafting of a bi-annual report. What are the results? The preparation of a first report for the 2004–2006 period was entrusted to an external consultant. It is difficult to obtain this report because, even if it was transmitted to the Council in October 2007, it has never been discussed or approved by the Commission as a working paper (External consultant, 2006).

The European Parliament proposes to reinforce monitoring. According to the Parliament, the Alliance needs 'effective reporting mechanisms, including detailed indicators of progress and follow-up schemes' (European Parliament, 2008b). It also proposes the establishment of a 'permanent advisory and monitoring body for Sustainable Development', which would include Member States and civil society representatives and would scrutinise the mainstreaming of the concept into EU policies and programmes, with a particular focus on development cooperation (European Parliament, 2006).

4. CONCLUSIONS

In 2001, the European Commissioner responsible for Development and Humanitarian Aid, Poul Nielsen, formulated a key question on European Development Cooperation policy and provided an answer:

> The first question is of course, 'why does Europe do this at Community level?' The individual 15 Member States do their part development cooperation, so why do we also do it together? The answer to this is that it is simply a projection of the values on which European cooperation has been founded. These are the same values we want to project into the world, and also our own self-perception makes it natural for us to do these things. There is a strong moral aspect in this, but there is also ideology and some enlightened self-interest; promoting stability cooperation and prosperity in the world is definitely also part of making it a better world for us to live and work in.[23]

The challenge of adapting to climate change inspired the European Union to formulate a genuinely ambitious policy. However, the EU's role is evolving: '[t]he EU's function as a distributor of development assistance is diminishing in importance, while its regulatory role as Policy setter is becoming more significant' (Orbie and Versluys, 2008, p. 88). The question now is whether the means will match its ambitions. The policy is still in a consolidation phase. Will the Alliance be more effective than the previous Action Plan? No doubt the present economic crisis dampens a positive outlook for the 'post-2012' negotiations.

NOTES

* Senior Lecturer, Paul Cezanne University, France.

** Senior Researcher, National Center for Scientific Research, Aix-en-Provence (France).
We would like to express our gratitude to Yves Le Bouthillier, Marion Lemoine, Vanessa Richard and Stepan Wood for their very helpful assistance in the translation of our contribution.

1. European Commission (2007), *Annual Report 2007 on the European Community's Development Policy and the Implementation of External Assistance in 2006,* http://ec.europa.eu/europeaid/multimedia/publications/publications/annual-reports/2007_en.htm, (verified 13 February 2009).
2. Kyoto Protocol to the United Nations Framework Convention on Climate Change (1997), Art. 10, http://unfccc.int/kyoto_protocol/items/2830.php (verified 2 June 2009).
3. The Environment Action Programme defines the priorities and the main goals of the European environmental policy. The First Programme was established in 1972. The current Programme, the Sixth, covers the period 2002–2012.
4. Consolidated version of the treaty establishing the European Community [2002] OJ C325/33.
5. Countries within Art. 310 of Rome Treaty, ibid., at 153; Art. 217 of the Lisbon Treaty, Consolidated version of the Treaty on the Functioning of the European Union [2008] OJ C115/144.
6. Ibid., Title XX,; ibid., Arts 208–211.

7. Ibid., Title XXI; ibid., Arts 212–213.
8. See http://unfccc.int/resource/docs/convkp/conveng.pdf (verified 24 February 2009).
9. OECD (2005), Paris Declaration on Aid Effectiveness, 2 March 2005, http://www.oecd.org/dataoecd/53/38/34579826.pdf (verified 15 December 2008).
10. Supra note 10, at 65.
11. Section 34.
12. Supra note 1.
13. Monterrey Consensus on Financing for Development, United Nations, 2003, http://www.un.org/french/esa/ffd2002/ (verified 2 June 2009).
14. Resolution adopted by the UN General Assembly, no 55/2, 'United Nations Millennium Declaration'.
15. European Council Conclusions, http//:europa.eu/european-Council/index_en.htm (verified 23 February 2009).
16. IP/06/1309, 6 October 2006.
17. M.W. Wara and D.G. Victor (2008), 'A Realistic Policy on International Carbon Offsets', PESD Working Paper No. 74, 18 April 2008, http://pesd.stanford.edu/publications/a_realistic_policy_on_international_carbon_offsets/ (verified, 16 December 2008).
18. See www.unitaid.eu (verified 23 February 2009).
19. S. Dearden (2007), 'The Reform Agenda of the EU's Development Policy', European Union Studies Association (EUSA), Biennial Conference, 2007 (10th biennial conference), May 17–19, 2007, Montreal, at http://aei.pitt.edu/7794/ (verified 15 December 2008).
20. Ibid., at 1.
21. Ibid., in 2
22. Supra note. 8.
23. P. Nielson (2001), 'Fighting Poverty and Promoting Development: the EU Strategy', European Commissioner Responsible for Development and Humanitarian Aid, speech of 23 January 2001, www.deljpn.ec.europa.eu/home/speech_en_Speech%2001/01.php, (verified 15 December 2008).

REFERENCES

Adelle, C., J. Hertin and J. Andrew (2006), 'Sustainable Development "Outside" the European Union: What Role for Impact Assessment?', *European Environment*, 16, 57.

Balleix, C. (2008), 'Atouts et défis de la politique européenne de coopération au développement à l'horizon 2015', *Revue du Marché commun et de l'Union européenne,* 519, 376.

Bretherton, C. and J. Vogler (2008), 'The European Union as a Sustainable Development Actor: the Case of External Fisheries Policy', *Journal of European Integration*, 30(3), 401.

Colombier, M., et al. (2007), 'Négociations sur le changement climatique. Propositions pour une nouvelle stratégie française', *IDDRI Synthèses,* No. 4.

Council of the European Union (1999), *Climate Change and Developing Countries, Council Conclusions*, 12308/99.

Council of the European Union (2004), *Climate Change in the Context of Development Cooperation, Council Conclusions*, 15164/04.

Council of the European Union (2005), *Joint declaration by the Council and the representatives of the governments of the Member States meeting within the Council, the European Parliament and the Commission on the development policy of the European Union entitled 'The European Consensus'* [2006] OJ C46/1.

Council of the European Union (2007), *Global Climate Change Alliance (GCCA) between the European Union and Poor Developing Countries Most Vulnerable to Climate Change,* Council Conclusions, 15103/07.

European Commission (1998), *Communication from the Commission to the European Council of 27 May 1998 on a partnership for integration: a strategy for integrating the environment into EU policies (Cardiff–June 1998)*, COM(1998)333.

European Commission (2000), *Commission Communication of 18 May 2000, Integrating Environment and Sustainable Development into Economic and Development Cooperation Policy – Elements of a Comprehensive Strategy,* COM(2000)264 final.

European Commission (2002), *Communication from the Commission on Impact Assessment,* COM(2002)276 final.

European Commission (2003), *Communication from the Commission to the Council and the European Parliament – Climate Change in the Context of Development Cooperation,* COM(2003)85 final.

European Commission (2004), *Integrating Environmental Considerations into Other Policy Areas – A Stocktaking of the Cardiff Process,* COM(2004)394.

European Commission (2007a), *EU Report on Policy Coherence for Development, Commission Working Paper of 20 September 2007,* COM(2007)545 final.

European Commission (2007b), *Green Paper from the Commission to the Council, The European Parliament, the European Economic and Social Committee and the Committee of the Regions, Adapting to Climate Change in Europe – Options for EU Action,* COM(2007)345 final.

European Commission (2007c), *Communication from the Commission to the Council, the European Parliament, the European Economic and Social Committee and the Committee of the Regions – Limiting Global Climate Change to 2 Degrees Celsius – The Way Ahead for 2020 and Beyond,* COM(2007)2 final.

European Commission (2007d), *Communication from the Commission to the Council and the European Parliament, Building a Global Climate Change Alliance between the European Union and Poor Developing Countries Most Vulnerable to Climate Change,* COM(2007)540 final.

European Commission (2008a), *Proposal for a Directive of the European Parliament and of the Council Amending Directive 2003/87/EC so as to Improve and Extend the Greenhouse Gas Emission Allowance Trading System of the Community,* COM(2008)16 final.

European Commission (2008b), *The EU – A Global Partner for Development Spending on Progress towards the Millennium Development Goals,* Communication from the Commission to the European Parliament, the Council, the European Economic and Social Committee and the Committee of the Regions, COM(2008)177.

European Commission (2008c), *Communication from the Commission to the European Parliament, the Council, the European Economic and Social Committee and the Committee of the Regions – Second Strategic Review of Better Regulation in the European Union,* COM(2008)32 final.

European Court of Auditors (2006), *Special Report No 6/2006 Concerning the Environmental Aspects of the Commission's Development Cooperation* [2006] OJ C235/1.

European Parliament (2006), *European Parliament Resolution of 2 January on Mainstreaming Sustainability in Development Cooperation Policies,* www.europarl.europa.eu/RegData/seance_pleniere/textes_adoptes/definitif/2007/02-01/0014/P6_TA(2007)0014_EN.pdf (verified 24 February 2009).

European Parliament (2007), *European Parliament Resolution of 20 June 2007 on the Millennium Development Goals – The Midway Point*, www.europarl.europa.eu/RegWeb/application/registre/searchResult.faces (verified 24 February 2009).
European Parliament (2008a), *Draft Report by Anders Wijkman (PE406.017v02-00) on Building a Global Climate Change Alliance between the European Union and Poor Developing Countries most Vulnerable to Climate Change, (2008/0000(INI)). PR\724676EN.doc,* www.europarl.europa.eu/RegData/commissions/deve/projet_rapport/2008/406017/DEVE_PR(2008)406017_EN.pdf (verified 24 February 2009).
European Parliament (2008b), *European Parliament resolution of 21 October 2008 on Building a Global Climate Change Alliance between the European Union and Poor Developing Countries most Vulnerable to Climate Change*, www.europarl.europa.eu/sides/getDoc.do?pubRef=-//EP//TEXT+TA+P6-TA-2008-0491+0+DOC+XML+V0//EN (verified 24 February 2009).
European Parliament (2008c), *Decision of the Committee Responsible, 1st Reading/Single Reading*, COD/2008/0013, www.europarl.europa.eu/oeil/resume.jsp?id=5588462&eventld=1033678&backToCaller=NO&language=en (verified 6 October 2009).
External consultant (2006), *EU Action Plan on Climate Change and Development: 1st Bi-annual Progress Report (2004-2006)*, Request for Services No. 2006/129304–Version 1, Final Report, Ecorys.
German Advisory Council on Global Change (2008), *World in Transition – Climate Change as a Security Risk* (Earthscan), www.wbgu.de/wbgu_ig2007_engl.pdf, (verified 16 December 2008).
High Representative and the EC (2008), *Climate Change and International Security: Paper from the High Representative and the EC to the European Council,* 14 March 2008, S113/08.
Intergovernmental Panel on Climate Change (IPCC) (2007), *Fourth Assessment Report: Climate Change* (IPCC).
Kempf, H. (2008), 'La conférence de l'ONU avance, sans conclure, sur le financement', *Le Monde*, 13 December.
Moliner-Dubost, M. (2004), 'Le mécanisme pour une développement propre: une nouvelle voie de coopération et de transferts Nord Sud?', *Revue générale de droit international public*, 108, 963.
Orbie, J. and H. Versluys (2008), 'The European Union's International Development Policy: Leading and Benevolent?', in Jan Orbie (ed.), *Europe's Global Role: External Policies of the European Union* (Ashgate).
Stern, N. (2006), *The Economics of Climate Change* (HM Treasury).
Streck, C. and J. Lin (2008), 'Making Markets Work: A Review of CDM Performance and the Need for Reform', *European Journal of International Law*, 19, 409.
United Nations Development Programme (UNDP) (2007), *Human Development Report (2007–2008), Fighting Climate Change: Human Solidarity in a Divided World* (UNDP).
United Nations Security Council, (2007), April 2007, 5663rd Meeting, SC 9000, S/PV.5663.
Vogler, J. and S.R. Hannes (2007), 'The European Union in Global Environmental Governance: Leadership in the Making?', *Environmental Agreements*, 7, 389.
Wara, M.W. (2007), 'Is the Global Carbon Market Working?', *Nature*, 445, 595.
Zito, A.R. (2005), 'The European Union as an Environmental Leader in a Global Environment', *Globalizations*, 2, 363.

Index